The Latin American Studies Book Series

The Latin American Studies Book Series promotes quality scientific research focusing on Latin American countries. The series accepts disciplinary and interdisciplinary titles related to geographical, environmental, cultural, economic, political and urban research dedicated to Latin America. The series publishes comprehensive monographs, edited volumes and textbooks refereed by a region or country expert specialized in Latin American studies.

The series aims to raise the profile of Latin American studies, showcasing important works developed focusing on the region. It is aimed at researchers, students, and everyone interested in Latin American topics.

Submit a proposal: Proposals for the series will be considered by the Series Advisory Board. A book proposal form can be obtained from the Publisher, Juliana Pitanguy (juliana.pitanguy@springer.com).

More information about this series at http://www.springer.com/series/15104

Victor Bravo · Nicolas Di Sbroiavacca

Oil and Natural Gas Economy in Argentina

The Case of Fracking

Springer

Victor Bravo
Bariloche Foundation
San Carlos de Bariloche, Argentina

Nicolas Di Sbroiavacca
Bariloche Foundation
San Carlos de Bariloche, Argentina

ISSN 2366-3421 ISSN 2366-343X (electronic)
The Latin American Studies Book Series
ISBN 978-3-030-65519-8 ISBN 978-3-030-65520-4 (eBook)
https://doi.org/10.1007/978-3-030-65520-4

This Springer imprint is published by the registered company Springer Nature Switzerland AG
The registered company address is: Gewerbestrasse 11, 6330 Cham, Switzerland

Introduction to the Problems of Hydrocarbons in Argentina and the Case of Fracking

This volume gathers four papers written by the authors of this volume: the oil engineers Victor Bravo and Nicolas Di Sbroiavacca.

The central axes are the application of the extraction techniques of two non-convencional hydrocarbons, oil and natural Gas, by means of the fracking techniques, which are massively used in the USA, and of which Argentina possesses important resources.

To develop these techniques has generated controversial discussions all over the World, due to the possibility of large environmental impacts and the significative requirements of huge capital investments.

Argentina is not out of this controversy since the national authorities of the country believe that this could be one of the main actions that could contribute to the payment of the outsized external debt, and from other sectors, researchers understand that it is not possible to develop the country only by means of the massive exploitation and export of the natural resources.

Consequently, the problem of hydrocarbons is a problem of Energy Politics.

This is the reason why there is merit to make a previous analysis of the Energy Politics of the recent governments of Argentina by means of the papers: "Analysis of the National Law 27,007," known as the "Hydrocarbon law" and "The Hydrocarbon Politics in Argentina of the period 2003 to 2014" and "Energy Politics in Argentina during the period 2014 to 2018," and then enter deeply in the problem of "Fracking," with two complementary analyses, one of them centrally referred to the technical aspects and prospective scenarios of Fracking, "Shale Oil and Shale Gas in Argentina: situation and perspectives," and another, "An opinion about Fracking," which discusses the impacts of this technique, particularly the environmental ones.

However, any of the papers is self-contained and they may be read separately.

Victor Bravo
Nicolas Di Sbroiavacca

Contents

About the Authors

Victor Bravo is a Chemical Engineering graduated from the Universidad Nacional del Litoral, Santa Fe, Argentina. Afterward, he obtained a Master's degree in Oil Resources, at the Universidad de Buenos Aires, Argentina. He is a Senior Researcher at the Bariloche Foundation, and a consultant at ECLAC and UNDP. He has been a Visiting Professor in many universities in Argentina, in Latin America and Spain. He is the Director of the "Latin American Course in Energy Economics and Planning" that the Bariloche Foundation offers annually since 1967.

Nicolas Di Sbroiavacca is a Petroleum Engineer, Master in Energy Economics and Environment (Scuola Enrico Mattei, Italy) and has a specialization in hydrocarbon management (National Petroleum Directorate, Norway). He is a Senior Researcher at the Bariloche Foundation and a consultant at ECLAC, UNFCCC, UNDP, among others international institution. He is presently the acting President of Bariloche Foundation and as Director of the Master's degree course, in collaboration with University National of Comahue, Argentina.

Chapter 1
Analysis of the Hydrocarbon Law 27,007 and the Hydrocarbon Policy for the Period 2003 to 2014

Abstract The present document, beyond representing only the author's opinions on the subject and not committing the Bariloche Foundation, aims to analyze the hydrocarbon policies applied in Argentina in the period 2003–2014 and in particular Law 27,007. To do so, first a very brief look is taken at the previous historical periods, since a large part of what is happening in the present is influenced by the past, which it improves or worsens and almost never leaves as it was. Afterward, the regulations that supported the policy for 2003–2014 are analyzed, and then we will look at some essential aspects of this policy. We will continue with the analysis of Law 27,007 of Hydrocarbons of 2014, which reforms the mother law 17,319 of 1967. The following is a criticism, which we believe is constructive, of this law 27,007. In order not to remain in the criticism, some ideas are presented regarding the contents of a new Energy Policy, perhaps an old one, because it gathers the successes of a generation of Argentines, including Mosconi, Huergo, Scalabrini Ortiz, Illia, Silenzi de Stagni, Juan Sábato and others who showed us and followed a path, perhaps, for us to continue.

Keywords Energy policy · Hydrocarbon law 27,077 · Privatization of YPF · ENARSA · Refining · Expropriation YPF · Liquefied petroleum gas

1.1 Introduction

One of the objectives of this work is to analyze the recently enacted National Law 27,007 called the "Hydrocarbons Law." The other is to analyze in particular the hydrocarbon policy of the Kirchnerist governments, i.e., the governments of Mr. Néstor Kirchner and his wife, Cristina Fernández, between 2003 and 2015.

Before this analysis of the Law, in particular and in general, we will mention which, in our opinion, are the reasons that led to its elaboration.

Without being exhaustive, we will refer to some of the antecedents (laws and decrees) that preceded it, essentially the National Law 17,319 of 1967, which this new law reforms, for which we will have to go back a little far in time.

But, first, a little bit of "politics".

V. Bravo and N. Di Sbroiavacca, *Oil and Natural Gas Economy in Argentina*,
The Latin American Studies Book Series,
https://doi.org/10.1007/978-3-030-65520-4_1

Since the beginning of the twentieth century, after the commercial discovery of oil in Argentina there have been two issues that have been the basis, for good or for bad, of the laws and decrees that have regulated hydrocarbon activity in Argentina:

(a) the role of the state and the role of the private sector;
(b) the national or provincial ownership of the hydrocarbon resources found in the subsoil.

Of course, everything is subordinated to a larger issue, the appropriation of the income and profits that this activity generates.

The economic value of oil and natural gas reserves, as of 2013, at December 2014 international prices, reached US $214 billion dollars, not including the non-conventional resources that are not yet proven.

We are not going to make a critical description of the oil industry in Argentina. For those who want to know more about the subject, I would like to refer to two texts that contain my thoughts, since there are other authors who present different conceptions:

"Legislación de los Hidrocarburos en Argentina": Víctor Bravo, Publication of the Bariloche Foundation, year 1971.

"The Policy of Oil Deregulation in Argentina, Background and Impacts": Roberto Kozulj-Víctor Bravo—with the collaboration of Nicolás Di Sbroiavacca—University Libraries Latin American Publishing Center 1993.

In our Latin American region, from the antecedents of the Law of the Indies, the mineral resources of the subsoil belong to the State and not to the surface owners (unlike what happens in the United States of America). The State's ownership of these resources has never been questioned, at least in law, but not in fact.

There have been differences in the ownership of these resources by the National State or the Provincial States, Argentina being, at least constitutionally, a Federal Republic.

Going through history, we find that these conditions have changed, firstly, to Provincial Property, then National Property and now again Provincial Property (corresponding to the Provincial State in which the resources are located) and to the National State (in the Continental submarine shelf).

Both concepts have their pros and cons.

Argentine history has always shown a tendency toward centralization and the provinces have felt disadvantaged.

First, the National Law 24,145 of 1992, then the Article 124 of the reformed National Constitution of 1994 and then the National Law 24,197 of December 2006, attributed to the provinces the original dominion of the natural resources existing in their territory. Manuel Rebasa, in an article published in the newspaper Página 12 (Buenos Aires) on June 1, 2014, distinguished between dominion and property, considering that property corresponds to the National State and dominion to the Provinces, making articles 41 and 75 of the National Constitution prevail, which would authorize the Nation to establish the general policy for hydrocarbons and the regulation of natural resources.

In my opinion, and beyond what the laws may say, the conception that grants the National State ownership of the hydrocarbon resources existing in the subsoil of the continent and of the Submarine Platform should prevail.

But this does not mean that the Nation appropriates all the profits generated by the exploitation of the resource. The provinces in whose territory these resources are located must participate in these benefits. This goes beyond the desire to compensate them only with royalties.

The Nation must, via Congress where all the provinces are represented, whether they are hydrocarbon producers or not, elaborate a serious "Hydrocarbon Law", not a patch like the one we are considering (a patch, moreover, that is very influenced by the situation, when a law must go beyond the short term) or an Energy Code, as the Mining Code exists, for which the National Congress is the ideal place to discuss it.

The Congress, that is, the Nation and the Provinces, must establish the sector's operating standards, which are uniform throughout the country, setting the essential role of the State through a public company (in this case, the State-owned YPF, not a corporation as it is now, which allows it to be governed by Chapter II, Section V of National Law 19,550, (among other things, not to be covered by the regime of access to public information) and also to recreate the State Gas Company.

The role of the State would then also be defined.

These companies would be in charge of executing the hydrocarbon policy and surely in this way the internalization of the benefits of the activity would be assured, leaving to the private sector the important task of providing the equipment and materials that the industry requires (that is to say, the investments of these public companies would be the source that feeds the activity of the private sector). For this purpose, they would have the purchasing power of the state-owned energy companies, which can vary between US $15 and 20 billion dollars a year. These figures this would arise from the energy plans that are drawn up.

The foreign private sector would be used to acquire knowledge and technology that the country is not in a position to provide.

The provinces, in addition to participating in the preparation of legal instruments, would form part of the boards of directors of State enterprises and would decide on and control the actions that these enterprises carry out.

The Provinces, because they are the owners of the domain and not the property of the hydrocarbons, would receive a proportional part of the benefits that the companies would obtain. Today, the benefits are received by the private partners of, for instance, YPF S.A., since they are owners of 49% of the shares, as a bonus for entering the areas of the state companies, and of course the percentage of royalties that correspond to the characteristics of the deposits.

In this way, the provinces would not become renting entities but partners.

The fact is that today, with the contracts that the provinces formalize with private companies, via concessions (which in the new National Law or Code, which we are herein proposing, should disappear as a contractual figure), they usually seek, almost as their only objectives, the amounts that the right to enter the areas implies, royalties and other taxes, and to that often subject the environmental conditions, the habitat of the surface areas (usually native peoples) and other facilities so that the offers of some provinces are more attractive than those of others, unhealthy competition that is detrimental to the country as a whole.

Unfortunately, the recent history of Argentina shows examples of this type of behavior.

1.2 Why It is Needed a Prompt Elaboration of the So-Called Hydrocarbon Law

The main reasons given by the proponents of the law are the following:

The National Law 17,319 that currently governs the activity is very old, dating from 1967, and it does not reflect the current situation. This is true.

Argentina is importing (the estimates are for the year 2014) almost US $12,000 million dollars per year of fuels, which represents almost 16% of the total exports of the country. The deficit of the balance of fuels is almost US $7000 million dollars which affects the total commercial surplus which is only US $9000 million dollars. In other words, high fuel imports in a country that has a great shortage of dollars or any other hard currency. This is true, but these consequences must be sought in the causes, essentially, the privatization of YPF, the historical state-owned oil company. During the decade of the 1990s, the reserves of the oil and gas deposits, which had been almost all of them discovered by the old YPF, were squeezed to the maximum, due to the application of an irrational export policy, without thinking on the future hydrocarbon supply of the country. Additionally, exploring for oil and gas was almost stopped, with negligible investment.

The existence, as estimated by international organizations such as the International Energy Agency in a report of June 2013, of considerable resources (not reserves) of hydrocarbons in the national subsoil, contained in non-conventional formations of shale and tight, which would multiply by 11 those reserves of oil and by 67 those of natural gas. See observations to this estimate in two works published by Fundación Bariloche:

"Shale oil and Shale Gas in Argentina. Estado de situación y Prospectiva", by Nicolás Di Sbroiavacca, Fundación Bariloche, 2013, Working Documents; in Spanish. In English in this volume.

"Una opinión sobre fracking", by Víctor Bravo, Fundación Bariloche, 2013, Working Documents; in Spanish. In English in this volume.

These documents indicate that the reserves, that is to say, what is really recoverable from these hydrocarbon resources, could reach in the case of the Vaca Muerta oil field (Neuquén Basin, northern Patagonia) only 3% of the current oil reserves and 0.2% of the natural gas reserves. Then, we should be careful. At the end of 2014, the oil production of the Vaca Muerta oil field represented about 2% of the total hydrocarbon production of Argentina.

The interest to invest in Argentina was awakened in international oil companies to put those resources in value. For this purpose, there is an Argentine company,

YPF S.A., with which they are associated or want to be so. This is the way to insist on having a profitable country which puts the natural resources, and foreigners that bring in capital and sometimes technology, but takes a great part of the benefits.

Since the investments required to valorize these non-conventional deposits are considerable and we are talking at least about hundreds of million dollars, the country is not in a position to contribute them on its own and the weight of fuel imports is growing, we must generate stable and attractive legal instruments for these international companies to come and invest in Argentina. This could lead to neglect the exploration and extraction of conventional hydrocarbons, which are cheaper than non-conventional ones, as it happened between 1993 and 2012, when there was almost no oil and gas exploration in Argentina.

YPF S.A.'s contract with Chevron Company to explore and exploit the Vaca Muerta oil field, already formalized, should be generalized and serve as an example for other similar types of contracts. This will require a framework law. See observations on this contract in a paper by Fundación Bariloche "Una opinión sobre el fracking", by Victor Bravo: Fundación Bariloche Working Documents, 2013, in Spanish. Also in English in this volume.

1.3 Some Legal Background and Other Circumstances Prior to the Full Privatization of the State-Owned YPF Company

1.3.1 From the "Origins" to the "Menemism" Period

We will not make a detailed description of these legal instruments but a synthetic presentation of those that we consider most important for the purposes of this work.

The legal framework for the hydrocarbon industry in Argentina begins with the Mining Code (which dates back to 1886 and has undergone several modifications since then). In it, it was considered that the subsoil resources were property of the Nation or of the Provinces according to the nature of the territory where they were found.

The National Law 12,161, of 1935, modified the Mining Code (YPF Sociedad del Estado had been created in 1922) and established an eclectic regime that allowed the State to compete, have private participation through concessions and participate in the constitution of mixed companies. This law confirmed the provincial ownership of the subsoil resources.

The National Constitution of 1949, which in its Article 40 established that the hydrocarbon fields were indispensable and inalienable goods of the Nation, as opposed to the "provincialist" position of the National Law 12,161 of 1935. This constitution was repealed by the Coup d'état of 1955, returning to the prescription of the 1935 National Law.

In 1958, during the government of Arturo Frondizi, a new "Hydrocarbons Law" was passed, named as 14,773, which established the national domain of these fuels and delegated responsibility for oil and gas activities to both the state-owned YPF and Gas del Estado, respectively. This was the time of the so-called oil contracts, later cancelled by the government of Arturo Illia (1963–1966).

The de facto government, which emerged from the 1966 coup until 1972, made a 180° turn in Illia's oil policy and in 1967, a new hydrocarbons law, 17,319, was promulgated. With various modifications, this Law is still in effect at the time of writing this paper. This law allows the operation of the state companies coexisting with the private companies in all the range of contractual figures: concession, association, services. In general, several articles clearly show the two conceptions that have oscillated supported oil policies in Argentina: one that tends to accentuate the predominance of private company interests and the other that tries to preserve the country's interests through state companies. In fact, during the de facto period, the interests of private companies fully prevailed. These interests would accentuate their predominance during the entire period of the 1976–1983 military dictatorship. With regard to the ownership of the resources of the subsoil, the provisions of National Law 14,773, replaced by National Law 17,319, remained in force, that is, the ownership of the resources was assigned to the National State.

The democratic period established from 1983 did not fundamentally modify the inherited legal instruments. As an example, the Decree 1445 of August 5, 1985, established the Standard Contract for the associations of private companies with YPF. Also, Decrees 1443/85 and 623/87 launched the Houston Plan and later the Huergo and Olivos Plans 1 (Decree 1812/87, in response to the failure of the risk contracts under National Law 21,778/78. With Mr. Rodolfo Terragno as Minister of Energy, the Petroplan was launched, which would be the precedent for oil deregulation that took place in Carlos Menem's times. These instruments were not successful in the modification of the functioning of YPF and in actually increase the production, despite the fact that the objective was to favor the state companies and to internalize more of the oil income thus generated.

1.3.2 *The "Menemist" Period*

We will analyze here the essential legal instruments and we will stop especially in the privatization of the state-owned YPF:

The government of President Carlos Menem, which began in July 1989, was going to provoke the greatest surrender of Argentina's energy heritage in history. They were going to use National Law 17,319 which would allow it to do so given its "ideological" scope. Menem's policy considered as anachronistic that hydrocarbons were a strategic asset. They said that the international oil market was open and that if there was any excess of oil, it had to be exported and if there was any shortage, oil had to be imported. What was significant was the Economic Value but not the Strategic Value. Thus, the concept that had prevailed in the country in general was left aside, in this matter, which was linked to a conservationist production policy that implied extracting only the crude oil and gas required to satisfy the needs of the internal market and to preserve future generations from shortages and the weight of imports, starting from the premise that Argentina "was a country with oil but it was not an oil-producing country". We are not going to describe here the consequences that the effects of these policies had on the country and its inhabitants (loss of oil and natural gas reserves, massive imports of both hydrocarbons and a very unfavorable impact on the balance of trade and payments and the disappearance of the state energy companies as a whole). No government had ever dared to do so. These effects are carefully analyzed in the book: "La política de desregulación petrolera argentina y sus Impactos" by Roberto Kozulj and Víctor Bravo with the collaboration of Nicolás Di Sbroiavacca, from the Centro Editoral de América Latina, 1993.

The main legal instruments to establish the policy of the "1990s" were of two types: one that we will call general instruments (National Laws and Decrees that implied the reform of the State: the National Laws 23,696 and 23,697 of 1989, and 23,928 of 1991 and the Decrees 1224, 1225 of 1989 and 1757 of 1990 and 2408 of 1991) and those instruments specific to the oil sector (Decrees 1055, 1212 and 1589 of 1989 and 2278 of 1990).

We are going to refer a little more closely to the so-called Three Decrees, since some of them will remain in force after the approval of National Law 27,007 of 2014, which we are considering in this volume.

Firstly, Decree 1055/89, which essentially referred to exploration and exploitation activities: it tendered under the figure of a concession the poorly named, Marginal Areas; it tendered under the form of an association with YPF the Central Areas; it established the free availability of crude oil, regulated aspects of foreign trade, established the role of the producing provinces and indicated that YPF should return areas that it had in its power to explore.

Then, Decree 1212/89 whose main objective was to replace the State with "market rules" as a mechanism for setting prices, margins and tariffs. To this end, YPF had to renegotiate contracts with private companies to which it released the import of crude oil from payment of duties and established free prices for the domestic market.

Finally, Decree 1589/89: Mainly allowed revenues for crude oil and derivatives dumped in the domestic market to have the same income as if they had been exported and natural gas would have a price equivalent to a percentage of the price of Saudi light oil; 70% of exports would be freely available and if sold in the domestic market, they would enjoy the same prerogative as if they had been exported.

Decree 2778 of 1990 transformed YPF from a State-owned company into a joint-stock company, reducing the book value of its assets; removing powers from the new YPF and paving the way first for its fractional sale and then for its total privatization.

1.4 The Privatization of YPF

We must remember the economic context, which we are not going to describe here, just to mention that in April 1991 Domingo Cavallo arrived at the Ministry of Economy and established the Convertibility Plan through which the Argentine "peso" was to have a "one-to-one" parity with the US dollar, and that parity would be guaranteed by the State. This parity would last for eleven years. It promised stability, growth and especially privatization of all state enterprises whenever this was possible.

These were the reasons given for the privatization of YPF. See the article by Víctor Bravo, "Why privatize YPF", Revista Realidad Económica No. 110, September 30, 1992.

The cited article mentioned five reasons that have been refuted by me one by one and I will refer to them here.

Chronic budgetary imbalance;
Business inefficiency;
Imperative need of the National State to obtain funds to balance the budget;
Fear of the State-owned company of competition from private companies;
Lack of need for the State to have an oil company.

Privatization was carried out in several stages and it started with the PETROPLAN during the Rodolfo Terragno administration in March 1988.

The strategy of eliminating YPF continued from 1992, in various stages. In the first stage, YPF was reduced in size through the sale of assets (privatization of new central and marginal areas; sale of distilleries; sale of oil pipelines, port terminals and storage plants; concession for the wholesale marketing of fuels; sale of drilling equipment; sale of the Technological Assets Department located in the town of Florencio Varela, Buenos Aires province, which was postponed and would later be closed). In the second one, what was left of YPF was privatized.

The National Law 24,145 of 1992 was called the Law of Federalization of Hydrocarbons and the Privatization of YPF. This law transferred the domain of the deposits to the provinces, which would take place upon the expiration of the legal terms of the then current concession and association contracts. This is more a way to attract provincial votes for the privatization of YPF than a real federalization of hydrocarbons. The process of selling YPF was not very clear (see in this respect what is indicated on pages 228–230 of the aforementioned book "The Policy of Oil Deregulation…").

The Statute of YPF S.A. was contained in National Law 24,145. A Decree 1108/93 regulated the acquisition of shares by the National State in such a way that it was practically impossible for the State to control the company.

In 1994, the new, reformed Argentine Constitution was enacted, which in its article 124 established that the provinces were responsible for the original control of the natural resources existing in their territory and which gave constitutional legitimacy to Law 24,145 on the Federalization of Hydrocarbons.

National Law 24,474 of April 11, 1995, modified the "Federalization of Hydrocarbons Law" so that the capital of YPF S.A was represented by four types of shares. The A shares belonged to the National State; the B shares, which could be acquired by the provinces, up to 39% of the capital; the C shares, which could be acquired by the company's personnel, up to 10%; and the D shares, which could be sold by the National State and the provinces to private capital. The State would assume all the existing debts of YPF Sociedad del Estado (S.E.) and YPF S.A. as of December 1990 and December 1991, but since the balance sheets of those years had been positive, the sale was carried out free of all liabilities. The provinces would have access to the shares by crediting the debts for hydrocarbon royalties. The company's personnel would not be able to access the percentage with their own resources.

In July 1993, another package of shares was launched for sale, in order to cancel debts of the pension system. The shares were "sold" to retirees in exchange for their consent and silence.

After the first stage of sale of YPF, the stock package was constituted as follows:

National State: 20%;
Provinces: 12%;
Retirees: 12%;
Personnel: 10%;
Private Investors: 46%.

The National Law 24,474 of March 22, 1995, initiated the second stage of the privatization of YPF S.A. It modified Article 8 of the National Law 24,145 of 1992, on the Federalization of Hydrocarbons and included details on the characteristics of the A, B, C, and D shares.

The National State, as long as it had a class A share, should be consulted by the shareholders. For example, if 51% of the shares were to be sold to third parties, or if exploration work was to cease, or if the company was to be dissolved.

Therefore, as of February 1996, the stock of YPF S.A. was distributed as follows:

Private: 65%;
National State: 20.3%;
Provinces: 4.7%;
Personnel (in custody per National State): 10%.

In 1996, the so-called Second State Reform proposed to sell the type A shares of YPF, still in the hands of the National State.

In 1997, through Decree 628 of July 1997, the sale of 10% (class C shares) in the hands of YPF personnel began, as a consequence of the Participated Property Program. The 85% would be sold to holders outside the country and the 15% would remain in the domestic market.

The remaining 20.3% was to be sold to the National Government. To facilitate this, Decree 857 of July 1998 would be issued, which would return to the National Government the right to receive dividends from class A shares. Some members of the government favored an atomized sale of these shares to avoid a concentration of power in a few hands. But the Spanish group REPSOL appeared in the scene offering better prices than other potential interested parties. Furthermore, the statute of YPF S.A. did not allow the State to divest itself of more than 15% of the shares in its possession. But the small shareholders would accept it and it would be possible to obtain 66% of the votes necessary to change the YPF statute. In any case, only 14.99% of the shares of the National State were sold and 15 companies would be selected, among them the Spanish REPSOL. The procedure had to be hurried because in October 1999 there would be a change of government and the Alliance Party, the main candidate to win the elections, did not agree much with these sales. In fact, the sale was carried out by the Menem government, not for regional development as promised, but just to cover the deep fiscal gaps. On January 20, 1999, REPSOL bought US $14.99 million dollars of the company, contributing with US $2010 million dollars. But REPSOL wanted to reach the whole control of the company, for which it aspired to hold 51%. The national government still had 5.3% and to facilitate the sale in July 1999 it would be authorized to divest itself of this percentage by Decree 666. At that time, REPSOL would present a hostile offer to take over the whole of YPF. S.A., including the shares in the hands of the provinces of Santa Cruz, Chubut and Mendoza. The National State kept only a golden share and a member of the Board of Directors that would allow it to give its opinion and oppose a hostile sale of REPSOL-YPF to third parties. In July 1999, REPSOL paid US $13,150 million for all the shares and thus it kept YPF S.A. REPSOL was able to do so by means of a huge loan which it obtained from two Spanish banks (which were also part of REPSOL) and four international banks and finance companies. Most of the money that the private shareholders achieved from the sale would be invested abroad.

The Alliance Government, in its short mandate (from December 10, 1999 to December 10, 2001), did not produce any further changes regarding Menem's energy policy.

1.5 The Period from 2003 to 2014

The economic and social context of the two years following the fall of the Alliance Government would be very difficult and would be strongly marked by two facts: the devaluation of the Argentine peso that would move from one parity with the US dollar, one to one, to another one of three, and by the country's declaration of the impossibility of complying with the payments of the external debt (i.e., Default).

Since the fall of the Alliance Government, there have been four presidents in Argentina, being Eduardo Duhalde one of them, who changed the parity between the peso and the US dollar.

In 2003, Mr. Néstor Kirchner took office, followed by his wife Cristina Fernández in 2007, with a second term for her, starting in 2012 and ending in 2015.

As we will see, the main milestones in terms of hydrocarbon policy would be: the creation of trust funds; the establishment of export rights to crude oil, its derivatives and natural gas; the creation of ENARSA, a State-owned energy company; the granting of more "rights to the oil provinces"; the Total Energy Plans; the Oil

Plus, Refining Plus and Gas Plus programs and the implementation of the so-called Hydrocarbon Sovereignty.

Collaterally, some measures were taken regarding liquid petroleum gas (LPG); renewable energy sources, especially those related to electricity generation and the rational use of energy.

It was not until 2012 that an important, although not essential, modification of the hydrocarbon policy of Menem's time was attempted, with the expropriation of REPSOL's shares in YPF and recover the control of 51% of this company by the National State; the exploitation of non-conventional hydrocarbon resources and the possibility of the massive entry of foreign companies, facilitated by National Law 27,007 in October 2014, known as the "Hydrocarbon Law", which we are analyzing here in this chapter.

When Néstor Kirchner took office in 2003, the country was a net exporter, mainly of crude oil and natural gas. The reserves of both hydrocarbons were falling but were not yet a concern for the new government. REPSOL had practically stopped exploring; from the almost 120–150 wells/year that YPF S.A. was drilling, REPSOL had dropped to an average of only 29 wells in the period 2001–2004. Their only concern was to produce in order to recover as soon as possible, through exportable balances, the US $15 billion dollars spent on the purchase of YPF. But oil production was severely decreasing and with it, the exportable balances.

In other words, there was no actual evidence of an "energy problem"; the economic and social crisis and the effects of the cessation of payments of the foreign debt were then the real problems.

1.5.1 The Trust Funds

This topic would provide by itself enough information to write a separate document. Therefore, we will only refer here to the general concepts.

Based on the Article 75 of the National Budget Law 25,565, the Trust Fund was created to subsidize residential gas consumption for the Patagonian Region, the Malargüe Department in the province of Mendoza and the Puna, an isolated region of northwestern Argentina. The fund was perceived by the zonal distributors of natural gas and LPG. The Fund was constituted with a surcharge to the cost of the cubic meter of natural gas to all types of users in the rest of the Argentine National Territory.

This type of subsidies had been in place for a long time in Argentina.

Later, and for this same purpose, other laws were enacted, such as the National Law 26,337, decrees (as 786 of 2002 and 180 of 2004) and resolutions (such as 5 of 2010 and 209 of 2014).

But it was essentially the National Law 26,095, of April 26, 2006, that created specific charges for the development of energy infrastructure for the expansion of the system of generation, transport and/or distribution of gas and electricity services, which became trust funds for the use of the resources created.

This type of mechanism became more widespread and the government assumed that it could replace the role played by state-owned energy companies before the privatizations of the 1990s. To this, and in a contradictory way, the freezing of natural gas and electricity tariffs were added, distressing the most genuine form that companies have to finance a relevant part of their investments. With the exception of electricity transmission, the results of this policy paralyzed or drastically reduced investments in electricity and natural gas infrastructure and were responsible for the problems of supply shortages of these energies that manifested themselves in much of the period from 2003 to 2014.

1.5.2 Export Duties

In the early 2000s, oil and natural gas exports were important. Thus, in 2002, oil production reached almost 44 million m^3 and exports stretched to 15.28 million m^3 (the peak was achieved in 1998 with 19.2 million m^3). In other words, almost 35% of what was produced was exported. Something similar happened with natural gas. Almost 15.5 thousand million m^3 were produced and 5.85 thousand million m^3 were exported (the peak was reached in 2004 with 7.35 thousand million m^3), which was equivalent to almost 13% of production.

In 2001, these combined exports represented almost US $2.6 billion dollars.

Consequently, it seemed appropriate and necessary that the impoverished National State decided to apply export duties to hydrocarbons.

The Decree 310 of February 2002 established the first export rights of the period, setting them at 20%, based on the Article 6 of National Law 25,561 of Economic Emergency and Decree 806 of May 2004 by President Eduardo Duhalde. The different regulations were essentially based on setting cut-off and reference values for crude oil and establishing aliquots from which the amount of the duties would be deducted.

A series of subsequent Resolutions (809 of 13 May 2002; 645 of 26 May 2004; 526 of 22 October 2002; 335, 336 and 337 of 11 May 2004) ratified the validity of these export rights.

In August 2004, Resolution 532 maintained the twenty-five percent export duty. In case the price of WTI crude was equal or higher than a certain amount in US $ dollars, additional rates would be applied. The additional fees could result in an export duty of 45% if the price was equal to or higher than US $45 per barrel (in August 2004 WTI crude was quoted at US $44.84 of that year).

Subsequently, a series of laws and decrees forced this mode of determination of the export rights to subsist, for example, National Law 26,217 of January 15, 2007.

Resolution 394 of November 15, 2007, was generated by the then Secretary of Domestic Commerce, Mr. Guillermo Moreno and led to exporters receiving US $42 dollars per barrel exported (as long as the international price exceeded US $60.9 dollars). But the price of WTI crude was close to US $100. This was not accepted by Pan American Energy, Argentina's main oil exporter. Resolution 1 of 2013 of the

Ministry of Economy updated the cut-off and reference values for oil exports and this improved somewhat the income of the exporters.

But from October 2014 onward, the prices of WTI crude began to fall and this led the Ministry of the Economy to issue Resolution 803 on October 22, 2014, which established a relationship between the international price and the rates more in line with what those international prices showed in October 2014. Thus:

When the international price was less than US $80, the rate to be applied would be 13%, when it was less than US $75.0, 11.5% and if it was less than US $70, 10%.

Between October and early December 2014, WTI prices in current dollars ranged between US $90 and 55, with a downward trend.

Before this resolution, exporters who previously received US $42 per exported barrel received around US $70 in 2014.

The prices on the domestic market for the Escalante crude (the type of Argentine oil exported by Pan American Energy), between July and August 2014, were close to US $68.7 dollars and the Medanito type close to US $84 per barrel.

In short, when oil exports were very high (between 1996 and 2002), export duties did not exist at all.

More recently, with crude oil exports not reaching 2 million m^3 (in the year 2013), compared to 19 million m^3 in 1998, export rights have become "a relic of the past". The problem now is not hydrocarbon exports but imports, especially of natural gas with almost 12 billion m^3 in 2013 (almost 25% of the total consumption).

The most serious thing is that crude oil was exported at less than US $22 in 1999 per barrel and it would be imported in 2014 at no less than US $60–70.

Then, the situation would be reversed and import quotas would be established (essentially gas oil and gasoline) that would not pay taxes on fuels or similar (Resolution 99 of 2013).

Argentina was importing (estimates are for the year 2014) almost US $10 billion dollars per year of fuel, which represented almost 14% of the country's total exports.

The fall in oil prices in November 2014 would have improved the import bill somewhat, but the domestic price of Argentine crude oil (about US $84.70 per barrel in the case of the Medanito type and US $68.5 for the Escalante type) was above the international price of WTI, which in the last week of November 2014 was slightly below US $65. As a result, the export rights would be at 10% and with a clear tendency to disappear.

At the end of December 2014, WTI crude was slightly below US $55 a barrel and the government decided to take some actions, which were the following:

Reduce the domestic price of crude oil by about US $7.0 per barrel;

Stimulate oil producers by giving them a bonus of US $3.0 per barrel, if they increased the hydrocarbon production;

Also a subsidy of US $3.0 per barrel if they increased exports.

These two movements essentially benefited YPF S.A. and Pan American Energy.

The mobile retentions to oil exports were established, as Resolution 803 of October 2014 did, according to the variation of international prices. In December 2014, they did so at about US $55 per barrel with a 10% withholding).

The price of motor oil and gas oil would be reduced by approximately 5%.

In order to facilitate the actions of the oil companies at the end of 2014, Resolution 1077 of December 29, 2014 and Decree 2579 of December 30, 2014 were endorsed.

Resolution 1077 repealed Resolution 394 of 2007 and established a new reference crude oil rate, the Brent oil, with a slightly higher price than WTI, which it replaced in order to establish the rates of export rights for crude oil and its derivatives. The cut-off value was 71 U$S/barrel and in fact the export duty became almost symbolic, close to only 1%, while the Brent values did not increase much above that cut-off value.

Decree 2579 reduced the values of the motor oil and gas oil taxes by 10% and the water tax by 20%.

1.5.3 ENARSA

Since the privatization of YPF, Argentina had been left without a state-owned oil company, which has not been the case in any of the Latin American oil-producing countries.

The promised benefits that the country was going to receive as a result of the mentioned privatization had not taken place, and on the contrary, oil production was decreasing and something similar was in sight for natural gas, which was causing a drop in exports of these hydrocarbons, especially oil.

The Kirchner government did not want or did not consider necessary or could not nationalize YPF in the hands of the Spanish REPSOL.

But something had to be done and it enacted National Law 25,943 on October 20, 2004, through which it created the Empresa Energía Argentina Sociedad Anónima (ENARSA).

This company was to act in all energy activities, not just those related to hydrocarbons.

As practically all existing hydrocarbon areas in the Argentine continent were in private hands or in the hands of the provinces, the Continental Submarine Platform was assigned to ENARSA, so that through exploration permits and exploitation concessions it could activate the potential in hydrocarbons that may exist there.

Another responsibility of the new company would be to avoid dominant positions in the business by the private sector.

But perhaps the most outstanding, due to its later consequences, was to grant it the management of the country's hydrocarbon database, which, in fact, was in the hands of the Argentine Institute of Oil and Gas, which sold the information gathered. In other words, this was not freely available as it had traditionally been before the Menemist era, when it was centralized by the Secretariat of Energy. With this purpose, ENARSA called for a tender from companies to take over the task. Awarding the call to a company, apparently linked to Mr. Cristóbal López, which in turn is involved in oil business, has given rise to questions, criticism and complaints.

"The explicit purpose of the initiative was to provide the State with a comprehensive map of oil and gas wealth. This large digital file had certain degrees of confidentiality. Argentina would thus imitate many countries that have systematized geological information to facilitate oil companies in the exploration and exploitation of the subsoil. But what happened was very different. Before the public sector received them, the digital records were in the hands of Cristóbal López, which gave him privileged information" (Source: La Nación newspaper, Buenos Aires, Thursday, November 27, 2014).

I must use this opportunity to pay tribute to Engineer Hector Piegari, who, without computer resources, which were not available in his time, as Director of Statistics at the Secretariat of Energy, had made Argentina the most advanced country in this area and published the Fuel and Electricity Yearbooks that made data from all the energy chains accessible to everyone.

ENARSA came out as a corporation with 52% of the shares in the hands of the National State, 13% in the hands of the Provinces and 35% traded on the stock exchange.

Ten years after the creation of ENARSA, we can confirm that it did not fulfill any of the objectives that motivated its creation.

It was not a company with "productive" activities; it did not have the personnel or the facilities to do so. In fact, it can be defined as a "business unit", which was associated with private third parties that were the operators and ENARSA was just a mere partner.

It tried to establish partnerships with some companies to explore the Continental Submarine Shelf but these partnerships did not lead to positive results and were diluted. Today, the National Law 27,007 of 2014 removed ENARSA's ownership of the Submarine Platform.

When it became necessary to import liquid natural gas (LNG), ENARSA was granted the management of such imports. In 2014, apparently due to an unclear and blurred management of the issue, the responsibility passed to YPF S.A.

We believe that with the creation of YPF S.A., ENARSA's subsistence is today totally unnecessary and probably never will be.

1.5.4 The Role of the Provinces

It has already been mentioned that in Argentina the ownership or dominion of hydrocarbons found in the subsoil has oscillated between a "nationalist" position (i.e., the predominance of the National State) to a Federalist or Provincialist one (where the Provinces enjoy dominion over the hydrocarbons contained in their subsoil and Nation over those existing in the Continental Shelf beyond 12 miles measured from the coast). Since the sanction of National Law 24,145 of October 1992, and more firmly since Article 124 of the National Constitution of 2004, the balance seems to have tipped toward the provincialist position.

But laws are one thing and facts are another.

Let us then briefly describe the trajectory.

The Decree 1955 of November 4, 1994, defined the exploration areas to be transferred to the provinces.

It was not until 2003 that Decree 546 of August 6 recognized the right of the provinces to administer their energy resources and to grant exploration permits and exploitation concessions.

The National Law 26,154 of 27 October, 2006, created a promotional regime for exploration and exploitation. The benefits would last 12 years in the existing areas in the Continent that were not in production; 10 years if they were in production and 15 years if they were in the Platform (in this case, the Application Authority would be ENARSA). In all cases, the new private permit holders and concessionaires, arising from the contracts granted by the provinces, were obliged to join ENARSA if they wanted to enjoy the benefits of the law. This law established different terms for the benefits. These benefits had quotas assigned so as not to affect the National Treasury too seriously much, and referred to special treatments regarding VAT; income tax; minimum presumed income; import duties on equipment and machinery.

In order to receive the benefits, the private partners had to comply with the committed investment plans.

The National Law 26,197 of January 2007 confirmed the provincialist thesis and the provisions of National Law 24,145 of 1992, repealing what was established, in the opposite sense, by National Law 17,319 of 1967. In addition, an agreement on the transfer of information on hydrocarbons between the nation and the provinces was established.

National Law 26,360 of April 8, 2008, created a transitory regime for the treatment of the sources of financing of the purchases of capital goods destined to the hydrocarbon activity to be carried out between October 2007 and September 2010. This regime established differential treatment to the known concepts of Income Tax; accelerated amortization; early return of VAT and set a tax quota, for these purposes of 1 billion Argentine pesos.

But all these actions leading to give a greater role to the provinces in managing the policy for hydrocarbons, did not give the expected result, both in terms of a significant contribution of capital, as to stop the decline in production of oil and natural gas, which were the objectives. Thus, oil production fell from 43.95 million m^3 in 2002 to 36.6 million m^3 in 2008, and natural gas production fell from 52.4 million m^3 in 2004 to 50.5 million m^3 in 2008, while exports of both types of energy decreased.

Once again it was verified that the ownership of the domain between Nation and Provinces was not the problem to be solved, as it will be seen later.

Once again, the fatal error of privatizing state-owned companies and the ineffectiveness of the ENARSA patch to replace them were being confirmed.

1.5.5 Total Energy Plans

The Resolution 459 of July 12, 2007, created the Total Energy Program, within the scope of the Ministry of Federal Planning, Public Investment and Services, which was to encourage the replacement of natural gas and/or electric energy consumption by the use of alternative fuels for different productive activities and/or electric self-generation.

As already mentioned, in 2007, natural gas represented almost 51% of the gross domestic energy supply; almost 63% of the energy used to generate electricity; almost 46% of energy consumption in the industrial sector and almost 63% of residential energy consumption.

That is to say, natural gas was the most consumed energy source, when on the other hand, local production was decreasing and exports and the level of proven reserves were falling. It was warned then that if the level of consumption of natural gas continued as it was, it would be necessary to import gas to satisfy the internal demand.

In order to comply with the objectives set forth in Resolution 459, 930 million Argentine pesos were allocated to compensate for the higher costs incurred by companies that replaced natural gas or electricity.

This program absurdly had a duration of 90 days (I say absurdly because the replacement of one energy source by another cannot be carried out in such short term).

Perhaps this is why Resolution 121 of 3 March, 2008, extended the validity of the Plan until 21 December 2008.

Decree 2067 of 27 November 2008 created a trust fund to attend to natural gas imports and any other necessary to complement the injection of natural gas required to meet national needs. The trust fund was to be made up of, among other things, charges on the tariffs that would affect certain users of natural gas and electricity.

Resolution 1451 of 12 December, 2008, would regulate the use of the trust fund mentioned in the previous paragraph. The trust body would be ENARSA and the trustee would be a bank authorized by the Central Bank of Argentina.

Provision 287 of December 19, 2008, regulated the Total Energy Plan, extended it until December 31, 2009, and generated a series of secondary plans such as the following: Provision of liquefied natural gas (LNG) and its regasification; provision of propane gas for the networks of non diluted gas; provision of imported natural gas (surely coming from Bolivia); to ensure the supply of the demand for liquid fuels such as LPG; motonaphtas, gas oil and fuel oil; for the replacement of the consumption of natural gas or electric energy by other energy sources. These tasks were entrusted to ENARSA.

These different secondary plans were in force so far as ENARSA was not removed from them as it happened with the LNG plan that was passed to YPF.

We believe that this plan, with its different extensions, did not comply with the main objective, which was the substitution of natural gas consumption, since it went from representing 51% of the total energy supply in 2007 to 54% in 2012.

On the other hand, the copiousness of plans and trusts continued, which did not contribute at all to the enactment of a coherent and effective energy policy to solve the problems of the sector.

1.5.6 The Oil Plus, Refining Plus and Gas Plus Programs

The objective of these plans, as can be deduced from the respective resolutions, was to enhance oil production in the Argentine basins, especially those currently in production, and to increase the processing capacity of the refineries.

A very brief diagnosis of the situation of the oil market in Argentina in 2008 led to the identification of the following problems:

The sharp fall in the level of proven reserves and in production.

The decrease in investments in exploration reflected in the drop in the number of exploratory wells.

The concentration of reserves and production in the hands of a few companies, where REPSOL and Pan American Energy stood out.

The fall in exportable crude oil balances that affected the resources of the Ministry of Economy as a result of the consequent fall in income from retentions, previously by quantity and now also by price, and this in turn affected the budget surplus and the possible conversion of Argentina into a net oil importer in the very short term.

But it should be borne in mind that although 19.2 million m^3 were exported in 1998, because prices were very low, total hydrocarbon export revenues were US $2161 million dollars. In 2007, 3.3 million m^3 of oil were exported, but total income from hydrocarbon exports reached US $3141 million dollars, as a result of oil prices that were four times higher in 2007 dollars.

ENARSA's role has become that of a business management unit, with losses, rather than an energy or oil company.

The appearance in the oil business of new actors with Argentine capitals that seemed to recreate what happened with the Bridas and Perez Companc companies.

The irruption of the Provinces in the management of oil exploration, without a policy that would unify the conditions that each one proposed to the possible permit holders and concessionaires.

The extensions granted by several provinces (Chubut, Santa Cruz and Neuquén) to their oil production concessionaires in spite of the fact that in some of them, as in the case of the province of Neuquén, the committed investments had not been made yet.

The extension contract of Pan American (PAE) in the province of Chubut, which put as one of the conditions, in order to make effective its investment commitments, that the net price of crude oil at the wellhead actually received by PAE would not be less than the average price received by PAE in the province during the first quarter of 2006, when at that time the WTI price was US $63.20 per barrel, since at that price the price of Cerro Dragón oil was linked, when afterward the WTI price would be below US $45 per barrel.

The lack of investment for more than 10 years in new primary refining capacity and in conversion units partly explained the shortfalls in gas oil and fuel oil for the domestic market.

The growing imports of gas oil and fuel oil;
The concentration of refining in the hands of only four companies;
The concentration of sales in the domestic market of derivatives and in exports;
The situation of captivity of the expellers with respect to the oil companies;
The price policy of derivatives for the internal market artificially "frozen" until the end of 2007.

The retentions on crude oil exports and derivatives resulting from Resolution 394 of November 2007 that would allow during 2008, at least until September, to obtain an important income to the treasury, that would largely contribute to maintain the fiscal surplus and would serve to feed the enormous chain of subsidies to the energy and transport sector.

While the international price of WTI crude (Pi) was above the reference values (US $60.9 per barrel), the retentions were high (e.g., for a Pi = 104.78, average of September 2008, the State was left with 62.78 and the exporters with 42, cut-off value of the decree).

But for the average Pi of November 2008 (57.30), as this was lower than the Reference Value of the Decree (60.9), by applying a fixed rate of 45% to obtain the respective income, the State would have had 25.78 and the exporters 31.52.

The latter value was below US $42 per barrel as the price between the producers and refiners in the domestic market.

But if the value of the Pi was below US $45, as would happen in some days in December 2008, the State would have 90 days to apply the new percentages of retention to be applied.

In any case, the income from these retentions to exports would be very low and the aforementioned income effect would almost disappear.

As the quality differences between national crudes were great, the Ministry of Economy set a price, to be received by the oil companies, of US $42 per barrel for heavy crudes and US $47 per barrel for light crudes.

To aggravate the situation, it was very likely that Argentina in the not too distant future would become an importer of crude oil.

A good part of these problems had their cause in the sale and privatization of YPF, which took away from the State the essential tool to carry out its policies in the oil area, with the consequent loss of income that in other times served to finance the expansion of hydroelectric, nuclear and gas works.

The unfortunate history of the behavior of the Argentine oil market from the 1990s to the present time exempts from further commentary.

Always, depending on the investments of the private sector, the governments implemented different measures tending to give facilities to these companies so that they could explore, for example, the so-called Argentina Plan of the 1990s and the National Law 26,154 of the end of 2006, known as Incentives for Exploration. Both granted different facilities such as: the early return of VAT, accelerated amortization, exemption from the minimum presumed income tax, free import of machinery and equipment. But neither of them gave positive results.

These incentives were not very different from those granted by the current National Law 17,319 of 1967, in its Sixth Section: special fiscal regime, and they did not motivate the private sector either.

The only company that seriously explored in Argentina was the state-owned YPF. The private companies such as REPSOL and PETROBRAS concentrated only on oil extraction in order to recover as soon as possible the disbursements made for the "purchasing" of the YPF and Perez Companc areas, neglecting to ensure the internal supply with national crude for the future.

For its part, the provinces, with the approval of National Law 26,197 in December 2006, up to the end of 2008 had put out to tender 101 areas (some in basins that were not productive at the time) with investment commitments of around US $1.4 billion dollars, between 3 and 5 years. The results would not be seen if they were given before 2012.

ENARSA and the national government had subscribed investment commitments, of modest magnitude, with PETROBRAS, REPSOL; PAE and SIPETROL, to explore the continental marine basin and it would take no less than 5 years to have some ideas about the hydrocarbon potential of the basin. This brings to mind that in 1973, YPF had developed a plan to train people and investigate the marine areas, even acquiring a modern platform. The plan was thwarted by the 1976 military coup and the platform was leased to a third party).

In 2008, it was already evident that ENARSA, created in 2004, had failed to increase oil and natural gas production, recover reserves and avoid the inevitable future imports. Between 2004 and 2008, the mentioned concepts evolved as follows:

Concept	Unit (m^3)	2004	2008
Proven oil reserves	10^6	394	400
Proven natural gas reserves	10^9	541	398
Oil production	10^6	40.6	36.6
Natural gas production	10^9	52.4	50.5
Oil exports	10^6	10.1	3.3
NG exports	10^9	7.4	1.0

Source Secretariat of Energy of Argentina

Consequently, the government would implement the Oil Plus, Refining Plus and Gas Plus Programs to achieve these objectives.

In addition, the capacities and structure of the refineries had remained unchanged for many years.

The Ministry of Planning was in charge of the policies and the Secretariat of Energy was downgraded to a very secondary role. This characteristic would remain throughout the Kirchnerist period.

The Petroleum Plus and Refining Plus programs were created by Decree 2014 on November 25, 2008, for the exploration and exploitation of oil, to encourage production and the incorporation of oil reserves and fuel production, respectively. Those who signed the projects arising from the Programs would be granted a series of tax benefits considering them as basic infrastructure works in order to apply the provisions of Law 26,360 of 2008 which created a regime to encourage these investments.

Decree 2014 was regulated by Resolution 1312 of December 1, 2008, which created a follow-up commission to grant the benefits that were essentially tax credits. The condition for Petróleo Plus was to increase production on a basis corresponding to the years 2003–2007. The additional production to that base would receive the benefits. The tax credit certificates would be granted for the difference between the local price and the export price net of export duties and would be used for the payment of export duties on goods. Companies that exceeded a certain reserve replenishment rate would be granted additionally, tax credit certificates for TWELVE PERCENT (12%) of the difference between the local price and the value of the International Price pertinent to such export.

The Refining Plus program was aimed at new refinery projects and/or their expansion, and in a first stage, it sought to increase the production of gas oil and naphtha. The case of small refineries was also considered (those with a capacity of less than 30,000 m^3 month).

These programs would be suspended in 2012, leaving them without effect for the large companies, as a consequence of the partial expropriation of YPF-REPSOL. The companies involved were Pan American Energy, Occidental, Total, Sipetrol, YPF and Petrobras. In Refinación Plus, ESSO, Petrobras and YPF were damaged, as well as essentially the small refineries located in the city of Neuquén, northern Patagonia, such as RENASA and Petrolera Argentina.

These programs had been created with an internal oil price of US $35 per barrel and in 2012 the price was US $70.

According to the official version, during this period the State had granted fiscal benefits of around 10 billion Argentine pesos. These programs would have allowed the incorporation of reserves of 130 million barrels, in addition to the replacement of the annual production of each company. Likewise, the oil production of the beneficiaries of the program would have experienced an accumulated increase between 2008 and 2011 of 17 million barrels, in addition to the important discoveries of shale oil (non-conventional oil), which would also have been possible due to the legitimacy of this program.

In the case of Refining Plus, "14 projects were carried out with investments of US $2 billion, from which by 2015 we will have expanded refining capacity by almost 2 million cubic meters/year for diesel, that is, 16%, and 2.47 million cubic meters for naphtha, which is equivalent to 37%".

But the cold numbers at the country level reflected another reality and highlighted the failure of the aforementioned plans.

Concept	Unit (m^3)	2008	2012
Proven oil reserves	10^6	400	374
Proven natural gas reserves	10^9	398	315
Oil production	10^6	36.6	33.15
Natural gas production	10^9	50.5	44.1

Source Secretariat of Energy of Argentina

As for the results of Refino Plus, there would have been practically no change in refining capacity, no new refinery would have been built and some modifications would have been made to plants located at the city of La Plata (Buenos Aires province), Luján de Cuyo (Mendoza province) and Campana (Buenos Aires province) refineries.

It would seem that once the partial expropriation of YPF-REPSOL was overcome, the Petróleo Plus and Refino Plus programs, which were created in 2008, would be reactivated.

Oil Plus offered export tax benefits on "new" crude oil added to production. These incentives had a fiscal cost of up to 2 billion pesos a year.

On March 6, 2008, Resolution 24 created Gas Plus with the objective of increasing natural gas reserves and production, as a consequence of investments in new exploitations that presume greater financial outlays in areas without exploitation, areas in exploitation with particular geological characteristics (tight gas) and/or those areas that are not in production since 2004 or that, being in production, would add to said production that corresponding to new fields.

The natural gas produced under the GAS PLUS program would not be considered as part of the volumes of the "Agreement with natural gas producers 2007–2011", and whose marketing value would not be subject to the price conditions set forth in the cited agreement.

It was assumed that with this program those who joined projects arising from this plan could obtain prices above US $2.5 per million BTUs, which governed the market, and close to US $4 and not exceeding US $7.5.

The Gas Plus Plan, at least until November 2014, did not increase the natural gas production, which, except for the efforts of YPF S.A., continued to decline and put the country's external accounts at risk, despite the fall in international prices.

Once again it is confirmed, in our opinion, that the only way to guarantee future self-supply is to recreate YPF with 100% of the shares in the hands of the National and Provincial States and to convert all the current concession contracts into association contracts with YPF Sociedad del Estado.

1.5.7 *Hydrocarbon Sovereignty*

1.5.7.1 The Instruments

That is how we got to the year 2012.

All the actions, laws, decrees and resolutions generated by the Kirchnerist government had not succeeded in increasing oil and natural gas production and improving the situation of the energy trade balance. Hydrocarbon production was falling, natural gas imports were increasing and this in a context of high international prices of these energy products.

Something else had to be done.

In May 2012, the government decided to do something more drastic and severe, and it set its aims on the YPF Company in the hands of the Spanish REPSOL Group.

REPSOL had practically 100% of the shares, considering that the Petersen group, financed with REPSOL's money, had nominally "acquired" 20% of the shares.

We will first describe the instruments and then make the corresponding comments.

On May 4, 2012, the government enacted National Law 26,741, known as the "Hydrocarbon Sovereignty" Law.

This law considered the achievement of self-supply of hydrocarbons to be of national public interest. Thus, it created the Federal Council of Hydrocarbons and declared 51% of the assets of YPF S.A. and Repsol-YPF Gas S.A. to be of public interest and therefore, subject to expropriation.

(a) It established the following principles of hydrocarbon policy
The promotion of the use of hydrocarbons and their derivatives as a factor of development and increased competitiveness of the various economic sectors and of the provinces and regions;
(b) The conversion of hydrocarbon resources into proven reserves and their exploitation and the restitution of reserves;
(c) The integration of public and private capital, national and international, in strategic alliances aimed at the exploration and exploitation of conventional and non-conventional hydrocarbons;
(d) The maximization of investments and resources used to achieve self-sufficiency in hydrocarbons in the short, medium and long terms;
(e) The incorporation of new technologies and management modalities that contribute to the improvement of hydrocarbon exploration and exploitation activities and the promotion of technological development in Argentina for that purpose;
(f) The promotion of industrialization and commercialization of hydrocarbons with high added value;
(g) The protection of consumer interests related to the price, quality and availability of hydrocarbon derivatives;
(h) Obtaining exportable hydrocarbon balances for the improvement of the balance of payments, guaranteeing the rational exploitation of resources and the sustainability of their exploitation for the benefit of future generations;

(i) It created the Federal Council of Hydrocarbons and defined its functions.

This Federal Council was composed of the Ministry of Economy and Public Finances, the Ministry of Federal Planning, Public Investment and Services, the Ministry of Labor, Employment and Social Security and the Ministry of Industry; the provinces and the Autonomous City of Buenos Aires.

To promote the joint action of the National State and the provinces, the Council was dedicated to the fulfillment of the objectives and to establish the Argentine hydrocarbon policy that emerged from the National Executive Branch.

It decided to expropriate YPF, then in the hands of REPSOL.

With this objective, the government declared that 51% of the assets of YPF Sociedad Anónima and REPSOL-YPF Gas S.A. were of public interest and therefore, subject to expropriation.

The expropriated shares were distributed as follows: 51% to the national government and the remaining 49% to be distributed among the provinces that were members of the Federal Organization of Hydrocarbon Producing States (i.e., the hydrocarbon producing provinces), taking into account the production levels and proven reserves of each one of them.

It considered the new company as YPF Sociedad Anónima and the directors would be in proportion to the holdings of the national and provincial states and one director representing the workers and employees of the company.

The National Congress would be the only one authorized, by two-thirds of its members, to transfer the expropriated shares.

The price of the expropriation would be assessed by the National Appraisal Court.

Finally, in order to comply with the purposes established in the cited law, YPF S.A. would be able to resort to external and internal sources of financing and to the arrangement of strategic associations, joint ventures, transitory unions of companies and all kinds of association and collaboration agreements with other public, private or mixed, national or foreign companies.

1.5.7.2 Critical Analysis of the Expropriation of REPSOL-YPF

A. **The analysis**

It was declared to be in the public interest of the Nation to achieve self-sufficiency in hydrocarbons, as well as in practically all the activities that made up the hydrocarbon production chain.

As it will be seen later, this declaration of public interest provided the legal basis for the expropriation.

Additionally, the expiration of the concession areas had been based on the non-compliance with the provisions of the Hydrocarbons Law 17,319 of 1967.

Article 6 of the law states that concessionaires may dispose of hydrocarbons as long as they operate on "reasonable technical–economic bases that contemplate the convenience of the internal market and seek to stimulate the exploration and exploitation of hydrocarbons". As it will be demonstrated later on, REPSOL did not make any efforts to stimulate exploration, and therefore, it lowered the production of oil and natural gas, compromising the internal market.

According to Article 31, "every exploitation concessionaire is obliged to make, within a reasonable period of time, the investments that are necessary to carry out the work required for the development of the entire area covered by the concession, in accordance with the most rational and efficient techniques and in accordance with the characteristic and magnitude of the proven reserves, ensuring the maximum production of hydrocarbons compatible with the adequate and economic exploitation of the field and the observance of criteria that guarantee the appropriate conservation of the reserves".

Of course, REPSOL did not make the required and expected investments.

Article 32 forces the concessionaire to request the approval of the Application Authority (in this case the Secretariat of Energy of Argentina) of the development programs and investment commitments on a periodic basis. In this case, the Secretariat of Energy was in debt because it approved development programs of REPSOL's activities and investments that they would later be consider insufficient as it will be shown below. The insufficiency arose from the data that the basis of the expropriation bill includes.

Article 80 indicates the causes for the expiration of the concessions. Thus, paragraph (c) clearly indicates as one of these causes "the substantial and unjustified non-compliance with the obligations specified in terms of productivity, conservation, investments, works or special advantages committed", in this case, by REPSOL (a company that had not complied with the requested productivity and investments, as it has already been mentioned).

Article 37 indicates that the expiration, which implies the reversion of the concession areas to the State, "will entail the transfer in its favor, without charge, of full rights and free of any encumbrance of the respective wells with the normal equipment and installations for their operation and maintenance and of the fixed or mobile constructions and works incorporated permanently to the exploitation process in the concession area". According to this article, no amount should have been paid for the expropriated facilities, given the existence of grounds for expiration.

That is to say, the expropriation should be carried forward, based on the declaration of public interest, of REPSOL's shares in YPF and the expiration of the concessions granted in the areas, without any additional cost in the latter case.

On this last point, some doubts and perhaps problems would arise.

If the companies that hold the concessions for the fields presented their work plans and the amount of investment, and the Secretariat of Energy approved them, and then did not control compliance, the cause for expiration would fall, and I fear that this had happened. Then, all that remains is the expropriation based on public interest.

On the other hand, why was only 51% of REPSOL's shares expropriated and the other shareholders' (the Petersen Group and other private funds) remained untouched, thus recovering 100% of the company?

Perhaps because by controlling 51% of the shares, in this case expropriated only to REPSOL, the company was controlled and litigation with other shareholders was avoided.

The fear is that by reverting YPF to the State, as a corporation, other operators will be granted the operation of areas and distilleries in exchange for a minority participation of the State in mixed companies and with the lure of initial cash contributions by the private partners, in the way that occurred with the extensions in time of the concessions in several provinces, for example with Pan American Energy in Chubut.

The National State would lead the process with the collaboration of the provincial states and national and international public and private capital.

Consequently, the provinces would be subordinated and this would, in principle, recover the Nation's decision-making capacity, which in fact had been transferred to the provinces (National Law 24,145/1992, called the "Provincialization of Hydrocarbons Law", Article 124 of the reformed National Constitution of 1994, and the National Law 26,197, so-called the Short Law. I agree with this modification.

But it is worrying to see the possibility of the competition of national and international private capital which, in my opinion, in the way it has happened, has been one of the causes of the serious problems upsetting the Argentine energy sector.

The following principles of hydrocarbon policy were enunciated:

To promote the use of Hydrocarbons.

In fact, a previous step to define a Policy for Hydrocarbons would be to define an energy policy, within which Hydrocarbons would have a certain role. The Argentine energy matrix was excessively dependent on these sources of energy, since together, they contributed 87% of the total energy requirements and almost 58% of the electricity generation (1). Given the quite short duration of the proven reserves of oil (barely 11.4 years) and natural gas (scarcely 7.1 years) (2), it was essential to properly replace them and use them rationally. For instance, substitute them by hydroelectricity, nuclear energy and wind power, reasonably by ethanol and biodiesel, and other sources of energy.

Using them rationally would mean saving energy that is, satisfying the same needs of the social and productive system with less energy consumption. An example of this is the replacement of common lamps and spot lights by low consumption ones.

These measures were absent from the objectives mentioned in the law, but they were supported by Decree 140 of 2007, which created the Energy Efficiency Program.

Incorporate more oil and natural gas reserves.

I agree.

Integrate public and private, national and foreign capital in the tasks of the so-called upstream (exploration and production).

This was worrying as it could imply the replacement of REPSOL's role by another similar company that would once again play the leading role in the activity. The policy applied by the current government confirmed these doubts. Especially when, as will be seen, transitory unions of companies and joint ventures between this new YPF and the aforementioned capitals were being promoted. If what was intended was to incorporate capital to finance the activities, ensuring operational control of the company, it would have to be made explicit.

Investing to achieve self-sufficiency in the short, medium and long terms.

Without additional essential precautions, it could be very dangerous to meet this goal in the short term. It was true that the production and reserves of hydrocarbons had decreased dramatically and were compromising the trade balance due to the increasing imports of natural gas, and oil derivatives, but if the extraction of hydrocarbons from the current scarce reserves were to be accelerated (technically it could be done) it would be at the cost of an irrational exploitation of the fields that would inevitably lead to further shortening the life of the current reserves and this would compromise more seriously the present generation in the medium term. The fact is that the necessary and essential actions to increase exploration and apply tertiary recovery techniques will not bear fruit before the next 3–5 years. Even if the so-called non-conventional resources, which were to be exploited by YPF in partnership with Chevron, were to be used, they would have to move from resources to reserves, analyze the decline over time and consider the environmental impacts.

The procedures to revert this situation should have been taken at least 10 years ago, when the decline of reserves and production was already evident and a recovery of economic activity that was going to demand more energy was observed. On the contrary, the management of the activity continued to be left in the hands of private companies, which were more concerned about exporting than exploring, depleting resources and taking the economic benefits to their headquarters. The State behaved like a profitable partner, it did not control the processes, it sponsored the entrance of "friends" to the business, it was believed that the creation of ENARSA, a business unit more than a company, would be enough. This was not as expected, and the responsibility of those who conducted the "energy policy" cannot be ignored, so to speak, because actually it was a non-policy.

Incorporate new technologies and improve management.

All right.

Promote the industrialization and commercialization of hydrocarbons with high added value.

This would imply, for example, adding value to raw materials, such as petrochemicals, and if there were exportable balances, neither exporting crude oil, nor gasoline or virgin naphthas, as it is currently the case, but other petroleum derivatives such as high-octane motonaphthas and lubricants.

To protect the interests of the consumer in terms of prices, quality and quantity of oil derivatives.

To make these measures effective, a distillery should be built as soon as possible with the capacity of the one at the city of La Plata, in Buenos Aires province, which is equivalent to 50% of the capacity of YPF and 30% of that of the whole country (2). This, if the decision had been taken in that year 2012, could have been operating in 2017 or 2018. But this new distillery would have to be integrated, i.e., have conversion units such as catalytic cracking or hydrocracking and catalytic reforming, capable of producing the quantities and qualities of naphthas and gas oil required by consumers. Unfortunately, government officials in the area did not see this situation so clearly announced for no less than 5 years, since it has been already almost 10 years since any major, new distillery was built in the country. The other aspect was returning to have uniform prices for oil products throughout the country as was previously the case, which is technically and economically possible. Prices should be set according to internal costs, including some financing of investments. The sales prices of oil derivatives to the final consumer would be at sufficient levels, at present, to allow integrated companies (i.e., those that produce and process oil and sell its derivatives) to achieve an adequate level of self-financing of their investments in the production and refining stages, provided that the companies were in the hands of the State. The total sales of oil derivatives in 2010 reached about US $21 billion dollars a year (8), and the profits obtained should not be distributed among the shareholders by the state companies. For the exploration stage, destined to incorporate oil and gas reserves, it could be convenient to formalize contracts with private companies, where the mining risk would be borne by the investor and if hydrocarbons were discovered, the private company would have to associate itself with the State Company, which in all cases should have more than 50% of the society, and the free availability of the produced hydrocarbons for the private sector would not be authorized. For natural gas, tariffs should be established based on cross-subsidies where the poorest residential consumers (in general those who consume the least) have lower unit rates than those who consume the most, and the commercial sectors also have higher unit rates than the industrial sectors. The important thing would be for the average tariff to cover the costs of the State-owned enterprise since there would be no distribution of benefits.

Creation of the Federal Council of Hydrocarbons composed of the heads of the Ministries of Economy, Planning, Labor and Industry, with the participation of the Provinces and the Autonomous City of Buenos Aires.

In this composition, neither the National Congress nor representatives of the opposition, nor of the workers and employees appeared, which certainly will not facilitate an operative dialogue.

The functions of the Council would be: the coordination of the action of the National State and the provinces; the fulfillment of the objectives of the law and the establishment and consideration of the hydrocarbon policy that emanates from the National Executive Branch.

This confirms what was said before, as the National State would seem to recover the leading role in determining the hydrocarbon policy and its implementation, and not each provincial jurisdiction on its own, at least with regard to exploration and production.

It is possible that the drafting of an energy code, a new hydrocarbons law to replace the current one 17,319 of 1967, and the total and not partial repeal of the three deregulation decrees of 1989 (1055; 1212; 1589) that practically transferred the free use of hydrocarbons to private concessionaires, which were the legal instruments that facilitated the loss of reserves, should have been promoted. Once again the current Secretariat of Energy appeared almost totally ignored.

The expropriation.

As already indicated, the expropriation of 51% of REPSOL-YPF's shares, when declared to be in the public interest, allowed the application of the Article 17 of the National Constitution and the Article 2511 of the Civil Code, which in both cases would require fair compensation. In other words, the public interest was a necessary condition for expropriation, and the recitals of the bill amply demonstrate this by considering the majority shareholder of REPSOL-YPF responsible: for the insufficient investments made in practically all the links of the productive chains, for the lack of exploration work and for the failure to expand the capacity of the distilleries in time. All of this had led to the gradual fall in the level of hydrocarbon reserves, the accelerated decrease in the production of oil and natural gas, the shortage of oil and natural gas derivatives, with the consequent increase in hydrocarbon imports, putting the positive commercial balance at risk, while the company sent abroad dividends that in some years came to represent almost the total net profits obtained. Of this situation, I repeat, the current national authorities would also be responsible for inaction, especially the State representative on the board of REPSOL-YPF, which except for the last years, always accepted the decisions of the company.

The shareholder composition of REPSOL-YPF at the time of the expropriation was as follows:

REPSOL: 57.4%;
Petersen Group: 25.4%;
Mutual funds and savers: 17.2%;
YPF's property would remain so immediately after the expropriation:
Estado Nacional: 26.0%;
Estados Provinciales: 25.0%;
REPSOL: 6.4%;
Petersen Group: 25.4%;
Investment funds and savers: 17.2%.

But until the form of distribution of the shares among the provinces was defined, the National Executive would exercise control over all the expropriated shares.

This should be decided as soon as possible in order to avoid discretionality on the part of this power.

REPSOL had a shareholder composition in which the Spanish groups SACYR (10.1%) and La Caixa (12.8%) were not the majority, since they held 22.9% of the shares, with Petrolera PEMEX holding 9.4%, US investors 42% and funds of other origin 25.7%. But the Spanish State did have a golden share that would allow it to play a decisive role in the possible hostile disposal of REPSOL.

Subsequently, REPSOL and the Petersen group would divest themselves of their shares and the ownership would remain as it was:

National State: 26.0%;
Provincial states: 25.0%;
Inbuisa: 5.69%;
Mson Capital Management: 4.0%;
Lazard: 3.99%;
Soros: 3.39%;
Loebs: 3.4%;
Perry: 1.2%;

Third Point: 1.0%;
Kyle: 0.9%;
Others: 17.44%.

The distribution of the shares among the provinces envisaged only the hydrocarbon producers according to their production and oil and natural gas reserves.

This would pose a problem. If the shares were distributed according to the reserves and production of oil and natural gas of the total of the country, the following 10 provinces would be included in the distribution: Neuquén, Chubut, Mendoza, Santa Cruz, Río Negro, La Pampa, Tierra del Fuego, Salta, Formosa and Jujuy.

But if they were to consider (since it is a question of the expropriation of YPF) only the reserves and production of oil and natural gas of that company, they should only consider Mendoza, Neuquén, Chubut, Santa Cruz and Río Negro, since REPSOL-YPF does not appear, at least as an operator, in the provinces of Tierra del Fuego, La Pampa, Formosa, Salta and Jujuy.

The other conflictive issue was the lack of access to the shares of the non-producing provinces, at the date of the expropriation, of oil and natural gas, but which have distilleries and storage plants and concentrate a good part of the consumption of oil derivatives and natural gas, despite the fact that they would be part of the aforementioned Federal Council of Hydrocarbons.

The project would privilege the availability of hydrocarbon reserves but would not take into account the other components of the productive chain (distilleries, storage plants, for example) as in my opinion it should be.

The directors of the part of the shares of the expropriated YPF (Public Limited Company) would be in proportion to their ownership and a representative of the workers would be incorporated.

That is to say that the National State, in order to impose its decisions, should unfailingly count on the support of the directors of all the provinces.

The Petersen group was a case apart. The money was "lent" by REPSOL itself, assuming that the group would pay it with the dividends generated by YPF-REPSOL itself. It seems that this was one of the reasons why almost all of the company's profits went into dividends.

There were always doubts that the Petersen group would be able to pay the amounts due.

The transfer of the expropriated shares (51%) would require the affirmative vote of two-thirds of the National Congress.

It seems right and this would make what happened when the privatization of YPF Sociedad del Estado (YPF S.E.) was approved difficult, since in order for the provinces to give their support, they were given part of the shares that were later sold by each one at the time they considered appropriate, as did the retirees and former employees of YPF S.E.

According to the provisions of the expropriation law 21,499 of 1977, the National Executive would be the expropriation body.

It seems logical, perhaps the accompaniment of the National Congress should have been included.

The price of the expropriated assets would be determined according to the provisions of Law 21,499 and the appraisal would be carried out by the National Appraisal Court.

The aforementioned law indicated that the compensation would not take into account hypothetical profits, nor lost profits, therefore it would seem that the value of the company assets should have been quantified, deducting: the corresponding depreciations, according to the useful life of those assets; subtracting the financial liabilities and environmental damages as well as the damages generated by the breach of the concession contracts returned to the provincial governments and the exorbitant profits.

In short, the cost of the expropriation, including interest on subsequent payments, was close to US $6.5 billion dollars.

As for market sales between 2003 and 2011, they had totaled US $81,927 million current dollars or an annual average of US $9103 millions (4).

Profits between 1997 and 2011 had reached US $17,456 million current or an annual average of US $1246 million per year (4).

The dividends distributed in the period 1997–2010 would have been current US $14,200 million with an annual average of US $1014 million dollars. (4) This gave a very high percentage, 83.5% of profits, when most international oil companies do not exceed 30% and allocate a significant part of profits to investment. In the case of REPSOL, if it had destined 70% of its profits to more investments, it would have invested US $8960 million dollars more between 1997 and 2010, which would have represented nearly 160 exploration wells per year, assuming a cost of US $4.3 million dollars per exploration well.

The unfortunate thing is that the National State had a director on the Board of RESOL-YPF who neither questioned the lack of investments, nor the loan to the Petersen group, nor the very high distribution of dividends.

The operational continuity of the company had to be ensured until the election of the new directors.

Unfortunately, the same person who was responsible for the situation of lack of control and the erratic actions of the failed energy "policy" applied by the country since 2003, which led to the situation that we want to remedy today, was appointed as YPF's controller. The same is true of the person appointed as controller of the YPF regional office in the crucial Neuquén Province, where the Vaca Muerta oil field is located.

In order to ensure legal continuity and the management of YPF, it was maintained as a corporation and was authorized to resort to internal and external sources of financing, to enter into strategic associations, transitory unions of companies and all types of agreements with other public, private or mixed, national or foreign companies.

I do not share at all the maintenance of the corporation as a legal figure and rather it should have been declared a State Company.

As a corporation, the intervention of the General Auditor of the Nation, which is, by law, in the hands of the opposition to the political party in power, and of the general Syndicate of the Nation, was not required. Both are empowered to report on the behavior of state-owned companies, on their managers, their operations and their balance sheets, and this ensures much more adequate control to avoid or make more difficult mismanagement and corruption, which are usually "temptations", than in state-controlled corporations.

The possibility of business agreements with third parties seems to be based on the belief that it is impossible to achieve an adequate level of self-financing of the investments required to return to the path of hydrocarbon self-sufficiency with one's own resources.

In the first place, I repeat, the expropriation should have been based, not only on the causes of REPSOL-YPF's malfunctioning, but, and perhaps more importantly, on a concrete and detailed plan of what to do in the future with the expropriated company. The latter was not made explicit.

Rather, it would seem that it was based on two budgets that would go together: the lack of amounts to invest and the lack of technological capacity to face new challenges.

Nobody knows clearly, because there has been no energy planning, what is the annual amount to be invested in hydrocarbons, to explore, produce, distil, transport and market oil and natural gas in the country in the next, at least, 10 years.

Between 2005 and 2011, according to data from the Secretariat of Energy, US $25,356 million current dollars of investment were budgeted, or an average of US $3622 million per year (1), of which REPSOL would have been responsible for between US $1500 and 2000 million dollars per year (5).

As it is clear that these amounts would not have been enough, since the total proven oil reserves of the country fell (from 416.7 million m^3 in 1997 to 401.3 m^3 in 2010), the total proven natural gas reserves of Argentina fell (from 683.8 billion m^3 in 1997 to 38.7 billions in 2010). The necessary number of exploration wells was not drilled (between 1970 and 1992 the state-owned YPF drilled 110 wells per year and between 1999 and 2011 the private company only drilled 50; nor were the distilleries expanded and modernized (the capacity of the distilleries remains practically at 103.000 m^3/day for the last 10 years (2); and self-supply was lost, since in 1995 US $809 million dollars of fuel were imported and in 2011, US $9397 million current dollars(4).

Today, it is estimated that annual investments should not fall below US $6– US $7 billion dollars per year, not including, in full, work on the continental shelf or the exploitation and confirmation of proven reserves of unconventional hydrocarbons.

On the other hand, REPSOL's net profits in the last 15 years reached the current US $17,456 million dollars or 1250 million per year (4).

Domestic sales of oil derivatives were in the region of US $21 billion dollars per year (8).

These few numbers suggest that the domestic market for hydrocarbons would be able to afford no less than 70% of the annual investments required, or between US $4 and 5 billion dollars per year, not counting non-conventional exploitation. But for this it would be necessary to modify the tax component that oil derivatives have today, bringing it closer to that which existed in the 1970s. Even by replacing the

10% tax on the production or distillation of crude oil, an additional US $1420 million dollars per year could be obtained (6).

For exploration tasks, one could think of formalizing contracts with private companies, where the exploration risk (mining risk) would be taken by the private investor and the state company would associate, keeping the majority of the shares, in the production stage without admitting the free availability of crude oil for the private investor.

Additionally, funds could be obtained from banks, such as the National Bank and/or the National Development Bank, which should be recreated, from equipment suppliers and from savers in general through the issue of bonds.

But this mechanism of obtaining sources of financing, external to the company itself, would make the existence of YPF Sociedad del Estado and not of YPF Sociedad Anónima necessary.

Regarding the technological capacity, the exploration and exploitation of non-conventional hydrocarbons would require the training of YPF personnel abroad, taking advantage of the knowledge we have of drilling horizontal wells and hydraulic fracturing.

Where the technological deficiencies are most serious is in relation to work on the continental submarine shelf in deep waters near the slope. In the years of the process, YPF left the possibility of carrying out this training when the submarine platforms it had acquired in 1974 were sold.

As for the formation of mixed companies, the corporate figure of PETROBRAS is given as an example. This company is part of a holding company called PEPSA where PETROBRAS, together with the National Bank of Economic and Social Development (BNDES) of Brazil, holds about 40% of the shares (ordinary and special), but with the capacity to set the company's policy by the Brazilian State (7).

But the most important thing is that, unlike YPF Sociedade Anônima, it only owns 21% of the proven oil reserves and 18% of the natural gas reserves, 34% of oil production and 23% of natural gas production)(2), a product of this bill, and of ENARSA, (which is not a company but a Business Unit), while PETROBRAS produces more than 90% of Brazil's oil, and usually wins most of the bids called by the Brazilian Petroleum Agency, to explore and exploit areas, especially in marine waters. Even private companies seek to partner with PETROBRAS to have a better chance of winning bids.

Likewise, the exploration and exploitation of the important oil reserves discovered by PETROBRAS on the seabed, called "Pre-Salt", led to the creation of a new state company, where the operator would be PETROBRAS and the Brazilian State, with the aim of internalizing most of the profits, would have a majority that could even reach 100% of the stock package.

The other fear that arises, and much more serious, is that REPSOL will be replaced by another or other international companies, which will be the operators of the areas and that YPF S.A. will become another quasi-business unit, as ENARSA is today.

The "new YPF" would have its strength in the distillation of oil and sale of derivatives, where it had almost 55% of the market, with the aggravating factor of being a buyer of the crude it refines. While in the stages of exploration and production of oil and natural gas it only had between 18 and 34% of the market. It is the leading company in terms of oil and the second, after the French TOTAL Austral, in terms of natural gas.

But even with these values, if the concession contracts for areas that are ending are not extended again, it could recover ground. This reminds us that when YPF was created in 1922, its production was lower than that of private companies, but there was a determined policy to control the hydrocarbon business in Argentina.

Another problem is the struggle that will take place between the provinces and the Nation and between the provinces themselves in order not to lose power in the company and to appropriate most of its benefits. Without an energy code, without a new hydrocarbons law and without the cancellation of the three oil contracts signed during the times of the Menem government, this will be very difficult to avoid.

B. **Summary and Conclusions**

Conceptually, I agree with any action that allows the recovery of the State YPF but the expropriation project deserves some disagreements and doubts that were based on the analysis already made in point A.

It is intended to ensure, among other things, the country's hydrocarbon self-supply but the current REPSOL-YPF, only had 21.4% of the proven oil reserves (with a duration of 11.3 years based on decreasing productions) and 18.3% of the natural gas reserves (with a duration of 7.6 years); produced only 34.3% of oil and 23% of natural gas; processed 55% of crude oil and sold 55% of oil derivatives on the domestic market and a number of other important assets. That is to say that in spite of being the first oil company of the country and the second gas company, it was far from having the totality of the markets and this implies, on the part of the State, a control and regulation of the action of the remaining actors to achieve the ends proposed in the expropriation project.

It seems to me that the idea that the hydrocarbon activity should be conducted by the National State together with all the provinces and not only the oil and natural gas producers is adequate to avoid the independent management that each provincial district has been doing of natural resources that belong to all Argentines.

It would have been necessary for the National Congress to participate in the Federal Council of Hydrocarbons, with representatives from the majority and minority parliaments.

I believe that a previous step to the decision of expropriation should have been the elaboration of an energy policy, with the consequent hydrocarbon component, and medium and long terms planning, which would make clear what is to be done with the company, and probably the creation of an energy code by the National Congress.

I do not accept the figure of a corporation for the expropriated YPF and I am inclined toward that of a state-owned company that ensures greater control by bodies authorized to do so, outside the company.

It would probably have been preferable to expropriate 100% of the company, considered a State-owned company, since it would facilitate, on the one hand, the execution of the hydrocarbon policy and, on the other hand, it would not be necessary to distribute dividends among the shareholders, allocating all of them to the expansion of the company.

We would have to be very careful with the possibility of incorporating national or foreign private actors through corporate figures such as strategic associations, transitory unions of companies and joint ventures. These figures have been used in the past, for example with ENARSA, and have not been profitable to ensure the use of energy resources for current and future generations of Argentines. The contribution of capital and technology, which is usually the basis for these corporate figures, must be achieved, in the first place, by adequately directing the income that these hydrocarbons generate in the internal market (changing the distribution of taxes to fuels) and resorting, for example, to private risk capital, destined to exploration, with companies where the State has a majority share and the private minority partner is not granted free availability of the hydrocarbons. On the other hand, as far as the incorporation of new technologies is concerned, YPF personnel have sufficient experience and knowledge base (they have been pioneers in the drilling of horizontal wells and are used to simple hydraulic fracturing) to incorporate them promptly, even training themselves to that effect abroad, especially in exploration and exploitation in marine waters.

The incorporation of new partners to YPF can constitute in facts a change of actors, where REPSOL is substituted by another similar company and the priority is to get fresh monetary resources as a condition for the entrance, the same way that has happened with the temporary extension of the concessions in provincial territories, for example with Pan American Energy in the province of Chubut.

It is not known how REPSOL-YPF will manage the significant debt of US $8879 million dollars as of December 2011, nor how it will act against the debt of the Petersen group (which has 25.4% of the stock package) with REPSOL itself and with banks that reach almost US $3380 million dollars.

In any case, a vigilant attitude is important, especially on the part of all of us who have supported the measure in general, and especially the opposition political parties that have agreed to it, in order to avoid a repetition of the mistakes that were made in the 1990s with privatization, and to seek the means to reach 100% ownership of YPF and convert it into a State corporation.

Notes and References

Most of the non-monetary information comes from the National Secretariat of Energy of Argentina.

(1) The Energy Matrix and the Argentine Energy Policy—Bariloche Foundation; Victor Bravo; October 2012.
(2) Website of the National Secretariat of Energy of Argentina—Market Information—Hydrocarbons Market.
(3) Distribution of oil and natural gas reserves and production by province.
(4) Expropriation bill; Argentina's Hydrocarbon Sovereignty: National Executive Power, April 2012.

(5) It is false that YPF has not invested enough in Argentina: Diario La Nación, Buenos Aires, REPSOL request, April 24, 2012.
(6) In the year 2010 with a production of 35.27 million m^3 of oil, at US $65 per barrel and applying a 10% tax, an additional US $1420 million dollars per year will be obtained.
(7) PEP SA's shareholding structure of Petrobras.
(8) Calculated for the sales of oil derivatives in the domestic market in 2009, multiplying the sales of each derivative by the average sales price of the same in the Province of Buenos Aires and affecting the sum by the exchange rate of the average dollar in 2009.

Reserves and production by province

Total companies

	Proven oil reserves	Proven natural gas reserves	Oil production	Natural gas production
Province	(%)	(%)	(%)	(%)
	2010	2010	2010	2010
CHUBUT	43.62	9.14	28.08	7.72
FORMOSA	0.26	0.07	0.33	0.06
JUJUY	0.04	0.01	0.05	0.01
LA PAMPA	2.45	1.41	4.42	0.99
MENDOZA	12.87	2.25	14.66	5.01
NEUQUÉN	11.93	40.01	21.56	48.12
RÍO NEGRO	4.59	1.66	6.44	2.69
SALTA	1.27	12.09	1.46	10.4
SANTA CRUZ	20.92	9.70	20.28	9.51
TIERRA DEL FUEGO	1.26	10.08	2.05	7.46
National state	0.78	13.57	0.67	8.03
Total country	100.00	100.00	100	100

Source Own preparation based on data from the Secretariat of Energy of Argentina

Reserves and production by province

REPSOL-YPF as operator

	Proven oil reserves	Proven natural gas reserves	Oil production	Natural gas production
Province	(%)	(%)	(%)	(%)
	2010	2010	2010	2010
CHUBUT	15.16	1.33	16.8	1.6
MENDOZA	33.26	10.39	30.2	21.4
NEUQUÉN	14.52	80.57	19.3	69.3
RÌO NEGRO	8.12	0.27	2.2	0.3
SANTA CRUZ	28.94	7.44	31.5	7.4
Total country	100	100	100	100

Source Compilation based on data from the Secretariat of Energy and the Argentine Petroleum Institute

1.5.7.3 Complementary Decrees to the Hydrocarbon Sovereignty Law

The expropriation law would be regulated by Decree 1277 of July 25, 2012.

Essentially, the Regulation included the following figures to achieve the objectives proposed in the law:

The National Plan of Investments in Hydrocarbons;

The Plan's Strategic Planning and Coordination Commission.

It created a National Registry of Hydrocarbons and partially repealed (that is to say, it did not repeal them in their entirety), the famous three decrees of the Menemist era to which we already referred in point 3 of this document, that is to say, numbers 1055, 1212 and 1589 of 1989, which essentially referred to the free availability of crude oil and 70% of the foreign currency on the part of the concessionary companies.

The National Investment Plan was established as strategic axes to increase investments in exploration, exploitation, refining, transport and commercialization of hydrocarbons.

The planning commission would have the objective of ensuring compliance with the plan. It was made up of three members representing each of the following three Secretariats: Economic Policy and Planning of the Ministry of Economy, Energy of the Ministry of Federal Planning, and Internal Commerce of the Ministry of Economy.

To this end, it created a registry of investments where companies wishing to carry out hydrocarbon exploration, exploitation, refining, transport and marketing activities must be included.

As regards self-supply and recovery of reserves, refining and marketing and transport of hydrocarbons, companies should present an annual investment plan whose compliance would be controlled by the planning commission.

In order to ensure reasonable commercial prices, the Planning and Coordination Commission would establish the criteria that would govern the operations in the internal market and would publish reference prices for each of the components of the costs and reference prices for the sale of hydrocarbons and fuels, which should cover the costs of production and the obtaining of a reasonable profit margin.

In case of non-compliance with its obligations by the companies, the Planning Commission would establish fines and penalties that could lead to the nullity of the concessions granted.

Since the regulations of the Hydrocarbon Sovereignty Law did not specifically mention natural gas, the Planning and Strategic Coordination Commission of the National Hydrocarbon Investment Plan issued Resolution 1 on January 18, 2013, and created the program to stimulate the surplus natural gas injection. Natural gas was the main energy source consumed in the country, its production was falling year after year, imports were growing and with them the impact on the trade balance.

In order to facilitate the process, the deadline was extended until October 2013 so that those subjects registered in the aforementioned national investment registry (Decree 1277) interested in participating in the "Program to stimulate the surplus natural gas injection" could present a "Project to increase the total natural gas injection according to the bases established in the resolution.

The interested companies had to present their projects for the increase of the natural gas injection to the commission and if they were approved, they would receive monthly compensation from the State for the surplus production consisting of the difference between the price of this production and that of a base price.

In some cases, this difference for the price of the surplus production could reach US $7.5 per million BTUs when the companies usually received US $2.5.

In order to get this benefit, the companies would have to fulfill their investment plans and of course obtain real surplus production. Everything would be decided by the planning commission.

The Program would last 5 years.

The Program to stimulate the injection of natural gas for companies with reduced injection was created by Resolution 60/13 which was regulated by Resolution 83/13. The main objective was to increase the injection of natural gas by producing companies that, for reasons linked, among other things, to their production scales and/or the geological characteristics of the fields on which they operate, had a reduced injection of the hydrocarbon; so as to continue, on the one hand, increasing, in the short term, the injection of natural gas, thus reducing imports and, on the other hand, stimulating investment in exploration and exploitation in order to have new fields that would allow the recovery of the horizon of reserves; in order to achieve energy self-sufficiency, companies to be considered in this category should have accredited natural gas production of less than 300,000 m^3/day and the surplus production would have a higher price than the one they received, which could reach US $7 per million BTUs.

1.5.7.4 The Chevron-YPF S.A. Contract

The terms of the negotiation emerged from three documents, drafted in English and categorized as "strictly confidential", which both oil companies signed in mid-2013. They were identified as "Letter of Agreement", "Guarantee" and "Terms and Conditions of the Exploration Investment Project". A fourth document, the most sensitive and relevant of all—"Project Investment Agreement"—was not known.

This agreement, being signed by two companies and neither of them owned by the State (a euphemism since, although YPF was a corporation, the State controlled 51% of the shares), was presented as an agreement between private firms and subject to "commercial secrecy" and therefore it was inconvenient to publicly disseminate its terms. This gave rise to numerous controversies.

YPF S.A.'s agreement with the Chevron Company is over 380 square kilometers of the Vaca Muerta oil field (it has a total of 30,000 km^2) and the investment commitment is US $1240 million dollars.

If this pilot project gives good results, and only in that case, more funds would be invested.

Lack of serious consultation with and information to the peoples living in the areas where the operations will be carried out, as indicated by the International Labour Organization (Convention No. 169) and the Argentine Environmental Act (No. 25,675). The fact is that there may be environmental problems during exploitation.

The fact that the Convention is secret and not of public knowledge, at least for the legislators, opens questions regarding which clauses are not beneficial for the country. They say it is because of the commercial secret.

The association of State companies with international or national private companies is now common practice in almost all countries (including Venezuela, Ecuador and Bolivia), and everything depends on how these associations are made and how the national interest is protected.

The company can keep 20% (in this case, after the fifth year of operation), which is better than what happened in the 1990s with Decree 1589 of 1989, which allowed 70% of the foreign exchange obtained to be transferred abroad and did not respect the condition that the country should be self-sufficient in energy petroleum before exporting.

If what is wanted is to stop importing by increasing the production it is necessary to take into account that today YPF does not have the totality of the country's deposits but only produces 35 percent of the country's energy petroleum and 23% of the natural gas.

With this agreement, assuming that the pilot plan on the surface of 20 km^2 satisfies the company, in the year 2048 a quantity of energy petroleum would be produced (10,000 m^3 per day) that would not be enough to compensate the normal fall in energy petroleum production between 2013 and 2048 which would be (being very conservative) of 16,000 or 20,000 m^3 per day. In other words, this investment of almost US $28 billion dollars between 2013 and 2048 would not even be enough to cover the natural decline of the current YPF deposits.

On the other hand, in the energy petroleum "non-conventional" fields, the production decreases year after year (e.g., wells start producing 350 barrels per day and after 5 years they produce 50 and in the natural gas ones they start producing 255,000 m^3/day and after 5 years they produce 11,300) and this forces, as it happens in North America, to drill many wells annually and to spend a lot of money. On the other hand, the "conventional" energy petroleum fields start producing 350 barrels per day and after 5 years they produce 205 barrels and the natural gas field start producing 255,000 m^3/day and after 5 years they produce 180,000. This requires spending less money and drilling fewer wells.

Investments are much higher for exploring "non-conventional" fields than for "conventional" ones. A "conventional" exploration well can range from US $2 to 4 million dollars. On the other hand, a "non-conventional" well can cost between US $7 and 12 million dollars.

This is an agreement fundamentally to produce energy petroleum and additionally some natural gas, when the fuel that the country needs most is natural gas and that represents, between what is bought from Bolivia by pipeline and what is brought in liquefied by ship, almost US $7 out of every US $10 dollars spent on importing fuels.

The knowledge to carry out the exploitation today is not available but can be "bought", hiring and paying foreign specialists until our personnel learns it.

In my opinion, given the characteristics of this type of exploitation, and this goes beyond the Chevron-YPF Agreement, those who really win are the companies that sell equipment and specialized services, all of them foreign.

1.5.8 Other Laws and Complementary to National Law 26,741

Decree 929 of July 11, 2013, always linked to the Hydrocarbon Sovereignty, created the investment promotion regime for the exploitation of hydrocarbons.

The decree set out the requirements and conditions for eligibility for this Scheme.

For example, companies should commit to an investment of no less than US $1 billion dollars over the next five years.

This entitles them to a number of benefits from the fifth year of work, such as:

To trade freely in the external market up to 20% of oil and natural gas production.

To have the free availability of foreign currency for the amount received by 20% of the production, whether the same was exported or marketed in the domestic market, since self-sufficiency has not been achieved.

Non-payment of Export Duties.

It dedicated a series of articles especially to non-conventional hydrocarbon exploitation concessions, understood as the extraction of liquid and/or gaseous hydrocarbons by means of non-conventional stimulation techniques applied to deposits located in geological formations of shale gas or shale oil, compact sandstone (tight sands, tight gas, tight oil), coal bed methane and/or characterized, in general, by the presence of low permeability rocks. Those registered in the respective registries would be entitled to obtain a non-conventional concession. The provinces may subdivide areas and consider part of them as non-conventional. The term of the new concession would be 25 years, plus 10 years of additional extension. If a non-conventional area has geological continuity with a contiguous conventional area, the non-conventional one can be extended to that contiguous area. Conventional tasks can also be carried out in non-conventional areas. The provinces would be the enforcement authority, but the commission would be the one to evaluate the behavior of the concessionaire.

Decree 1277 was regulated by Resolution 9 of July 15, 2013, and established the requirements and conditions for the presentation and subsequent approval of investment projects for the exploitation of hydrocarbons. These projects had to be evaluated, approved and controlled in their execution by the planning commission.

It essentially established all the geological and engineering information on reservoirs that the companies included in the register had to submit. In particular, the commitment to invest one billion dollars during the first five years (later, the National Law 27,007 of October 30, 2014, would reduce the amount to US $250 million dollars) and the corresponding investment disbursement curve.

Decree 927 of July 8, 2013, specifically referred to capital goods included in tariff positions, indicating that, given the importance for the country of investments related to the hydrocarbon investment plan, certain capital goods would be subject to extra-zonal import duties.

1.5.9 Regulation of the Liquefied Petroleum Gas (LPG) Market

LPG in cylinders is the fuel consumed by the poorest sectors of the population (it would be around 40% of the families, or about 4.5 million people) who usually do not have access to natural gas.

It is that the price of the LPG calorie, that the users of the "social carafes" pay (these people are those that acquired to 16 Argentine pesos the 10 kg container) is 8 times more expensive than the calorie paid by the residential users of natural gas. But if they do not get the "social carafe" the price rises to almost US $40 and this is almost 20 times higher than network natural gas. This is truly a clear inequality.

This situation tried to be improved by government decisions.

Thus, the National Law 26,020 of April 7, 2005, established the regulatory regime for the industry and commercialization of LPG (Resolution 792 of June 28, 2005, set regional reference prices) and Decree 1539 of September 19, 2008, regulated it, creating the national program for residential consumption of bottled LPG, whose objective was to establish the conditions for the cylinders to be acquired by residential users at a differentiated price (the values set were for containers of 10, 12 and 15 kg) for example for the 10 kg, it was set at 16 pesos of those time, almost 40% of the market price. The first is that the price was fixed at the point of sale and the buyer had to take it home (generally on foot or by bicycle) and the second is that the quotas assigned for these "social bottles" did not meet the requirements of consumers, who often had to and must acquire them at a market price of between 30 and 40 Argentine pesos.

Resolution 1083 of October 1, 2008, approved the regulation of the national program for residential consumption of bottled LPG, which would be repeatedly extended, while the parties in the LPG chain made a price stability agreement.

Resolution 532 of May 22, 2014, extended the price agreements until December 31, 2014, and established quotas and subsidies to distributors, breakers and producers to try to meet the requirements of consumers of social bottles with the program "Carafes for All".

This plan will receive in 2015 about 2400 million pesos to cover the difference between the costs of the bidders and the subsidized prices.

As of October 2014, the producer will receive a subsidy of US $450 per ton and the fractionators and distributors will receive US $1000 pesos each, also per ton. The market price ranges from 1600 to 4000 pesos. Those who manage the supply say that the price of 16 pesos for the 10 kg social carafe would barely cover 30% of the cost.

We believe that the solution would be to analyze in detail the productive chain of costs of the different links. Establish cross-subsidies among natural gas consumers, who should pay more for the m^3 they consume, finish the Northeastern Argentina (NEA) gas pipeline, increase natural gas coverage for users who are currently not supplied, resorting when necessary to "virtual pipelines", and essentially increase the local supply of natural gas by giving a decisive boost to the tasks of a State YPF.

1.5.9.1 Renewable Energy Sources

It is clearly known that the Argentine energy matrix is highly distorted.

As for primary energy, between oil (33.2%) and natural gas (52.2%), they monopolize 85% of the supply (and this has been repeated for almost 20 years), while renewable energy contributes almost 10% (hydroelectrical 4.2% and biomass 5.7%.

If we observe the generation of electricity, the hydrocarbons absorb 70.3% (the oil 17.4% and the natural gas 52.9%) while the renewable ones hardly supply 15.8% (essentially by the hydroelectric one with 13%).

The evolution of the energy matrix allows us to draw the following conclusions:

Liquid and gaseous hydrocarbons have always represented the substantial part of the energy consumption.

It is remarkable the participation of natural gas, which since the time of the privatizations, has become the main energy source.

Natural gas has replaced oil (considered as its derivatives) in practically all consumer sectors. In the residential sector, in a first period together with the LPG, displacing kerosene; in the transport competing with the gasoline; in the industry replacing the fuel oil; in the generation of electricity, not only the fuel oil and gas oil, but becoming the foundation of the thermal power stations in decline of the expansion of the hydraulic and nuclear energy sectors.

The mineral coal, whose reserves are important, (they would last 640 years with the production levels required for the thermal power station at Río Turbio, Santa Cruz province, southern Patagonia) appears with a very low participation. The failure of the exploitation of the Río Turbio field and the availability of gaseous and liquid hydrocarbons practically excluded this energy source from the energy matrix. On the other hand, its contribution was never significant because unlike the central countries, Argentina went directly from biomass (firewood) to liquid and gaseous hydrocarbons.

Another particularity of Argentina among its Latin American peers is the low relative weight of biomass (mainly wood and agro-industrial waste). This shift in the consumption of firewood, especially from the 1940s and 1950s, has a lot to do with the policy implemented by Gas del Estado state company regarding the dissemination of LPG in cylinders, which made Argentina a pioneer in this area and opened the way for the dissemination of natural gas from the gas pipeline in the 1950s.

In the generation of electrical energy, until the beginning of the years 2000s, the displacement of the petroleum derivatives takes place mainly by natural gas, partly by hydroelectricity and to a lesser extent by nuclear fuels. This strategy is interrupted from the years 2000s and natural gas, in spite of some oil recovery, continues with its high relative weight, due to the practical abandonment of hydroelectricity and nuclear energy sources.

As the demand for electrical equipment is behind, the contribution of the combined cycles is intensified, in general with a lower investment cost and a shorter installation time, to the detriment of hydroelectricity (it has been losing participation since the 1980s) and nuclear energy.

In summary, both at the level of the energy matrix and the electricity generation matrix, natural gas accounts for more than 50% of the respective consumption, and in the first case, if oil is added, it can be stated that Argentina is a dependent hydrocarbon country.

If the availability of renewable and non-renewable energy is analyzed, the conclusion is immediate as shown below:

Non-renewable energy	"Duration"
GN	7.8 years
Oil	11.0 years
Uranium	33 years
Coal	640 years
Renewable energy	
Hydroelectricity	2.5 times the installed hydroelectric power in 2011
Biofuels	10–15% of the demand for naphtha and gas oil

(continued)

(continued)

Non-renewable energy	"Duration"
Wind	344 times the installed wind power In 2011
Solar	The solar radiation incident on the national territory would make it possible, theoretically, to satisfy all the needs of electrical energy, but, among other things, at costs between 3 and 5 times those of other sources of generation

This panorama should be kept in mind as a justification for the coming out of National Law 26,190 on January 2, 2007, which created the "Régimen de Fomento" (a promotion regime) for renewable sources destinated to the generation of electricity.

The objective of the law was to reach 8% of electricity generation with renewable sources in 10 years, that is to say, in the year 2017.

To this end, the well-known benefits were granted for 10 years, in terms of investments, VAT, income tax, minimum presumed income (which would not include the goods destined for the purposes of the law) and a trust fund was created, consisting of a US $0.3 per MWH tax to be paid by the distributors and a 0 remuneration was recognized. Energy at US $0.15/kwh (kilowatt/hour) was generated by wind, geothermal and hydroelectric means of less than 30 MW. The Senate of the Nation, on December 5, 2014, gave half an endorsement to a project to bring it to 50 MW). In turn, photovoltaic generation would receive US $0.9/kwh generated.

The Decree 562 of May 20, 2009, regulated the National Law 26,190.

In any case, the objectives proposed by National Law 26,190 on the promotion of renewable sources to reach an 8% participation of these sources in the generation of electricity by 2017 was very delayed in 2014 (only a little less than 2% would have been reached) partly due to the slowness in applying the regulatory decree and the inconveniences presented by the establishment of the trust fund to promote the activity. The delay in awarding the projects tendered, which would have only 15% implementation, was also questioned.

One of the sources that would have the greatest potential for electricity generation is wind power, not only because of the quality of the resource (speed and permanence of the winds in the "roaring fifties") but also because of the maturity of the local industry for the production of wind turbines, despite the fact that a large part of those planned for 2014 were of foreign origin.

Previously, National Law 26,093 of May 12, 2006, established the promotion regime for the production and sustainable use of biofuels. This regime defined ethanol, biodiesel and biogas as biofuels and, in addition to granting the well-known benefits, authorized the mixture of up to 5% of biodiesel with gas oil and 5% of ethanol with motor oils.

On January 1, 2007, National Law 26,334 established the ethanol production promotion regime and encouraged the formation of value chains by integrating sugarcane and ethanol production.

Resolution 1125 of the Secretariat of Energy of December 30, 2013, authorized an increase in the biodiesel content of mixtures with gas oil from 5 to 10%, setting a limit of 5% by the end of 2015.

Subsequently, the tax rate on biodiesel exports was reduced from 21.7 to 11% and finally to 9.3%.

National Law 26,942 of June 17, 2014, exempted biodiesel from the payment of 19% of the fuel tax that was levied on the gas oil component in the mixture.

Another incentive for the diffusion of biodiesel was the setting of attractive prices to producers, although some argue the non-application of the methodology of calculation of the same, apparently agreed by the authors.

The Secretariat of Energy published monthly prices for ethanol, both sugarcane and corn, e.g., Resolution 44 of September 16, 2014, set them at US $8616 per liter for November 2014.

Resolution 170 of 28 November, 2014, modified the parameters for calculating the prices of corn and sugarcane ethanol and raised the content of this biofuel in blends with motor gasoline to 10%.

In the case of biodiesel, the Secretariat of Energy established a calculation methodology that included the following parameters: soybean oil costs, methanol costs, labor costs and a 3% capital recovery.

The production of these biofuels, especially biodiesel, generated resistance, which is understandable, especially because of the environmental impacts that soybean production would have generated in the country, for example deforestation and the displacement of native populations from their fields.

1.5.9.2 Rational Energy Use Programs

Possibly one of the most mentioned energy policy measures to try to reduce unnecessary energy consumption is the so-called rational use.

It is that by operating rationally from the consumption side and from the supply side, it is possible to satisfy the energy needs of the socioeconomic system by consuming less energy.

In Argentina, in 2007, a situation of energy deficit in the medium term, evidenced by the information we have provided repeatedly in this document and the objective of mitigating the greenhouse effect of gases, which contribute precisely the energy, prompted the government to take on December 21, 2007, the Decree 140 for the program for rational and efficient energy use.

This project approved the general guidelines of the program.

It mentioned the development of a series of actions for the short, medium and long terms.

It also called for the development of a series of short-, medium- and long-term actions:

a massive education campaign;
the replacement of incandescent lamps with energy-saving ones;
carefully labeling of energy-consuming appliances, where efficiency is stated;
improving the energy consumption of public lighting.

For the medium and long terms, it was advocated:

the realization of energy audits in the industries;
actions and procedures for the commercial and services sectors;
incorporation of the topic of rational energy use into the curricula;
introducing co-generation;
extend labelling not only to appliances but also to energy-consuming equipment;
regulate consumption;
insist on improvements in public lighting;
analyze possible improvements in the transport sector.

The same was proposed for housing construction;
contemplate the problem of climate change;
It also listed a series of actions to be developed.

On 15 January 2008, Resolution 24 established the structure and composition of the commission for support, monitoring and control of compliance with the measures of the aforementioned program and regulated what would be called the National program for rational and efficient use of energy (PRONUREE).

As concrete measures, we can mention, for instance:

Resolution 8 of January 24, 2008, which referred to the replacement of between 15 and 20 million incandescent lights with fluorescent lights and between 500,000 and 2 million incandescent lights with LEDs. In both cases, it happened in 2 years. Cammesa was entrusted with the purchase of the devices to be replaced.

It also verified the delivery of energy-saving lamps in exchange for other incandescents without charge to users and established a deadline for the cessation of incandescents.

In 2004, the program for the rational use of electrical energy (PUREE) was created and was replaced by Resolution 745 in 2005, which became known as the 2005 version.

The idea of this program was to lower the electricity consumption of certain users, especially residential ones, through a system of prizes and penalties. Thus, those who consumed more than they had in a base period would be punished, and those who consumed less would receive a bonus.

As an example of the results of the PUREE, we mention what is included in Resolution 324 of 2 December, 2014, for users of EDENOR and EDESUR, energy distribution companies. The result was negative and a total of 282,000 mwh more had been consumed in the reference month shown in the resolution.

We believe as a conclusion that the intentions of the REU plans were good, but not the results obtained.

It is that one of the problems that are generated when partial energy policy measures are applied and not those that should include all the problems of the sector is that partial measures fail.

In the case of electricity and natural gas, the low level of tariffs discourages any REU measure.

On the other hand, if we look at the consumption behavior of motorcycles, which we believe to be excessive, the increase in prices in real terms has led to a fall in consumption, especially in the case of those with the highest octane rating.

In other words, taking REU measures without taking into account tariff schemes and values does not make sense.

1.6 The Hydrocarbon Law 27,007

Once the contract between YPF S.A. and Chevron Company was formalized, the government considered it convenient to generalize the provisions so that other companies could also have access to the benefits established in the contract.

This led to the enactment of National Law 27,007 in October 2014.

The law consists of 32 articles that essentially modify the same number of articles of the National Law 17,319 of 1967.

1.6.1 Specific Analysis

Article 1

It replaces Article 23 of National Law 17,319.

It refers to the basic deadlines for the exploration stage.

It distinguishes between conventional, non-conventional and continental submarine shelf objectives, when National Law 17,319 was not making this distinction.

In general, it shortens the deadlines and reduces them to two periods.

The shortening is 3 years for conventional ones, one year for non-conventional ones and increases by one year for the Continental Submarine Shelf, always in relation to the provisions of National Law 17,319.

The extension is maintained for up to 5 years.

The desire to shorten deadlines is noted, when in 1967 3D seismic was not used and this could require at least maintaining them.

Article 2

It replaces article 25 of National Law 17,319.

The difference lies in the fact that it now sets no limit to the exploration, permitting that the same persons (natural or legal) may have. Previously the limit was five.

It does not seem prudent to tend to the concentration of areas in a few private hands in a business that naturally tends toward oligopoly or monopoly. It is interesting to mention that shortly after the oil discoveries in Comodoro Rivadavia (province of Chubut) and Plaza Huincul (province of Neuquén), in order to avoid monopolistic actions, the areas surrounding the discoveries were declared a state reserve zone.

Article 3

It replaces the 26th of the National Law 17,319.

This law is more demanding in terms of the deadlines for returning the remains of areas under exploration.

The general objection mentioned in Article 1 of this law is therefore valid, but can be accepted.

Articles 4, 5 and 6

Replaces article 27 of National Law 17,319 and adds articles 27a and 27b.

The concept of a non-conventional exploitation concession appears in these articles. It defines what is understood by non-conventional exploitation (shale gas or shale oil geological formations; compact sands-tight sands, tight gas; tight oil-coal methane) in general characterized by the presence of low permeability rocks.

What is "new" is that unconventional areas can be subdivided and give rise to new unconventional concessions. Furthermore, if adjacent to the areas, the concessionaire holds another, pre-existing, exploitation concession, it can unify both areas, but the remaining area can continue to be exploited under the previous conditions. As if this were not enough, in the non-conventional areas it may also carry out conventional exploitation activities.

I believe that this distinction between conventional and non-conventional activities will give rise to more than one problem if the Authority of Application exercises its control obligations properly.

Especially considered are the tertiary production projects (e.g., polymer-enhanced recovery type), extra heavy oil (API less than 16 and Offshore). This does not make sense here as it is an unambiguous location.

In these cases, the royalties can be reduced by up to 50%.

These benefits are reminiscent of others requested by private concessionaires, in another era, where they requested extensions or better conditions in their operations promising, for example, tertiary recovery, and then continued to produce by "slightly" primary methods.

I do not agree with this halving of the royalty rate. Let them cut their costs and prorate their profits. It is safer for the country and the provinces.

Article 7

It replaces the article 29th of the National Law 17,319.

Exploitation concessions will be granted by the National or Provincial executive branches who may also grant them on proven areas.

Although the existence of hydrocarbons in such areas is not guaranteed, it is not convenient to grant concessions to private or mixed companies over this type of areas that should be reserved to the state company that I insist to propose that they should exist.

The reason is obvious, someone made the exploratory effort that led to the discovery and that should be used by the National and Provincial State.

Article 8

It replaces article 34 of National Law 17,319.

It preserves the maximum area for a new exploitation concession.

But in my opinion, it is not wise to limit the maximum number of areas granted to a single concessionaire to five.

It does not seem prudent to tend to the concentration of areas in a few private hands in a business that tends toward oligopoly or monopoly, as was mentioned for the case of exploration areas in article 2 of the law under analysis.

Article 9

It replaces article 35 of National Law 17,319.

It refers to the validity of exploitation concessions. For conventional concessions, they will be for 25 years (as established by National Law 17,319); for non-conventional ones, they will be extended to 35 years, including the 5-year Pilot Plan, and for those located on the Continental Submarine Shelf, the term is 30 years.

They can request up to 10 years of extension; with 1 year in advance of the expiration. National Law 17,319 indicated 6 months.

The most striking thing, and in my opinion unacceptable, is that the concessions already extended may ask for a new extension of 10 years at the end. In fact, this benefit almost converts the concession into hydrocarbon property since the period can reach almost 50 years in these cases, and it is assumed that during that time the reserves of the area would have been exhausted.

Of course, as National Law 17,319 had stated, they must have fulfilled their investment plans.

The time periods for private concessionaires are exaggerated because 35, 45 or 40 years, depending on the case, is more than enough to exhaust the reserves, and if the actions of the private concessionaires are not carefully and permanently monitored, areas held by these types of actors are being "frozen" in economic terms, almost until the end of the useful life of the deposits.

Article 10

It replaces article 41 of National Law 17,319.

It simply makes the transport concessions, which can be granted to the exploitation concessionaires, equivalent to the duration and extension of the respective exploitation concessions.

Article 11

Replaces article 45 of National Law 17,319.

It refers to the awarding of concessions, giving priority first to the right that, according to the modification introduced by Article 5 of the law under analysis, is granted to pre-existing concessionaires or those who become holders of a non-conventional concession, for the remaining concessions there will be a bidding process.

In my opinion, any conversion of the type of concession should have been submitted to a public bidding process and surely this would have given the State the possibility of obtaining better conditions from third parties than from the former holders of conventional concessions. In other words, the process would have been more transparent.

Article 12

It replaces article 47 of National Law 17,319.

With respect to the original Article 47, the amendment establishes the preparation of a model bidding document for tenders.

This document would be prepared by the provinces and the National Energy Secretariat, but does not mention the special advantages that bidders would be willing to grant to improve their bids.

Later on, in Article 29, a period of 180 days was given (that is, until 30 April 2015, approximately) to prepare the above-mentioned request for proposals, which may be reviewed and updated periodically.

Article 29 gives some very general guidelines for the preparation of the bidding documents, such as the bid bonds, the scope of the investments and the income that the applicant would grant to the granting authority, as well as the special conditions that the contracts must have depending on whether they are in conventional or non-conventional areas or on the Continental Submarine Shelf.

Let us wait until we know this Base Document to give a more informed opinion.

As in all the content of the law under analysis, the State is very careful to establish advantages for the possible awarders of contracts.

Article 13

It replaces article 48 of National Law 17,319.

It establishes certain criteria for the award, in addition to the generic ones, such as the greatest investment or the greatest exploratory activity.

Unfortunately, it does not say anything about the quality of the work, about taking care not to depredate the resource, or about the transfer of technology or the employment of local labor.

This is supposed to be stated in the specifications.

Nor does it seem appropriate, if there is only one bidder who can be awarded the tender. On the contrary, the tender should be called again, but it seemed that the legislator was in a hurry.

Article 14

It replaces articles 57 and 58 of National Law 17,319 and incorporates the Section "Canon and Royalties".

In the same way that National Law 17,319 establishes the payment of canons for exploration and exploitation.

The increases in the exploration canon for the extension were relatively greater in National Law 17,319 (they were increased by 50% and here they are increased by 25%).

The amounts are fixed in local currency when they should have been indexed to the costs of some operation related to exploration and exploitation.

There is also no distinction, and should be, as to whether activities are carried out in conventional, non-conventional or Continental Submarine Shelf areas. Again, care is taken with the interests of the companies.

Article 15

It incorporates an article 58 bis to National Law 1731.

Here appears the famous bonus of extension that seems to have been, until the present times (the Cerro Dragón case, for example), the "true" motivation of the provincial governments to grant them and whose non-inclusion in this law gave rise to strong discussions between the Nation and the Provinces.

In the end, a compromise was reached.

The payment of the bonus has a maximum and is established for exploitation concessions, we are talking about pre-existing concessions, as well as complementary activities of non-conventional exploitation to be carried out. The amount is fixed as 2% of the remaining value of the reserves of the respective hydrocarbons of the area until the end of the concession period taking as a price the hydrocarbons of the basin in the 2 years prior to the granting of the extension.

In other words, it is the internal price of hydrocarbons and since in the years 2014 and 2013 prior to the enactment of the law these were lower than the international prices, the concessionaires benefitted from this.

The other aspect refers to who quantifies the remaining hydrocarbon reserves in the areas, because if what happened before the privatization of YPF is repeated, the values could be below the real ones.

Article 16

It replaces article 59 of National Law 17,319.

It refers to the hydrocarbon royalties.

The exploitation concessionaire, both for oil and natural gas, will pay 12% of the value of the respective hydrocarbons (the value is in wellhead, minus the corresponding freight and is declared by the permit holder or concessionaire). If the authority of application does not consider this price acceptable, it can formulate objections. Article 17, modifying 61 of the National Law 17,319) states that if the State requests payment in kind, it must assure the concessionaire that this criterion has a certain permanence.

The granting authority may, at its discretion, reduce the royalty up to 5%.

In case of extension, there is an additional 3%, in each successive extension but up to a maximum of 18%.

Non-conventional concessions may pay an additional 3% royalty, up to a maximum of 18%.

Article 28 of the complementary and transitory provisions established that the granting power may reduce up to 25% the amount of the royalties applicable to the corresponding productions during the 10 years following the completion of the pilot project for non-conventional concessions that request it within 36 months from the effective date of this National Law 27,007.

The amount of the royalties should have been set at a minimum value of, for example, 15% and leave a higher amount to the bidder's offer. The province of Neuquén, for example, had already obtained 18% in previous negotiations.

It is also not acceptable to limit the amount of royalties to the only income that can be obtained by provincial jurisdictions. This should also have been left to the offers of potential bidders in the tenders.

The reason for this additional removal of non-conventional concessions is not understood. Again, another undue advantage.

Article 18

It is incorporated as article 91 bis to the National Law 17,319.

This article is absolutely inadmissible.

It prohibits the National State or the Provinces from reserving oil field areas in favor of public or State-owned entities or companies.

It also limits the provinces that have State enterprises with reserved areas at the time the law is enacted to analysis and that do not have association contracts with third parties to associate only with a participation in the development stage proportional to the investments they make in them.

Thus, in the future, all sedimentary basins on land or on the Continental Submarine Shelf will only be able to be explored and exploited by means of the respective bidding process, without reserving particularly attractive areas for state-owned companies, as is the case in Brazil with Petrobras, for the use of the National State or the Provincial States.

This is an unspeakable commitment and I cannot understand what reasons of general interest may have existed for such a decision.

Articles 19–22 are part of the so-called Title II, Investment Promotion Regime for Hydrocarbon Exploitation.

The objective of this title is to maintain harmony with Decree 929 of 2013, which we analyze in point 5.8 as part of the so-called Hydrocarbon Sovereignty described in point 5.7.

We can say that the so-called Hydrocarbon Sovereignty was regulated by National Law 26,741 of May 2012 which essentially created the Federal Council of Hydrocarbons, enacted the National Plan of Hydrocarbon Investments and implemented the Expropriation of YPF-REPSOL.

Decree 1277 of July 2012 created the Strategic Planning Commission of the National Hydrocarbon Investment Plan.

Resolution 1 of 2013 created the Program to stimulate the injection of surplus natural gas; Decree 929 of July 2013 established the investment promotion regime for the exploitation of hydrocarbons, which was regulated by Resolution 9 of July 2013, and finally Decree 927 of July 2013 referred to the levies for the import of capital goods destined to these activities.

Article 19

As Decree 929 incorporated into the investment promotion regime for the exploitation of hydrocarbons those that exceeded one billion US dollars to be invested during the first five years, this provision seemed excessive to some companies. Then, it was reduced to 250 US $ million in the first three years. This implies a reduction of 60%.

The decree also granted the benefit of being able to export and free up foreign currency for 20% of hydrocarbon production as of the fifth year and exempted them from the payment of export duties. The same benefits were granted if exports were not authorized and the hydrocarbons were sold in the domestic market.

This article 19 reduces the above-mentioned period to three years and extends the benefits to 60% if the oil comes from offshore locations exceeding 90 m depth, between low and high tides.

These benefits imply a return to what the decrees of Menem's time provided for, which the Sovereignty Law erroneously proclaimed to have repealed.

Since foreign currency is a scarce good in Argentina, this provision is established to attract foreign investors at the expense of general welfare.

Again, we identify the search to conform the foreign investor at any cost.

Article 20

The rights granted to investment projects for the exploitation of hydrocarbons approved before the law under analysis is recognized.

This implies those arising from the failed Oil Plus, Refining Plus and Gas Plus program.

Article 21

Bearing in mind that the approval of National Law 27,077 had caused significant friction with the producing provinces that considered that the National State was subjugating their rights, this article grants them two special monetary benefits.

A 2.5% of the initial amount of the project (ranging from US $6.25 to 25 million dollars) to be contributed by the companies directed to something called the corporate responsibility.

An amount, although not determined, by the National State to finance social infrastructure works.

The objective was to "pacify" the rage of the producing provinces.

Article 22

It indicates that the investments in capital goods of the companies involved will pay the import duties provided for in Decree 927 of 2013, already mentioned.

Here is a Title III of complementary and transitory provisions.

Article 23

As the operations of the hydrocarbon industry are considered unfriendly to the environment, and especially those destined for fracking, which has provoked resistance from both residents near working locations and groups of citizens, this article promises to establish specific environmental legislation.

Argentina has an excellent law, the National Law 25,675, called the General Law of the Environment, in addition to at least 16 provincial laws, the first ten principles of which stand out for the preservation of the environment, especially those of prevention and precaution.

I believe that the environmental impact studies and the public hearings should be previous, obligatory and conditioning for the hydrocarbon activities.

That is to say, overcoming environmental impacts should be a restriction for the beginning of any activity in this sector.

Unfortunately, these precautions have not been taken ex ante and all that remains is to have "faith" in the good practice of the companies (something that they have not demonstrated up to the present). There are permanent examples of environmental "accidents" in the hydrocarbon fields.

Article 24

It establishes uniform taxation in each of the territories where this National Law 27,007 applies.

It seems appropriate since one of the problems of the atomization of the provisions in this area is to generate undue competition among the provinces to attract investors by offering each one better benefits than the other.

The solution is to return to the State companies such as YPF or Gas del Estado and thus with the participation of the provinces in them to avoid these problems of multiple taxation.

Article 25

It incorporates the program of stimulus to the surplus natural gas injection and the one of stimulus to the natural gas injection for companies with reduced injection (Resolution 60/13) to the benefits of the one of the National Law 27,007 and it entrusts the administration to the strategic commission of the national plan of hydrocarbon investments.

There is a proliferation of commissions and plans that will generate, as it has already done, more than one problem. The simple is always better than the complex in matters of legislation and administration.

Article 26

Correctly, on paper, it tries to unify procedures.

The overlapping of tasks and controls has been a characteristic of the current government's energy policy, and this has generated conflicts between ministries and secretariats that are of public knowledge.

Article 27

It repeals article 62 of National Law 17,319 that referred to the royalties that natural gas should pay.

This issue is supposed to be resolved by article 16 of this law, which is passed on Article 59 of National Law 17,319.

Article 28

It was dealt with in the analysis of Article 16.

Article 29

It was dealt with in the analysis of Article 12.

Article 30

The transfer of the exploration permits and exploitation concessions to ENARSA will not be effective and these will be transferred to the National Energy Secretariat. If there are association contracts with ENARSA, these will become exploration permits and exploitation concessions.

Two things are evident here.

One is the uselessness of ENARSA for the new stage of the government's hydrocarbon policy.

The other is that it proceeds exactly the opposite of what would be advisable. That is to say, association contracts can be converted, supposedly with the State, into permits and concessions, contractual figures more favorable to the private investor than the Association.

My opinion is diametrically opposed. I believe that it would be necessary to recreate the state company YPF and negotiate with the private permit holders and concessionaires the transformation of these contracts into associations with the state YPF (this has been done in large part by Bolivia) in order to have a secure handling of the hydrocarbon policy which, of course, would be another very different to that applied by the Kirchnerist government.

Article 31

This refers to cases in those provinces that have had previous extension processes. They are given 90 days to conclude them and then adapt them to the provisions of this National Law 27,007 in its article 9 which replaces 35 of National Law 17,319.

This is consistent with the idea of bringing all contractual figures into line with the new provisions.

1.6.2 General Analysis and Conclusions

The law understudy is a patch of the old National Law 17,319. This does not imply having a new law but adopting the very old one, from 1967, and which gives for everything, the "new" situation.

Throughout its articles, the desire to attract foreign private investors essentially to exploit non-conventional reservoirs is evident. As we Argentinians are aware of a large part of these new technologies and it does not seem convenient to risk being exporters of a new crude material. There is no oil-exporting country in the world that is an example to imitate for Argentina, which must be industrialized indeed), we do not see the need to grant concessions to foreign private companies.

For the technologies we do not have, we should have applied the same procedure used in 1974 when a marine platform was bought so that Argentine technicians could learn the new techniques in situ. Of course, the 1976 military dictatorship sold the marine platform instead of using it.

As for the amounts of investment, if one goes "slowly", one does not neglect conventional exploration, and one takes into account the environmental impacts of non-conventional exploitation, with one's own resources, that is to say, with state companies and prudent associations with third parties one can get ahead. The idea would not be to export hydrocarbons but to conserve them for future generations and use them for industrialization.

Practically the concessionaires of exploitation, in any of its forms (of conventional areas; non-conventional or of the Submarine Platform), will be able to become in fact the owners of the reserves of their concessions, since the system of successive extensions makes it possible. This is equivalent to the handing over of the country's hydrocarbon assets. As there is no limit to the number of concessions that a single company can have, the handing over is aggravated by the presence of strong oligopolies.

By limiting the rate of royalties (they cannot exceed 18% and on the contrary can be reduced to 5%) the provinces and the National State cease to receive an important part of the only source of appropriation of the hydrocarbon income allowed by the cited Law, since the income and gross income taxes are not used as an income.

The prohibition for state companies (at present there are several provincial ones) or where the State may have participation (such as in mixed companies like YPF) to reserve areas in their favor, implies that all of them must be granted to private companies. This is very serious because there may be geological areas of very low exploratory risk and high probability of containing important reserves that could be convenient to pass on to the State and not be put out to tender. In Brazil there was a time when this type of area was reserved for PETROBRAS. In our country, the law does not even allow them to be reserved for the mixed company YPF S.A.

The extension bonus, a gift to be given to the provincial states, is related to the magnitude of the areas' reserves and it is not established who will carry out the corresponding audition of reserves. It should be the State, since when YPF was privatized the task was entrusted to a foreign consultant who reduced them to reduce the value of the company.

It is clear that the intention is to attract foreign companies by lowering the requirement of the minimum amount of investment to be made to a quarter of what is established in Decree 929.

The percentage of hydrocarbons and foreign currency freely available to foreign investors has been maintained, although it has been reduced, which implies accepting the criteria of the three decrees from the Menemist era, whether or not exports are made.

ENARSA's uselessness is recognized by removing the association it had made with companies to act in the Submarine Platform.

A firm, clear and forceful defense of the environment is left aside; there is only a promise to do so, when the handing over of areas to be explored and concessions to be exploited have already been authorized without these procedures having been submitted to environmental impact studies and public hearings.

The proliferation of commissions continues, which will do nothing but generate conflicts between different State and Provincial States.

We believe that what should have been done is to elaborate a new Hydrocarbon Law that would replace the old and patched up 17,319 of 1967, where the execution of the hydrocarbon policy was in the hands of State companies, recreating YPF Sociedad del Estado and Gas del Estado, (with authentic participation of the provinces in the management) at the same time that all the exploration and exploitation concessions permits were converted into service or association contracts with State majority. Partnerships with private companies should be made to obtain technology transfer where not available in the country and human resources training.

The requirements of the investment plans should arise from the recreation of the tax system that taxed liquid and gaseous fuels in the 1970s before the military dictatorship, plus the contribution of funds from a National Development Bank and a Latin American Bank, such as Banco Sur. The idea would be to preferably develop the conventional hydrocarbon resources; to be very careful with the non-conventional ones without launching into massive exploitation plans, the same as in the maritime platform, associating through risk contracts in the exploratory stages to conclude, if discoveries are made, in association contracts with the majority of the State and without free availability of the hydrocarbons for the exploitation stage. In this way, the internalization of the hydrocarbon income and the internal supply for the present and future generations of Argentineans would be ensured, using these energy resources for an adequate industrialization of the country.

1.7 Conclusions on the Hydrocarbon Policy for the Period 2003–2014

Firstly, we will present some comparative data on concepts in the hydrocarbon industry, in general between 2002 and 2013.

This period covers, approximately, the governments of Néstor Kirchner (May 2003–December 2007), the first presidential term of Cristina Fernández de Kirchner (his wife; December 2007–December 2011) and her second term (December 2011–December 2015).

We thought it was convenient to present, first, these few data that give an idea of the "results" of the policy to be applied in the sector we are analyzing.

Of course, the energy sector cannot be separated from the political, social and economic sectors, but we will stop at the first aspect, which is what is of interest for this work.

We will then draw up a report on the results of this work, and I repeat, on the hydrocarbons sector.

Some Figures

Concept	Unit	2002	2013
Proven reserves oil	10^6 m^3	472.7	370.3
Proven reserves natural gas	$10^9 m^3$	777	329
Oil production	10^6 m^3	43.95	32.4
Natural gas production	$10^9 m^3$	46.5	41.7
Exploration wells decade of 70	N°	115	
Exploration wells 2003–2013	N°		51
Oil imports	10^6 m^3	0.28	0.41
Natural gas imports	$10^9 m^3$	0.1	17.2
Oil exports	10^6 m^3	15.7	2.3
Natural gas exports	$10^9 m^3$	5.85	0.0
Oil production in refineries	10^6 m^3	30.2	30.7
EP's share of internal primary energy supply	(%)	40.8	32.6
NG's participation in internal primary energy supply	(%)	46.9	54.1
Gas oil price at pump	$/liter	1.44 April 2006	11.20 August 2014
YPF PE production	10^6 m^3	19.8	16.8
YPF's share of total PE production	(%)	46	36.8
YPF NG production	$10^9 m^3$	16.8	10.42
YPF's share of total NG production	(%)	33.2	25.0
Price of NG in Buenos Aires City	$/m^3	0.291 July 2007	1.13 August 2014

Source Secretariat of Energy of Argentina

As you can see, practically all the concepts included show negative results.

Proven reserves and production of oil and natural gas fell significantly.

Exploration wells, an indicator of the risk investments to replenish and increase reserves, decreased, on average, to 44% of those made by YPF in the 1970s.

Imports of natural gas increased drastically and the company went from being an exporter of hydrocarbons to an importer and in increasing magnitudes over time.

The penetration of natural gas (the most scarce energy source) followed, partly replacing oil derivatives, and this can be seen with the increase in the participation of natural gas in the country's gross domestic energy supply.

As there were no expansions or changes in the structures of the refineries, the crude oil processed remained stable, even in absolute values, and fuel oil and gas oil had to be imported to replace natural gas in the generation of electricity, and naphthas and gas oil to supply the reduced supply from the refineries to meet the demand of the transport sector and agriculture. Even in some years there were problems with the supply of liquid derivatives and natural gas, so it became customary to restrict the supply of natural gas.

The participation of YPF, the main Argentine hydrocarbon company, until 2012 in the hands of REPSOL, in the production of oil and natural gas decreased almost 9 points and here lies one of the causes, not the only one, of the drop in the production of both hydrocarbons and of the imports of natural gas.

Natural gas rates were practically frozen, which caused an excessive increase in consumer subsidies, especially to residential consumers, and the appearance of the trust funds in 2002. This led to the cessation of work on the companies' distribution networks. As at the same time, the price of natural gas in wellheads was practically frozen, private suppliers stopped exploring and exploiting it, and this also contributed to the increase of liquid natural gas imports (with a cost of the calorie almost 6 times higher than the value recognized in wellheads). But problems of transport infrastructure to bring it and lack of supply in Bolivia restricted the natural gas from this origin). With the Gas Plus Plans that, until now, have failed, the government tried to give an incentive by increasing the price of gas in wellhead up to three times. In the case of natural gas for residential users, the increase in tariffs between August 2014 and July 2007 would be 40% in constant currency, but there are legal resources that are making the increase difficult. In contrast, oil derivatives, especially since the appearance of YPF S.A. between August 2014 and December 2006, increased much more than the inflation rates (7.8 times in current currency and almost 2 times in constant currency, in the case of gas oil or naphthas).

Let us now turn to the fundamental objections to Kirchnerism's hydrocarbon policy.

The management, which we are analyzing, was defined and is defined as "progressive" or "reformist." It might have been so in other areas, such as human rights, but in no way in relation to the management of the energy sector.

In 2004 it seemed that with the creation of ENARSA, which changed the course, but in fact it was desired by many, to replace the State YPF, and became a business unit without adequate technical personnel and without a real presence in the productive activities.

Recognizing the failure, an attempt was made with a proliferation of plans such as Oil Plus, Refining Plus, Gas Plus and the use of trust funds to reverse the decline in the production and export of hydrocarbons and the increase in imports.

These plans also failed, and in 2012 the expropriation of 51% of REPSOL-YPF's shares was used to try another path. We have already analyzed this fact further back and at the end of 2014 it was a little early to judge the viability of the measure.

We can see that a person with vast knowledge of the subject was put in charge of the company, a public limited company, and not of the State. But since YPF, as already mentioned, only has a reduced part of the reserves and participation in the production of hydrocarbons, despite the good performance, in the second aspect, and which is illustrated in the following tables, it has not even been able to retake the participation it had when it was owned by REPSOL.

Oil production

Thousands of cubic meters

January to October each year

Concept	2010	2011	2012	2013	2014	Variation (%) 2010–2014
Pan American energy	5416	5300	4872	4641	4748	−12.33
YPFSA	10.150	8987	9444	9615	10.471	3.16
Total	28.955	26.569	26.745	26.077	25.701	−11.24
YPFSA's share (%)	35.05	33.8	35.3	36.9	40.7	5.65 profit

Source National Secretariat of Energy of Argentina

Natural Gas Production

Millions of cubic meters

January to October each year

Concept	2010	2011	2012	2013	2014	Variation (%) 2010–2014
TOTAL Austral	11.110	11.440	11.119	10.328	9735	−12.38
YPFSA	9851	8867	8653	8689	9754	−0.98
Total	39.588	37.967	37.031	34.751	34.515	−12.81
YPFSA's share (%)	33.1	23.4	23.4	25.0	28.3	4.8 loss point

Source National Secretariat of Energy of Argentina

With regard to natural gas imports (the most serious quantitative problem), the following table shows the evolution between 2010 and 2014:

Year	Amount	Volume
	Millions of U$S	Millions of m^3
2010	826	3286
2011	2461	6398
2012	4015	8913
2013	5134	10.095
2014	5273	11.180
Variation 2014–2010	6.4 times	3.4 times

Source National Secretariat of Energy of Argentina

In other words, YPF S.A. will not be able to supply the country with oil and natural gas on its own and reduce the balance of imported natural gas.

The "occurrence" of non-conventional hydrocarbon resources, perhaps the cause of the National Law 20,007, which the International Energy Agency publicized, motivated first the agreements between YPF S.A. and the Chevron company, and then with other companies (some of them are still under discussion) and other private companies independently of YPF S.A. (such as EXXON, SHELL; Wintershall; Total Austral), all of them in Neuquén province in the Vaca Muerta oil field. We are not going to refer here to this subject and for further information you can consult the following works:

"Shale oil y Shale Gas" Nicolás Di Sbroiavacca, Working Documents Fundación Bariloche, August 2013; in Spanish. In English in this volume.
"Una opinión sobre el fracking", Víctor Bravo, Working Documents Fundación Bariloche, September 2013; in Spanish. In English in this volumen.
"Hydraulic fracturing 101", George E. King Apache Corporation September 2012.
"Noble gases identify the mechanisms of fugitive gas contamination in drinking-water wells overlying the Marcellus and Barnett Shales" by Thomas H. Darraha, Avner Vengosha, Robert B. Jackson, Nathaniel R. Warner and Robert J. Poredae, edited by Thure E. Cerling, University of Utah, Salt Lake City, Utah, U.S.A., and approved August 12, 2014.

We will only say that the risk assumed is undoubtedly great, since it has not yet been demonstrated that the initial acceptable levels of high productivity of non-conventional wells are maintained at those levels over time; that if we were to try to export oil of this origin again we would make the same mistake of the 1990s, aggravated by the possible negative environmental impacts of these techniques, which are not fully analyzed when Argentina must preserve the hydrocarbons for present and future generations; that in general we are essentially looking for oil, when as we have seen, the main hydrocarbon problem is natural gas; that in the last 10 years, exploration has practically stopped in conventional areas, which are less expensive than non-conventional areas and from which not only oil but also natural gas can be extracted; that the tasks of exploration and exploitation must be accompanied by the consequent construction of the trunk gas pipelines for evacuation (in this sense we have the bad examples of the delay in the contribution of gas from the area of the Southern Sea due to the delays in the construction of the gas pipelines in time, as well as the delay in the Northeastern Argentina (NEA) gas pipeline which would make it possible to bring gas from Bolivia, which is cheaper than imported liquid natural gas and at the same time would supply the population and productive activities of that area which today lacks supply.

The proliferation of regulations (Laws; Decrees and Resolutions) sometimes replacing or duplicating functions (more than 60 are mentioned in this document) and overlapping activities of different secretariats and ministries. It was vox populis that the Secretariat of Energy of Argentina was practically subsumed by the Ministry of Planning, and the Minister of Planning was actually the acting Secretary of Energy.

The policy of freezing electricity and natural gas tariffs.

The energy sector tariff policy has been totally distorted, and this has helped to generate a very large weight on the country budget. The generalization of indiscriminate subsidies that under the pretext of not wanting to affect the lower-income sectors or to moderate inflation, has unduly benefited medium and high-income sectors that could pay more for natural gas and electricity. In economics, one principle used is that of cross-subsidies, where higher-income sectors pay more per unit of energy than lower-income sectors and productive sectors of final consumption than intermediate consumption (that is, more expensive unit rates for businesses than for industries.

On the contrary, the prices of oil derivatives (gasoline, gas oil) have been constantly rising above the inflation rate.

The impact on the trade balance of Argentina has been motivated essentially by the imports of gaseous and liquid natural gas, gas oil and fuel oil that, although they still do not turn the balances negative, it prevents to destine funds equivalent to other productive uses. We already mentioned the causes of these hydrocarbon imports (fall of the production of oil, natural gas and derivatives of oil).

Investment was stopped to increase refining capacity and modernization of the refineries. It has been more than 20 years since a large oil refinery was built in the country. This is reflected in the fact that the production of naphthas and gas oil does not manage to satisfy the demand, beyond the decrease in the production of light crudes of the Medanito type.

Investment in trunk and distribution pipelines to supply the internal market has been stopped and there are still regions of the country (e.g., the NEA, Northeastern Argentina region) that do not have natural gas and there families must be supplied with liquefied petroleum gas (LPG), which is much more expensive than natural gas.

Investment in LPG recovery plants for natural gas was stopped, among other reasons because of the drop in natural gas production. This generates in some regions the overuse of electricity for caloric purposes (heating, cooking and water heating) facilitated by tariff structures. It is cheaper to use electricity for caloric uses than LPG.

1.8 What Should Be Done

After the observations made on the hydrocarbon policy for the period 2003–2014, it seems necessary to make some alternative proposals.

A previous budget.

Without an integrated and sustainable development plan that gives an idea of the type of country that Argentines want in the medium and long terms, all energy policy will cease to have any support.

As energy, at least in Argentina, is a satisfactory need of the socioeconomic system, its quality, quantity and diversity will respond to the country model that is decided.

Otherwise, the market or, better yet, the banks will take over.

Recreate energy planning as an indicator of the most appropriate allocation of energy resources and which has as its main objective to ensure the supply of energy in quantity and quality required by the socioeconomic system not only for present generations but also for future ones. These plans must be independent of the periods of government and must be updated annually or biannually.

The Secretariat of Energy, or the Ministry to be created, should be given a higher profile and be responsible for implementing the policies of the energy plan, rather than the tremendously short-term non-planning that has been carried out by the Ministry of Planning or the Ministry of the Economy during many years of the period analyzed.

Recreating the energy information systems, for example, by recovering the annual fuel and electricity bulletins, which have, of course, been improved and made available electronically, thus greatly facilitating the tasks of energy scholars.

Preparation of an energy code and a new Hydrocarbons Law, which will govern activity throughout the country. The fact is that the constitutional reform transferred natural resources to the exclusive domain of the provinces and this is generating, to be mild, imbalances and asymmetries in the management of these resources and makes it very difficult to design and apply a coherent energy policy. This code should be drafted by the National Congress.

A strong role for energy regulators. As long as private energy companies exist, their annual and medium-term plans should be monitored for compliance by these entities, ensuring that they are compatible with energy plans and, in particular, that they comply with investment plans in a timely manner. In this way, the supply of all energies would be assured in terms of quantity, quality and time. The public energy companies should also be subject to the rules established by the corresponding entities.

In general, massive energy exports will not be encouraged, except in the case of interconnections and agreements with Latin American and Caribbean countries or exceptional discoveries. In this last case, a policy similar to that of Norway should be followed, not that of Nigeria, and the surplus resources should be used to improve the quality of life of the population, ensure the supply of future generations and industrialize the country, that is, break with the policy of exporting raw materials and importing industrialized goods.

A necessary condition for the implementation of an energy policy that contemplates the present and future interests of the country and its inhabitants is the recreation of the State Energy Companies, each one with its sphere of influence and coordinated through the Secretariat of Energy.

But these companies must have management autonomy, their management personnel must be selected by competition and not by political cronyism, they must present their plans annually in the National Energy Plan, they must account for their actions in terms of compliance with the goals contained in the plan, not only those of production and be federal, that is, the provinces must join their boards and participate not only in decisions but also in the "benefits" that the companies produce.

Perhaps the model, adopted to our reality, is that of companies like "Electricité de France" and its plan contracts.

The Provinces should be integrated into the directories of companies so that they are not just recipients of royalties.

In this way, the natural resources existing in provincial territories will now belong to the provinces and to the nation, and the same should happen with the energy resources existing in marine waters.

> As today there are no national energy companies and in the case of hydrocarbons, the private companies in fact exercise control over the existing resources in their concessions, it seems important to draw up a new hydrocarbons law to replace the inadequate one. For the purposes proposed here, a detailed analysis of the current Hydrocarbon National Law 17,319, reformed in 2007, will be carried out, especially with regard to compliance with investments and good practice, and in the event of non-compliance, the resources will be returned to the state's assets.

In this way, it would be necessary to suspend the extension of the concession contracts that the provinces are making, review those granted and convert the exploration permits and these exploitation concessions into service or association contracts with the majority of the state YPF and without free availability of foreign currency or oil and natural gas.

The following two points should be analyzed in the following context which will give an idea of the magnitude of the hydrocarbon "business" in Argentina.

The sale of oil derivatives plus exports and minus imports would generate in 2014 no less than US $33 billion dollars per year at the official exchange rate.

The economic value of oil and natural gas reserves, at international prices, would reach US $214 billion dollars.

But how will the implementation of the energy works be financed?

Years ago, a large part of the financing of the energy works came from the taxes on fuels that were assigned to build hydroelectric, nuclear and conventional plants and to construct the country's gas pipeline networks.

During the 1990s, everything was privatized and the State delegated the investments to private companies because "the market was a good allocator of resources" (sic).

This policy continued with subsequent government administrations.

But the private entrepreneur must respond to the interests of his shareholders and these do not necessarily coincide with what the country needs in terms of the type and opportunity of the works.

The objectives of making the market more transparent and more competitive were practically not met and the consequences are clear.

In addition to good management and resources from fuel taxes, via the Energy Fund, resources can be obtained from equipment suppliers' credits, from multilateral banks, from a regional bank like Banco Sur, from the recreation of the National Development Bank (i.e., redirecting the current banking system) and to associate through bond issues to Argentineans.

But what is the role of the private entrepreneur in this scheme?

In a very general way, it can be said that the equipment of the energy sector in Argentina requires between 5 and 8% of the GDP, depending on the delay of works and the growth of the demand, which as it was said is intended to be attenuated by the energy saving plan.

Then the private sector that wants to be linked to the energy business will have to reconvert, leave the operation of the fields and the power plants, gas pipeline, etc., to the state companies and dedicate itself to the manufacture and installation of the equipment that the activity requires.

For this purpose, it will have the purchasing power of the state-owned companies, which can vary between US $15 and 20 billion dollars a year (this will emerge from the Energy Plan that will be drawn up).

It seems that this role of entrepreneurs may be more useful to the country than that of the current Argentine so-called "entrepreneurs", participating in YPF S.A.

This "Argentinization" of YPF is quite similar to that of YPF's old contractors, when some of them created their empire with virtually no business risk.

Energy Prices and Tariffs Policy.

The most genuine way for a company to finance the capacity of its productive structure is throughout the income it receives from the sale of its products.

In the case of the energy sector, it is not usual for a public or private company to self-finance all its investments. The fact is that often large investments have to be made that mature several years later.

For this reason, bank and supplier credit and sometimes company capitalization are often used.

In any case, it is not advisable that the company's own income is not enough to cover a more or less important part of the investments.

The country is applying an erroneous policy of prices and tariffs for energy products, especially for electricity (it is said that tariffs barely cover 20% of costs) and natural gas, whose sales values to the final consumer are too low.

The case of oil derivatives is different and the current prices, measured in constant dollars per liter, for gas oil and super naphthas are today above those in effect during the convertibility.

The low tariffs for electricity and natural gas are "compensated", in part, to the companies through a policy of global subsidies that has generated a situation whose exit can be very traumatic. The fuel subsidies took about 26 billion Argentine pesos in 2010 and imports totaled almost US $7 billion dollars.

Trust funds have been another instrument for private companies to take over ownership of the power plants.

A gradual increase in average electricity and natural gas rates would therefore be appropriate in order not to repeat the traumatic departure from the convertibility plan.

In terms of average tariffs, cross-subsidies are being proposed between different types of users so that those who consume most pay more per unit consumed than those with low consumptions and businesses more than industries.

In other words, the tariff structures must be reviewed. This should not be a policy for the economy but should be the country pricing policy.

Of course, if the operation of the sector were in the hands of state-owned companies, the implementation would be simpler, because the values should cover the costs and an adequate expansion of the productive capacity and not allocate amounts to the shareholders who own the private companies.

In summary, it is needed:

A comprehensive national plan for Argentina sustainable development.
Energy planning.
Recreation of the decision-making power of the Secretariat of Energy.
Preparation of an energy code and a new hydrocarbons National Law.
Recreation of national and federal state energy companies.

Recreation of the Energy Fund from the taxes on oil and natural gas derivatives (which are non-renewable energies and are putting a lot of pressure on the Energy Matrix structure) to finance part of the expansion of productive capacity, including the mining risk involved in exploration.

Assign the private sector the role of supplier of the equipment and inputs required by the companies' investments.

Recreate the National Development Bank as a genuine source of financing for part of the sector's investments.

Reformulate another price and tariff structure to gradually emerge from the current quasi-freezing and also serve as a tool for modifying the Energy and Electricity Generation Matrix.

Finally, it will not be possible to implement an energy policy that serves the country's present and future generations if the State Energy Companies are not recreated.

Annex: Text of National Law 27,007 of 30 October 2014

National Law No. 17,319. Amendment.
Sanctioned: October 29, 2014
Enforced: October 30, 2014

The Senate and the Chamber of Deputies of the Argentine Nation, meeting in Congress, etc., sanction with the force of law:

Title I

Amendments to National Law 17,319

SECTION 1—Section 23 of National Law 17.319, as amended, shall be replaced by the following text:

Section 23: The terms of the exploration permits shall be set in each tender by the Authority of Application, according to the objective of the exploration, as follows:

Basic Term:

Exploration with a conventional objective:

1st period up to three (3) years.
2nd period up to three (3) years.
Extension period: up to five (5) years.

Exploration with non-conventional objective:

1st period up to four (4) years.
2nd period up to four (4) years.
Extension period: up to five (5) years.

For exploration on the continental shelf and in the territorial sea, each of the periods of the Basic Exploration Period with a conventional objective may be increased by one (1) year.

The extension provided for in this article is optional for the permit holder who has fulfilled the investment and other obligations under its charge.

The partial transformation of the area of the exploration permit into an exploitation concession carried out before the expiration of the Basic Term of the permit, in accordance with the provisions of Article 22, authorizes the addition to the term of the concession of the period not elapsed of the exploration permit, excluding the term of the extension.

At any time, the permit holder may renounce all or part of the area covered by the exploration permit, without prejudice to the obligations prescribed in Article 20.

Article 2—Replace Article 25 of National Law 17,319 and its amendments with the following text:

Article 25: Exploration permits shall cover areas not exceeding one hundred (100) units. Those granted on the continental platform shall not exceed one hundred and fifty (150) units.

Article 3—Replace Article 26 of National Law 17,319 and its amendments, by the following text:

Article 26: At the end of the first period of the Basic Term, the permit holder shall decide whether to continue exploring in the area, or whether to return it to the State in full. The permit holder may maintain the entire area originally granted, provided that it has complied with the obligations arising from the permit.

At the end of the Basic Term, the permit holder shall return the entire area, unless it exercises the right to use the extension period, in which case such return shall be limited to fifty percent (50%) of the remaining area before the expiration of the second period of the Basic Term.

SECTION 4—Replace section 27 of National Law 17,319, as amended, with the following text:

Article 27: The exploitation concession confers the exclusive right to exploit the hydrocarbon deposits existing in the areas covered by the respective concession title during the term set forth in Article 35.

The holders of hydrocarbon exploration permits and/or exploitation concessions shall have the right to apply to the Authority of Application for a non-conventional hydrocarbon exploitation concession, under the terms set forth in Article 22 or Article 27 bis, as appropriate.

SECTION 5—The following shall be incorporated as section 27 bis of National Law 17.319, as amended:

Article 27 bis: Non-conventional hydrocarbon exploitation is understood as the extraction of liquid and/or gaseous hydrocarbons by means of non-conventional stimulation techniques applied in reservoirs located in geological formations of shale gas or shale oil, compact sandstone (tight sands, tight gas, tight oil), coal bed methane, and/or characterized, in general, by the presence of low permeability rocks. The exploitation concessionaire, within the concession area, may require the subdivision of the existing area into new areas of non-conventional hydrocarbon exploitation and the granting of a new non-conventional hydrocarbon exploitation concession. Such request must be based on the development of a pilot plan which, in accordance with acceptable technical-economic criteria, aims at the commercial exploitation of the discovered field. The national or provincial Authority of Application, as appropriate, will decide within sixty (60) days and its validity will be calculated as established in Article 35.

The holders of a non-conventional hydrocarbon exploitation concession, who in turn are holders of an exploitation concession adjacent and preexisting to the first one, may request the unification of both areas as a single non-conventional exploitation concession, provided that the geological continuity of such areas is reliably demonstrated. Such a request must be based on the development of the pilot plan provided for in the previous paragraph.

The concession corresponding to the area duly granted and not affected by the new non-conventional hydrocarbon exploitation concession shall remain in force for the terms and conditions previously existing, and the granting authority shall readjust the respective title to the extension resulting from the subdivision. It is hereby established that the new concession for non-conventional hydrocarbon exploitation shall have as its main objective the non-conventional exploitation of hydrocarbons. However, the holder of the concession may carry out complementary conventional hydrocarbon exploitation activities, within the framework of the provisions of Article 30 and related articles of this law.

Article 6—The following article shall be incorporated as article 27 ter of National Law 17,319 and its amendments:

Article 27 ter: Those projects of tertiary production, extra heavy oil and offshore that due to their productivity, location and other unfavorable technical and economic characteristics, and that are approved by the Authority of Application and the commission of Planning and Strategic Coordination of the National Plan of Hydrocarbon Investments, may be subject to a reduction of royalties of up to fifty percent (50%) by the provincial or national Authority of Application, as applicable. Tertiary production projects are considered those production projects in which enhanced oil recovery techniques are applied (enhanced oil recovery—EOR or improved oil recovery—IOR). Extra heavy oil projects are considered those requiring special treatment (crude oil quality below 16 degrees API and with viscosity at reservoir temperature above 1000 centipois).

SECTION 7—Section 29 of National Law 17,319, as amended, shall be replaced by the following text:

Article 29: Exploitation concessions shall be granted, as the case may be, by the national or provincial executive branch to the individuals or corporations exercising the right agreed upon in Article 17 by complying with the formalities set forth in Article 22.

The national or provincial executive branch, as appropriate, may also grant exploitation concessions on proven areas to those who meet the requirements and observe the procedures specified in Section 5 of this Title.

This modality of concession does not imply in any way that the existence of commercially exploitable hydrocarbons in such areas is guaranteed.

The national or provincial executive branch, as appropriate, shall also grant non-conventional hydrocarbon exploitation concessions in accordance with the requirements set forth in Articles 27 and 27 bis.

Article 8—Article 34 of National Law 17,319 and its amendments shall be replaced by the following text:

Article 34: The maximum area of a new exploitation concession granted after the term of the present law and not arising from an exploration permit shall be two hundred and fifty square kilometers (250 km^2).

SECTION 9—Section 35 of National Law 17,319 and its amendments shall be replaced by the following text:

Article 35: According to the following classification, exploitation concessions shall have the terms set forth below, which shall be counted from the date of the resolution granting them, plus the additional ones resulting from the application of Article 23:

(a) Concession for conventional hydrocarbon exploitation: twenty-five (25) years.
(b) Concession for non-conventional exploitation of hydrocarbons: thirty-five (35) years. This term shall include a Pilot Plan period of up to five (5) years, to be defined by the concessionaire and approved by the Authority of Application at the time the concession begins.
(c) Exploitation concession with the continental platform and in the territorial sea: thirty (30) years.

The holders of the exploitation concessions (whether or not they have been extended as of the effective date of this amendment) and provided that they have complied with their obligations as exploitation concessionaires, are producing hydrocarbons in the areas in question and submit an investment plan consistent with the development of the concession, may request extensions for a period of ten (10) years each.

The respective app lication must be submitted no less than one (1) year prior to the expiration of the concession.

It is established that those exploitation concessions that at the date of sanction of the present law have been previously extended will be governed until the exhaustion of the terms of such extensions by the existing terms and conditions. Once these extension periods have expired, the holders of the exploitation concessions may request new extensions, and must comply with the extension conditions established in this law.

Article 10. Substitute Article 41 of the National Law 17,319 and its amendments, by the following text:

Article 41: The concessions that the present section refers to will be assigned and extended for time periods equivalent to those given to the concessions of exploitation related to the transport concessions. Once these periods have expired, the facilities and equipment would be transferred to the dominion of the National or Provincial states, as corresponding, without charges or taxes of any kind and of full rights.

Article 11. Substitute Article 45 of the National Law 17,319, and amendments, by the following text:

Article 45: Without affecting what has been established in Article 27 bis, the permits and concessions regulated by this law will be adjudicated by means of tenders in which offers could be presented by any physical or legal person, which would exhibit the conditions established in Article 5 and fulfill the conditions requested in this section.

The concessions that result from the application of the Articles 29, first paragraph, and 40, second paragraph, will be adjudicated complying with procedures established in Title II of the present law.

Article 12. Replace Article 47 of National Law 17,319, as amended, with the following:

Article 47: Once the call for bids has been issued in any of the procedures considered by Article 46, the Authority of Application shall prepare the respective tender document, based on the Model Tender Document prepared between the Authorities of Application of the provinces and the National Secretariat of Energy, which shall include, by way of illustration and with an indication of its origin, the available information concerning the submission of tenders.

The tender documents shall also contain the conditions and guarantees to which the bids must conform and shall set forth the fundamental bases that will be taken into consideration to assess the suitability of the bids, such as the amount and terms of the investments in works and projects that are committed. The call for bids shall be disseminated during no less than ten (10) days in the places and by national and international means considered suitable to ensure its widest knowledge, seeking the greatest possible competition, and the Official Gazette shall necessarily be included among these. Publications shall be made at least sixty (60) days prior to the date indicated for the start of the reception of offers.

SECTION 13. Section 48 of Act 17.319, as amended, shall be replaced by the following text:

Article 48: The Authority of Application shall study all proposals and may require from those bidders that have submitted the most interesting ones, the improvements it considers necessary to reach satisfactory conditions. The award shall be made to the bidder that has submitted the most convenient offer, rather than to the duly founded criteria of the national or provincial executive branch, as appropriate, in particular proposing the largest investment or exploratory activity.

It is the duty of the National or Provincial Executive Power, as appropriate, to reject all the bids submitted or to award the contract to the only bidder in the tender.

Article 14. Include in Title II of National Law 17,319 and its amendments, Section VII "Fees and Royalties" is incorporated to Title II of National Law 17,319, as amended, which shall comprise sections 57, 58, 59, 60, 61, 62, 63, 64 and 65, and sections 57 and 58 of National Law 17,319, as amended, are replaced by the following texts:

Article 57: The holder of an exploration permit shall pay annually and in advance a fee for each square kilometer or fraction thereof, according to the following scale:

(a) Basic Term:

1st Period: two hundred and fifty pesos ($ 250).
2nd Period: one thousand pesos ($ 1000).

(b) Extension:

During the first year of its validity, it shall pay in advance the amount of seventeen thousand five hundred pesos ($ 17,500) per square kilometer or fraction thereof, increasing said amount by twenty-five percent (25%) annually. The amount to be paid for this concept corresponding to the second period of the Basic Term and the Extension Period may be readjusted by compensating it with the investments actually made in exploration within the corresponding area, up to a minimum fee equivalent to ten percent (10%) of the fee corresponding to the period per square kilometer that will be paid in all cases.

Article 58: The exploitation concessionaire shall pay annually and in advance a canon for each square kilometer or fraction thereof covered by the area of four thousand five hundred pesos ($ 4500).

Article 15. The following shall be incorporated as article 58 bis of National Law 17,319 and its amendments:

Article 58 bis: The Authority of Application may establish for the extension of exploitation concessions, the payment of an extension bonus whose maximum amount shall be equal to that resulting from multiplying the remaining proven reserves at the end of the concession term by two percent (2%) of the average basin price applicable to the respective hydrocarbons during the two (2) years prior to the granting of the extension.

In the case of complementary activities of conventional hydrocarbon exploitation, from the expiration of the period of validity of the concession duly granted and within the non-conventional hydrocarbon exploitation concession, the Authority of Application may establish the payment of an exploitation bonus whose maximum amount shall be equal to the result of multiplying the remaining proven reserves associated with the conventional exploitation of hydrocarbons at the end of the period of validity of the concession duly granted and by two percent (2%) of the average basin price applicable to the respective hydrocarbons during the two (2) years prior to the granting of the non-conventional hydrocarbon exploitation concession.

SECTION 16. Section 59 of National Law 17,319, as amended, shall be replaced by the following text:

Article 59: The exploitation concessionaire shall pay the grantor a monthly royalty on the production of liquid hydrocarbons extracted at the wellhead, in the amount of twelve percent (12%). The same percentage of the value of the volumes extracted and actually exploited shall pay monthly the production of natural gas, as a royalty. For the payment of this royalty, the value of the gas shall be fixed in accordance with the procedure indicated for crude oil in Article 61.

In both cases, the National or Provincial Executive, as the case may be, as the granting authority, may reduce this royalty up to five percent (5%) taking into account the productivity, conditions, and location of the wells. Likewise, in the event of an extension, an additional royalty of up to three percent (3%) shall be payable with respect to the royalty applicable at the time of the first extension and up to a total maximum of eighteen percent (18%) of royalty for the following extensions.

In the case of the exploitation concessions referred to in the last paragraph of Article 35, a total royalty of not more than eighteen percent (18%) shall be payable.

For the performance of the complementary activities of conventional exploitation of hydrocarbons, referred to in Article 27 bis of the present National Law, as of the expiration of the period of validity of the concession duly granted and within the non-conventional hydrocarbon exploitation concession, the Authority of Application may also set an additional royalty of up to three percent (3%) over the current royalty up to a maximum of eighteen percent (18%) as appropriate in accordance with the mechanism established in Article 35.

The rates of royalty provided for in this article shall be the only mechanism for income from the production of hydrocarbons to be received by the jurisdictions holding title to the hydrocarbons in their capacity as background.

Article 17. Article 61 of National Law 17,319, as amended, is replaced by the following text:

Article 61: The cash payment of the royalty shall be made in accordance with the value of crude oil at the wellhead, which shall be declared monthly by the permit holder and/or concessionaire, subtracting from the amount established in accordance with the rules set forth in paragraph I, section c) of Article 56, the freight of the product to the place that has been taken as a basis to establish its commercial value. When the Enforcement Authority considers that the sales price reported by the permit holder and/or concessionaire does not reflect the actual market price, it shall raise the objections it considers relevant.

Article 18. The following is incorporated as Article 91 bis of National Law 17,319, as amended:

Article 91 bis: The provinces and the National State, each in relation to the exploration and exploitation of the hydrocarbon resources in its domain, shall not establish in the future new reserve areas in favor of public or state-owned entities or companies, regardless of their legal form. With respect to the areas that have been reserved by the authorities to date in favor of provincial entities or companies with State participation, whatever their legal form, but which to date do not have association contracts with third parties, association schemes may be carried out, in which the participation of such entities or provincial companies during the development stage will be proportional to the investments committed and actually made by them.

Title II

Investment Promotion Regime for Hydrocarbon Exploitation

Article 19. The National State shall incorporate into the Investment Promotion Regime for Hydrocarbon Exploitation, created by means of Decree 929/13, those projects that imply the realization of a direct investment in foreign currency of no less than U$S two hundred and fifty million dollars (US $250,000,000) calculated at the time of the presentation of the "Investment Project for Hydrocarbon Exploitation" and to be invested during the first three (3) years of the project.

The benefits provided for in said decree shall be recognized as from the third year counted from the implementation of the respective projects.

The percentage of hydrocarbons for which the benefits provided for in Articles 6 and 7 of the cited said decree shall be applied, shall be the following:

(a) Conventional exploitation: twenty percent (20%).
(b) Conventional exploitation: twenty percent (20%).
(c) Offshore exploitation: sixty percent (60%).

Within paragraph (c) above, those offshore exploitation projects in which the drilling of wells is carried out in locations where the distance between the seabed and the surface, measured at the location of the well, on average between the high and low tide exceeding 90 m, shall be included. Any other offshore exploitation project not meeting those requirements shall fall under subparagraph (a) or (b), as appropriate.

Article 20. The conditions for access to the Investment Promotion Regime for Hydrocarbon Exploitation provided for in Article 19 shall be in force as from the entry into force of this law, recognizing the Investment Projects for Hydrocarbon Exploitation approved previously, the investment commitments and the promotional benefits committed at the time of their approval.

Article 21. Within the framework of the Investment Projects for the Exploitation of Hydrocarbons that are approved in the future by the Strategic Planning and Coordination Commission of the National Plan for Hydrocarbon Investments, referred to in Article 19 of the present law, the following contributions are established for the producing provinces in which the investment project is developed:

(a) Two point five percent (2.5%) of the initial investment amount of the project, aimed at Corporate Social Responsibility, to be contributed by the companies.
(b) An amount to be determined by the Strategic Planning and Coordination Commission of the National Hydrocarbon Investment Plan, depending on the magnitude and scope of the investment project to finance infrastructure works in the producing provinces, to be contributed by the national State.

Article 22. The capital goods and inputs that are essential for the execution of the Investment Plans of the companies registered in the National Registry of Hydrocarbon Investments, whether imported by such companies or by those who prove to be their service providers, shall be subject to the import duties indicated in Decree 927/13 or the regulations that replace it. This list may be extended to other strategic products.

Title III

Supplementary and transitional provisions

Article 23. The National State and the Provincial States, in accordance with the provisions of Article 41 of the National Constitution, shall promote the establishment of a uniform environmental legislation, which shall have as a priority objective to apply the best environmental management practices to the tasks of exploration, exploitation and/or transportation of hydrocarbons in order to achieve the development of the activity with an adequate care of the environment.

Article 24. The National State and the Provincial States shall encourage the adoption of a uniform fiscal treatment that promotes the hydrocarbon activities foreseen in the present law to be developed in their respective territories.

Article 25. The National Executive Branch, through the Commission for Strategic Planning and Coordination of the National Plan for Hydrocarbon Investment, shall administer the Program for Stimulating the Injection of Surplus Natural Gas created by Resolution 1/13 and the "Program for Stimulating the Injection of Natural Gas for Companies with Reduced Injection" created by Resolution 60/13, in both cases of the Commission for Strategic Planning and Coordination of the National Plan for Hydrocarbon Investment, and the plans which it establishes for the purpose of stimulating the surplus production of natural gas in the future.

Article 26. The national and provincial Application Authorities, as appropriate, the National Energy Secretariat and the Commission for Strategic Planning and Coordination of the National Plan for Hydrocarbon Investment shall promote the unification of procedures and records aimed at fulfilling their respective competences and the exchange of information for that purpose and with the fulfillment of the self-supply objectives provided for in National Law 26,741.

SECTION 27. Section 62 of National Law 17,319, as amended, is hereby repealed.

SECTION 28. The National or Provincial Executive Branch, as the case may be, as the Granting Authority, may reduce up to twenty-five percent (25%) of the amount corresponding to royalties applicable to the production of hydrocarbons and during the ten (10) years following the completion of the pilot project, in favor of companies requesting a non-conventional hydrocarbon exploitation concession, pursuant to the terms of Article 27 bis of National Law 17,319, as amended, within thirty-six (36) months as of the effective date of this law.

Article 29. The provincial enforcement authorities and the National Secretariat of Energy shall prepare within one hundred and eighty (180) days as from the effective date of this law the Model Bidding Form established in Article 47 of National Law 17,319 and its amendments, which may be reviewed and updated periodically according to the opportunity and convenience of the bids. Said Model Bidding Document shall contemplate the general terms and conditions applicable to the bids, including, among others, the guarantees to which the bids must conform, the scope of the investments and the income that may eventually correspond to the respective background authorities. Likewise, the model bidding terms and conditions shall contain the special conditions applicable to awards whose object is the conventional exploration and/or exploitation of hydrocarbons, non-conventional exploitation, offshore, extra heavy oil, exploration in border areas and other situations that may be contemplated by said application authorities.

Article 30. Section 2 of National Law 25,943 is hereby repealed, and for such purpose, all hydrocarbon exploration permits and concessions for exploitation of national offshore areas shall be reverted and transferred to the National Energy Secretariat, with respect to which there are no association agreements entered into with Energía Argentina Sociedad Anónima within the framework of National Law 25,943.

Except for such reversion to the exploration permits or exploitation concessions existing at the time, this law comes into force that was granted prior to National Law 25,943.

The Executive Branch is authorized to negotiate by mutual agreement, within a term of six (6) months, with the holders of association contracts signed with Energía Argentina Sociedad Anónima under National Law 25,943, the conversion of such association contracts to exploration permits or exploitation concessions under National Law 17,319, as amended.

Article 31. When on the date of entry into force of this law any province has already initiated the process of extension referred to in Article 35 of National Law 17,319, as amended, with respect to concessions granted by the national state, and provided that such process has established certain conditions precedent in accordance with the will of the province and the srespective concessionaire and the laws in force, the province shall have a period of ninety (90) days to conclude the process of extension by issuing the necessary administrative acts in charge of the Provincial Executive. The extensions thus determined will subsequently be treated as provided for in Article 35 of National Law 17,319 and its amendments.

Article 32. Communicate to the National Executive Branch.

GIVEN IN THE SESSION HALL OF THE ARGENTINE CONGRESS, IN BUENOS AIRES, ON THE TWENTY-NINTH DAY OF OCTOBER OF THE YEAR TWO THOUSAND AND FOURTEEN.

—REGISTERED UNDER N° 27,007—AMADO BOUDOU.—JULIAN A. DOMINGUEZ.—Lucas Chedrese.—Juan H. Estrada.

Bibliography

Books, Documents and Journal Articles

Bravo V (1971) Hydrocarbon Legislation in Argentina. Working Documents, Fundación Bariloche. San Carlos de Bariloche, Argentina

Bravo V (1992) Why Privatize YPF. Economic Reality IADE N° 110

Bravo V (1993) YPF S.A.: What now? Economic Reality IADE N° 117

Bravo V (2012) The Argentine Energy Matrix. Fundación Bariloche, Working Documents

Bravo V (2013) "Una Opinión sobre el Fracking", Fundación Bariloche, Working Paper, San Carlos de Bariloche. In Spanish; in English in this volumen

Darrha TH et al (2014) Noble gases identify the mechanisms water wells overly the Marcellus and Barnett Shales. University of Utah, Salt Lake City, Utah, U.S.A., 12 Aug 2014

Di Sbroiavacca N (2013) Shale oil y Shale Gas en Argentina. Fundación Bariloche, Working Documents. In Spanish; in English in this volume

King GE (2013) Hydraulic Fracturing 101. Apache Corporation, September 2013

Kozulj R, Bravo V, with the collaboration of Nicolás Di Sbroiavacca (1993) Oil Deregulation Policy in Argentina. Antecedentes e Impactos. University Libraries, Latin American Publishing Center

Laws, Decrees and Resolutions

National Law 17,319/1967

Mining Code 1886

National Law 12,161/1935

National Constitution 1949

National Law 14,773/1958

Decree 1445 5/8/1985

Decree 1443 1985

Decree 633 1987

Decree 1812 1987

National Law 23,696 1989

National Law 23,697 1989

National Law 23,928 1991

Decree 1224 1989

Decree 1225 1989

Decree 1757 1990

Decree 2408 1991

Decree 1055 1989

Decree 1212 1989

Decree 1589 1981

Decree 2278 1990

National Law 24,145 1992

Decree 1108 1993

Argentine Constitution, reformed, 1994

National Law 24,474 11/4/95

Decree 628 July 1997

Decree 857 July 1998

National Law 25,565 6/3/2002

National Law 26,337 28/12 2007

Decree 786 2002
Decree 180 2004
Resolution 5 2010
Resolution 209 2014
National Law 26,095 2006
Decree 310 February 2002
National Law 25,561 2002
Decree 806 May 2004
Decree 809 13 /5/2002
Decree 645 26/5/2004
Decree 526 22/10/2002
Decree 335 11/5/2004
Decree 336 11/5/2004
Decree 337 11/5/2004
Resolution 532 August 2004
National Law 26,217 2007
Resolución 394 15/11/2007
Resolution 1, 2013
Resolution 803 22/10/ 2014
Resolution 99 2013
National Law 25,943 10/10/2006
Decree 1955 4/11/2004
Decree 546 6/8/2003
National Law 26,154 27/10/2006
National Law 26,197 2007
National Law 26,360 2008
Resolution 459 12/7/2007
Resolution 121 3/3/ 2008
Resolution 1451 12/12/2008
Decree 287 19/12/2008
Decree 2014 25/11/2008
Resolution 24 6/3/2008
National Law 26,741 4/5/2012
National Law 21,499 1977
Decree 1277 25/7/2012
Decree 1 18/1/2013
Resolution 60 2013
Resolution 83 2013
National Law 25,675 27/11/2002
Decree 929 11/7/2013
Decree 1277 2013
Resolution 9 15/7/2013
Decree 927 8/7/2013
National Law 26,020 7/4/2005
Resolution 792 28/6/2005
Decree 1539 19/11/2008
Resolution 1083 1/10/2008
Resolution 532, 22/5/2014
National Law 26,190 2/1/2007
Decree 562 10/5/2009
National Law 26,093 12/5/ 2006
National Law 26,334 1/1/2007
Resolution 1125 30/ 12/2013

National Law 26,942 17/6/2014
Resolution 44 16/9/2014
Resolution 179 28/11/2014
Decree 140 21/12/2007
Resolution 24 15/1/2008
Resolution 8 24/1/ 2008
Resolution 745 2005
National Law 27,007 October 2014
Resolution 1077, 29 December 2014
Decree 2579, 30 December 2014

Chapter 2
Argentine Energy Policy During the Period 2014–2018

Victor Bravo

Abstract This paper analyzes the Energy Policy applied in Argentina during the period 2014–2018, which includes practically all of President Mauricio Macri's administration. This analysis is necessary to place the case of the Fracking de Vaca Muerta in its energy context. This is how the following points will be developed:-Energy planning, if any, and its scope-Crude oil and its derivatives-Gaseous and Liquid Natural Gas-Liquefied Petroleum Gas-Especially the situation in the Vaca Muerta sites-Electric power-Nuclear power-Coal-Renewable energies, especially wind, solar and biomass energy-What are Private Participation Contracts applied to the energy field?-How distributed energy is considered in the electrical system.-The efficient use of energy—The problem of natural gas and electricity tariffs, particularly the effects of the so-called Tariffs—The constitution and structure of fuel taxes—The price of crude oil and its derivatives, as well as biofuels-Special paragraph deserves the issue of energy subsidies—The panorama closes by describing the composition of the Argentine Energy Matrix—The document concludes with general suggestions for an Alternative Energy Policy to those in force during the last two governments.

Keywords Energy planning · Crudeoil · Natural gas · Liquefied petroleum gas · Vaca muerta · Electric power · Nuclear power · Coal · Renewable energies · Distributed energy · Tariffs · Energy subsidies · Argentine energy matrix · Alternative energy policy

This work is dedicated to all friends of the Bariloche Foundation, academics, technicians and administrators. An extraordinary human group.

2.1 Why This Document

Energy policy is an exciting subject, and it is appropriate, having lived through several different versions, to analyze the current one in 2019.

V. Bravo and N. Di Sbroiavacca, *Oil and Natural Gas Economy in Argentina*,
The Latin American Studies Book Series,
https://doi.org/10.1007/978-3-030-65520-4_2

It seemed necessary to leave a document from this period that could be useful to those who are concerned about the fate of Argentina.

The same was done, as it is done here, with what happened in Argentina since the 1960s, and the analysis made with Robert Kozulj and Nicolás Di Sbroiavacca in the 1990s was detailed. Then, the author only, wrote of what happened in the 2000s until 2015 and now.

Some say that this is a time of great change. But really great changes were those of the 1990s. There they made a 180-degree turn and now they are back on track.

Of course, in this document, there is an explicit and implicit ideology, which the author drank from the Gospel, learned in the Conade (National Council of Development) in 1965 and confirmed it afterward in the Bariloche Foundation.

Of course, nothing the present author says or fails to say here compromises the Bariloche Foundation.

Everything in this paper is uniquely the responsibility of who is writing this paper.

2.2 Summary

This paper analyzes the energy policy applied in Argentina during the period 2014–2018 which includes practically all of President Mauricio Macri's administration.

This analysis is necessary to place the case of fracking in the Vaca Muerta oil field in its energy context.

This is how the following points will be developed:

Energy planning, if any, and its scope;
Crude oil and its derivatives;
Gaseous and liquid natural gas;
Liquefied petroleum gas;
Particularly, the situation in the Vaca Muerta oil field;
Electric power;
Nuclear power;
Coal;
Renewable energies, especially wind, solar and biomass energy;
What are private participation contracts applied to the energy field?
How distributed energy is considered in the electrical system?
The efficient use of energy;
The problem of natural gas and electricity charges, particularly the effects of the so-called tariffs;
The constitution and structure of fuel taxes;
The price of crude oil and its derivatives, as well as biofuels;
Special paragraph deserves the issue of energy subsidies;
The panorama closes by describing the composition of the Argentine Energy Matrix;

The document concludes with general suggestions for an alternative energy policy to those in force during the last two governments.

2.3 Comments

While the comments are contained in the various paragraphs of the text, usually after the numerical information, it has been thought appropriate to summarize them here for those who will not venture to read the extensive text.

In the end, something will be said about alternative policy guidelines.

Firstly, it is necessary to situate oneself in the socioeconomic context of the country in the period under analysis.

A justification of this situation would, in itself, merit a new document to be prepared by, for example, economists or political scientists, altogether.

In this regard, some statements will be made, it is believed, based on numerous social and economic indicators available, for example, in the National Institute of Statistics and Census (INDEC).

The economic and social situation is certainly not easy. It includes the fall in economic activity, deindustrialization, "primarization" of the economy, unemployment, inflation and deficit trade balance; increase in poverty, indigence and marginalization, heavy foreign debt and flying capital; fall in the purchasing power of salaries and pensions, and more unregistered employment; and decline in support for education in science and technology.

Of course, not all of them appeared under the current government, some come from before, but they would have been accentuated.

The comments that follow are statements that are supposed to be based on the hard core of the text.

2.3.1 Energy Planning

In Argentina, today, there is no energy planning or energy plan. There are energy scenarios and each area of the current Secretary of Energy and Mining would seem to behave as a watertight department with its own objectives and strategies without much coordination with the rest.

This also implies the need for socioeconomic planning, which serves as a framework for energy planning, not as an immovable corset but as a guide that points toward the direction to be followed.

The so-called energy plan to 2030 seems to be designed to offer external investors business opportunities as if the country were at the top and Vaca Muerta oil field was the "grandmother's jewel," still available to be sold.

2.3.2 *Crude Oil and Its Derivatives*

(i) **The audited reserves**

There is evidence today of an almost abandonment of the search for "conventional" oil, and much effort has been put into confirming the non-conventional oil and gas resources of the Vaca Muerta oil field.

The bidding of areas for exploration in the Argentine Sea (offshore) did not privilege the national oil company (YPF S.A.), nor can it serve to train Argentine technicians in this little known technology in Argentina, with the aggravating factor of the large extension of the submarine platform. That is to say, the possible resources of marine hydrocarbons would be handed over to foreign companies and only royalties, income tax, and some not very qualified employment would remain.

(ii) **Investments**

Investments in production and exploration decreased between 2015 and 2017 and rebounded in 2018 almost exclusively due to the Vaca Muerta oil field.

(iii) **Production**

Production has fallen sharply, especially in the non-conventional fields, and almost all the effort has been put into the non-conventional ones, neglecting the reactivation of "mature" fields that did not necessarily cease to be suitable for production.

(iv) **The oil trade balance**

Both in volume and in value, the balance has deteriorated and the upturn in 2018 is only due to the fall in domestic consumption.

(v) **Refining and import of derivatives**

There have been no investments to modernize and expand the refineries. Companies attribute this to the fall of the domestic market and, on the other hand, prefer to import the derivatives. This has been verified especially in the naphthas and also by the changes of flag produced in several companies.

(vi) **Pipelines**

The lack of coordination within the sector (in reality it is managed by private companies at their best convenience) has been evidenced in the lack of capacity of the pipelines to transport the growing production of Vaca Muerta oil field, and it is only at the end of 2018 that the concern to expand OLDEVAL is noted.

(vii) **Something about the prices of derivatives**

One of the first actions was to modify the reference crude oil by changing the traditionally used WTI for Brent, which has become almost 15% more expensive.

The release of prices in the domestic market (it now depends on the variation of the international price of crude, the value of the US dollar, the modification of the fuel tax and the prices of bioethanol and biodiesel) is very convenient for companies, not for consumers.

Furthermore, in Argentina, ironically, the prices of oil derivatives are inelastic to the fall of the international price and elastic to the rise.

(viii) **Changes in the companies**

Downstream Exxon (now AXION) and Shell (now RAIZEN) have withdrawn from the market and together with YPF S.A. oligopolize the domestic derivatives market, accounting for more than 90% of sales.

That is to say, the oligopolistic characteristic, by nature, of this market is now maintained with other actors.

(ix) **Tax and customs benefits**

They are granted for what he calls new ventures in the activity related to hydrocarbons. These activities are the extraction of crude oil and natural gas; services related to the extraction of oil and gas, except prospecting activities; and geological and prospecting services.

This provision would not appear to be necessary. The oil companies are making good profits (it can be seen in the balance sheets), and it can be stated that a good part of the activities in Vaca Muerta oil field already has profits and one of them is the subsidized price of natural gas.

2.3.3 Natural Gas

(i) **Argentina is a "gas-bearing" country**
(ii) **Proven reserves**

It is worth mentioning that oil reserves have fallen for the same reasons.

To change this situation, almost all the efforts have been put into the resources of Vaca Muerta oil field and the efforts to increase the conventional ones have also been neglected. This especially affects the Argentine northwest which depends almost entirely on imported Bolivian natural gas.

We will return to this when commenting on the Vaca Muerta oil field case.

(iii) **Production**

But, just as the participation of natural gas in the energy matrix has increased, production has fallen and Argentina, from being a net exporter until 2006, has become a net importer on gas, and in growing quantities, especially of liquid natural gas.

In the last two years, the price effect has decreased consumption in all sectors except for electricity generation, which is responsible for the slight increase in natural gas sales.

These facts, together with the lack of investment in exploration in conventional fields, have led to a fall in production between 2006 and 2017.

For this reason, it would be advisable not to neglect exploration in non-conventional areas.

(iv) **The "Gas PLUS" plan**

During the administration of the former Minister of Energy and Mining, the first of the Macri administration, the "Gas Plus" Plan originated in the government of Cristina Kirchner was continued in order to increase gas production, which, as mentioned, was in sharp decline.

To this end, Mr. J. Aranguren established a subsidy essentially for the production of non-conventional gas, based on recognizing prices of US $7.5/MMBTU in 2018; 7.0 in 2019; 6.5 in 2020; and 6.0 in 2021. Then, the subsidy ranging from US $4.0 to US $3.0 would disappear and producers would receive the price set by the "market," which would be between US $3.5 and 4.0/MMBTU.

These actions proved once again the lack of coordination between the policies of the different ministries and secretariats.

With the mentioned incentives, essentially Tecpetrol, in the Fortín de Piedra field, strongly increased the production of natural gas, taking it to 11 MMm^3/day by middle 2018 and hoping to reach 19 by the end of that year or in the first semester of 2019.

These values implied a high disbursement of the government in subsidies.

But the economic and social crisis in Argentina is "declared," and it is decided to go to the International Monetary Fund as a lender and this required an adjustment plan.

Consequently, the government decided to reduce the subsidy to the production of non-conventional gas.

The 2019 budget proposal limits the amount of this subsidy to US $700 million, which would be equivalent to a production of about 17 million m^3/day, much less than what Vaca Muerta oil field would deliver (only Tecpetrol would keep all the subsidy). But it seems that a trust fund would be created with the idea of contributing resources above the mentioned US $700 million if the production to be subsidized would be increased.

At present, there are six concessions recognized as deserving the subsidy and eleven concessions more with the specifications approved by the province of Neuquén, pending ratification by the National State.

In other words, a real problem is: What criteria should be used to distribute the now limited and scarce amounts of subsidies? How to avoid those considered "acquired rights" by the companies? How to "attract new investors," in the face of permanent changes in the rules of the game? How to make this compatible with the government's hope that the exports of Vaca Muerta oil field and the harvests will take the country out of the crisis?

It has been estimated that the total cost (development cost plus operating costs) at the wellhead of non-conventional gas is no more than US $1.7/MMBTU (according to the Ministry of Energy, this corresponds to the production of Orejano and Fortín de Piedra localities), but it is linked to very high investments that some estimate at no less than 10–15 billion dollars a year. It is estimated that US $3.5/MMBTU would be enough. What would have happened if the exploitation of Vaca Muerta had been left in the hands of the totally state-owned YPF?

In any case, it would seem wise, in order not to repeat the history of the 1990s, to reserve the existing natural gas, even that of the non-conventional fields, for future generations, and to industrialize it in the country by adding value to it, for example, through petrochemical plants, the elaboration of fertilizers and the demand that an industrial country would require.

The situation can even cause the reduction of natural gas production from Vaca Muerta oil field, and YPF S.A. has informed that with US $4 /MMBTU the fields are not profitable to extract the tight gas and would allocate more resources to oil production, and the Total Company would be willing to curb its existing investment in Vaca Muerta oil field. However, this issue of YPF S.A. seems to contradict the announcements of contracting a Belgian company to install a ship in Bahía Blanca harbor that liquefies the natural gas of your property to export it later.

(v) **The Argentine domestic market**

The current size of the Argentine gas market is very small compared to existing potential resources.

Observing the market, it can be seen that 40% of the population does not consume natural gas from the network. In other words, there would be a relatively interesting additional potential market at the local level.

The Northeastern Argentina (NEA) pipeline, having been halted its construction nowadays, could have absorbed, at best, some 4 million m^3/day of the gas production.

But not all households, which are currently not supplied, have homes suitable for the installation of network gas.

YPF S.A. has expressed interest in building a large petrochemical project in Bahía Blanca using gas from the Vaca Muerta oil field, and it is looking for a partner to do so.

The Down Company with which YPF S.A. is developing the El Orejano area in Vaca Muerta oil field has been mentioned as a possible partner. Down has a very bad industrial record as a real petrochemical producer in Argentina, and it seems that its projects have always been aimed at blocking Argentine petrochemical development, which led to the installation of the state-owned pole in the city of Bahía Blanca.

This is why it would be necessary to build other trunk pipelines and expand the distribution networks in many urban areas of the country.

To do this, the gas pipelines in the Neuquén Basin would have to be expanded, and possibly, the NEUBA II pipeline that reaches Bahía Blanca has currently idle capacity.

In other words, if one does not have a view on the future of Argentina as an industrialized country, the alternative of exporting appears to be the only viable one. This should not be the case, and if a serious socioeconomic development plan were drawn up for the medium and long term, perhaps the Argentine gas market would not be so small.

(vi) **Imports**

Imports reached almost 30% of the requirements of the domestic market.

The most important part is constituted by the liquid natural gas, the most expensive and responsible for a good part of the subsidies received by the former ENARSA.

For geopolitical reasons, it would seem convenient, despite the expected surpluses, to maintain gaseous natural gas imports from Bolivia, but this would imply an open foreign policy toward Latin America that does not seem to be among the priorities.

(vii) **Natural gas exports**

Those markets confirmed to receive the Argentine gas exports, until the end of November 2018, were Chile and Brazil.

The objective was not to harm the companies producing in Vaca Muerta oil field and to generate an immediate market for them: that is, the Chilean market.

In middle August 2018, the new Secretary of Energy and Mining stated that Vaca Muerta Gas to be exported to Chile would not receive a subsidy for it.

Four types of contracts were established with the characteristics of supply interruption and non-interruption of supply.

Basically, companies that have the subsidy of the Gas Plan (which starts as mentioned above with a payment of US $7.5 per million BTUs) will not be able to export that gas, but they will be able to export the non-subsidized one.

For the moment, the amount of natural gas exported has been limited to 6 million m^3/day.

The idea is to export up to 22 million m^3/day in the future, firstly through the Del Pacífico pipeline the amount of 12.5 million m^3/day, which has been extended by Gas Fenosa, and through the Andes pipeline 9 million more.

At the end of August 2018, the first three contracts for exports to Chile that were not for Vaca Muerta natural gas were signed.

The gas is destined to the METTHANEX plant in Chile and comes from the Santa Cruz Austral Basin (Compañía General de Combustibles—CGC), from the Marina Austral Basin in Tierra del Fuego and from the Argentine State (Pan American Energy and Total Austral).

The price for this transaction was agreed at US $4.08/MMBTU, and the successive ones will have a floor of US $3.45 and a ceiling of US $4.50, in all cases far from the US $7.5 received by the companies benefited by the Gas Plan, while the price of natural gas imported in winter from Chile would have been around US $10.

The guidelines set a maximum shipment of 750,000 cubic meters each, and the three shipments can reach a maximum export of 2,250,000 cubic meters per day and extend it to 2020.

In October 2018, approval of new orders to export from the Neuquén Basin to Chile, with YPF up to a maximum of 1.5 million m^3/day and with Exxon up to 400,000 m^3/day, was imminent. In both cases, the buying company would be INNERGY and the selling price of the gas is US $4.20/MMBTU.

Pan American Energy requested and was approved in October 2018, 1.5 million per day to supply thermal power plants in the Metropolitan Region and the V Region of Chile, at a price of US $4.05/MM of BTU.

Wintershall requested and was approved in October 2018 to export 1.5 million per day to Nehuenco and Candela thermal power plants at US $4.20/MM BTU.

CGC requested and was approved in October 2018, for the same destination as Wintershall, another 1.5 million per day at US $4.20/MM BTU.

The natural gas would be injected into the Gas Andes pipeline.

Other companies are still waiting for the approval of their orders.

Due to the problems of lack of evacuation capacity and the fall in demand due to high tariffs, around 15 million m^3/day of natural gas would have been left unplaced.

For this reason, YPF S.A. and Tecpetrol would have as an alternate for the storage of gas in "depleted" fields such as Los Bastos in the Neuquén Basin.

They would inject 20% of the surplus production or 2.5 million barrels/day during 5 to 8 months in summer and would take it out in four months during winter.

As of November 2018, there were 11 export contracts: seven from the Neuquén Basin to Bío and Central Chile; one from Salta to Antofagasta, via the North Andean Gas pipeline; and three from Santa Cruz and Tierra del Fuego. The contracts were agreed at prices between US $4.2 and 4.3/million BTUs with a retention of 4 Argentine pesos per dollar.

Finally, also in November 2018, Wintershall was authorized to export to Brazil.

The idea of exporting liquid natural gas to the rest of the world seems to be an expression of desire and an incentive to attract investors to Vaca Muerta oil field.

The relative advantage that Argentina has is its location in the Southern Hemisphere, so that when in the Northern Hemisphere it is winter and there is a shortage of gas, in the South, it is summer and there would be a relative gas "surplus." Its main competitors would then be: Australia, Nigeria, and Indonesia and the Asia-Pacific market. To this, it should be added Argentina's relative distance from those markets, with the consequent higher freight rates.

(viii) **Toward the immediate future**

It is very likely that self-sufficiency in natural gas will be achieved soon.

Vaca Muerta oil field would be the main supplier, but the Southern Basin should not be neglected.

With the mentioned subsidy of US $7.5 per million BTUs that is being granted, especially to the so-called Gas Nuevo, essentially from Vaca Muerta oil field, the incentives to invest and extract are among the highest in the world.

After the year 2022, the "new natural gas" would be paid the price that "sets" the market.

The idea of the government, or part of the government, is that in the future the distributors and Cammesa bid the prices of natural gas that they acquire with the idea that prices will fall. This is because the values are dollarized and are affected by devaluation processes, in a market with a floating dollar, and this affects what users pay, who are condemned to permanently increasing rates, because even if the price in dollars drops, the peso will continue to lose value.

The budget adjustment proposals cast doubt on continuing to subsidize the production achieved by the Gas Plan.

The fiscal impact, foreseen by the previous Energy Minister, was to drop from US $600 million in 2018 to US $400 million in 2019. But the increase in production to be subsidized would leave that figure far behind, bringing it up to almost US $1.1 billion in 2019.

One thing was a subsidy of US $3 per million BTUs on 13 million m^3/day, and another situation on 21 million.

Then, it was thought to limit the volume of subsidized natural gas, and this would affect the considerable profits of Tecpetrol (Fortín de Piedra oil field) and CGC (Santa Cruz Basin).

This situation shows again the lack of coordination between the different ministries and secretariats and the lack of planning. Although the previous government had chosen to establish the subsidy, up to paying US $7.5/million BTU, to encourage the natural gas production essentially in Vaca Muerta oil field, the current government did not put a limit to the volumes reached by the measure and consequently favored extraordinary profits of the preferential companies.

Then, the government decided not to extend to new exploitations the benefits of the Gas Plan.

How different would the situation be if the state-owned YPF had been in charge of developing the "New Gas"?

(ix) **Debt and commitments to companies for natural gas**

There are two types of debt.

One type is for the so-called Accumulated Daily Difference (ADD) which allowed extra charges, at the end of each semester, to be made to users in the face of significant differences in the exchange rate agreed for acquisition contracts and the actual exchange rate. This debt would be paid in installments during the years 2019 to 2021 and would amount to approximately US $350 million.

The main companies that would be benefitted would be Metrogas, Naturgy (ex Gas BAN), Camuzzi and Ecogas. The DDA's nonsense comes from the laws and decrees of deregulation of the gas business of 1992. So as that the claims of the distributors not to be repeated, it would be necessary to repeal that legislation.

The other debt is for the "Gas Plan" that ended on December 31, 2017, and the creditors are gas-producing companies. The State would pay it from the beginning of 2019 until June 2021. The State would settle for US $1583.2 million with YPF, Pan American Energy and Total Austral, among others.

As for the "Gas Nuevo" in 2019, subsidies would be given to companies in Vaca Muerta oil field and the Austral Basin for US $700 million.

(x) Something about the tariff increases

The increases in "normal" rates, which were expected to start in October 2018, will be greater for users with lower consumption. At the national level for households in category R1 (up to 500 cubic meters of gas per year), the increases range from 60 to 63% and for households with higher consumption, R3 4 (from 1801 cubic meters of gas per year) increases range from 19 to 21%. For instance, in the Buenos Aires metropolitan area (the most densely populated area of the country), where 70% of households belong to the lowest consumption categories, the monthly increase would be around $400 Argentine pesos. Since the beginning of the current government's administration, the increase in natural gas rates in the AMBA area (the Metropolitan area including the city of Buenos Aires and 40 counties in Greater Buenos Aires), it was an average of 1848%.

This structure makes no logical sense and shows the lack of social sensitivity from the government.

To a great extent, the dollarization of the tariffs has considerably aggravated the situation.

(xi) Gas pipelines and treatment plants

In the Neuquén Basin, especially in the Vaca Muerta area, the proliferation of producers and the relatively rapid increase in production have generated a bottleneck due to the lack of sufficient capacity to evacuate production. This has motivated some concessionaires like Tecpetrol from Fortín de Piedra oil field to project the construction of a gas pipeline to the city of Rosario area, at the Río Paraná.

TGN and TGS would join to use a new gas pipeline that would connect Vaca Muerta oil field with the town of San Nicolás (Buenos Aires Province).

That is to say, what has already been mentioned for the subject of oil pipelines is repeated, and the proposal of the authorities is the same. This means that any company, not only the transporters and distributors of natural gas, could build pipelines (also treatment plants) and rent the service to the producers. This situation is also a legacy of the privatization of Gas del Estado in the 1990s since this company operated the pipelines very efficiently and coordinated with YPF S.E. the construction of the pipelines from the production fields. The fact is that this type of policy underlies the modality of the business units and the "atomization" of the activities linked to the hydrocarbon industry.

At the Mega Company's Treatment Plant in Loma de la Lata, province of Neuquén, natural gas, ethane, propane, butane and gasoline are extracted, but the increase in production at Vaca Muerta oil field has filled the plant's capacity and the companies that own it (YPF, Dow and Petrobras are studying its expansion).

(xii) The institutional aspects

At the institutional level, the majority shares (70%) of the gas distribution company Metrogas, the largest in the market, are currently for sale to YPF S.A. The process is being carried out by Citibank. It is argued that as YPF S.A. is a producer of natural gas, according to the National Law 24,076, it could not also be the controller of a distributor. It would be expected that the same criteria would be applied to other companies that act in various links of the natural gas chain.

Four companies are interested: CGC, the oil company of Corporación América, the holding company headed by Mr. Eduardo Eurnekian; Integra, a firm headed by Mr. José Luis Manzano; Camuzzi, controlled by Disvol, of Alejandro MacFarlane (Camuzzi is also a distributor); and Gas BAN, the distributor of the Spanish group Gas Natural Fenosa.

It is interesting to mention that Metrogas notably improved its operational and economic performance. Thus, in 2017, it registered a net profit of 775 million pesos due, among other things, to tariff increases.

It is also announced that Distribuidora Gas del Centro (active in Córdoba, Catamarca and La Rioja Provinces) will absorb Gas de Cuyo (working in Mendoza, San Juan and San Luis Provinces), both constitute Ecogas in the hands of Techint and serve almost 18% of natural gas users in the whole country.

(xiii) **The profits of the companies of the natural gas chain**

The gas companies have recorded large annual profits and even some of them have registered in the first half of 2018 profits higher than last year's total.

In total, last year, the gas complex, which can be analyzed with public access balance sheets, recorded a global gain of 29,476.4 million Argentine pesos.

These enormous profits of the companies, product of the great tariff increases, would not reconcile with the lack of investments in works like Ecogas that distributed profits in the fiscal year of 2017 by a total of 617.1 million Argentine pesos and only invested in different works 150.7 millions, when it should have contributed 278.2 millions to the expansion of networks.

2.3.4 *The Vaca Muerta Oil Field*

Something about the Vaca Muerta oil field has already been mentioned in the paragraphs on oil and natural gas.

We will refer to Vaca Muerta oil field because it is the Argentine geological formation where, since 2010, activities of the so-called non-conventional hydrocarbons are being carried out, especially those located in the province of Neuquén. The resources of the Neuquén Basin also extend to the provinces of Río Negro, La Pampa and Mendoza. In this last province, there is a strong popular resistance against the possible environmental impacts of the application of the fracking techniques.

But questions and concerns arise about what Vaca Muerta oil field and non-conventional hydrocarbons, in general, may mean for the development of Argentina.

The aim is to demonstrate that Vaca Muerta oil field will save the country by providing an annual growth rate of between 0.3 and 0.4% for Argentina as a whole and an accumulated growth rate of between 4.5 and 4.8% until 2030, according to a joint analysis by the former Ministries of Energy and Mining and of Production.

It has been said, and the numbers seem to confirm it, that Vaca Muerta oil field is essentially important in gas resources.

The production of both oil and natural gas, despite the recovery in 2018, especially in gas, is not, at present, sufficient to supply the requirements of the domestic market.

If the pace of investment by the companies working in Vaca Muerta oil field continues and intensifies, there is no doubt that gas self-sufficiency will be achieved within the next five years. Oil can also be achieved, but not only by Vaca Muerta oil field but by the exploration and development of conventional fields, with substitution and rational use.

The government's idea is subsidizing only the gas production until 2021, and then, the prices would be those of the "market," that is, lower than those in force in 2018. This could play against this objective. This could reduce the rate of investment of all or some of the companies that have not yet managed to receive the subsidies, despite the fact that costs are much lower than the prices of US $4.5 or 5 per million BTUs that the market would have after 2021.

In any case, for reasons of the magnitude of resources and because of production subsidies, companies, many of them foreign, are more interested in gas than in oil. But they are afraid that because of the adjustment plans the government will end the subsidies for the companies that have not yet obtained this benefit.

As the Argentine market is small, despite the fact that almost 60% of the families do not consume natural gas, and so is the Chilean market, there remains the expectation of exporting to the rest of the world. But here the disadvantage, beyond the strategy of some of the foreign companies that are in Vaca Muerta oil field, is the current almost saturation of the world market by the appearance of liquid natural gas from the USA, competing with Qatar and Australia and with the gas by networks of Russia. Thus, the panorama of exporting liquid natural gas does not appear as the additional outlet for Vaca Muerta's surpluses, let alone consider this venture to be the country's "salvation."

The investments required for the development of these fields are enormous (the decline of the wells is very high and this requires, as the experience of the USA shows, the drilling of a large number of wells to maintain and expand production), and despite the strong presence of YPF S.A. (associated in turn with multinational companies), it would seem that the amounts needed would be provided by this last type of company. Then, their strategies will prevail, which do not necessarily, but rather the opposite, coincide with those of the country.

Then, we should be aware of the so-called Dutch disease, which has always been suffered by two large exporters such as Nigeria and Venezuela, and it seems that in Argentina there would be no Norwegian strategy.

The Dutch disease originated in the 1960s in Holland due to the appearance of large gas resources in the North Sea and the Dutch thought that this would bring prosperity to the country. This was not the case. The great increase in their external income caused the appreciation of their currency, the destruction of their industry and agriculture, and the impoverishment of vast sectors of the country. Since then, the phenomenon of the expansion of foreign exchange income due to booms in the export of natural resources (oil and minerals) which, without the appropriate measures, generates revaluation of the local currency and contraction and loss of jobs in other productive sectors and finally, negative effects on the economy as a whole, has been called the "Dutch disease."

Norway, unlike, for example, Nigeria, in the face of the same fact, created a compensation fund with revenues that were essentially regulated by the State and its state oil company.

The case of Nigeria is tragic; since beyond being one of the most corrupt countries in the world, largely due to the actions of multinational companies, it exports oil and imports its derivatives for domestic consumption.

It has already been said about the presence of the multinational companies in Vaca Muerta oil field and that their strategy of development of the resources will be the one established by the parent company, since they will have freedom to dispose of the production according to their convenience and this can be that of REPSOL in the 1990s, to export up to the maximum possibility, without worrying about the depletion of the resources or the supply of future generations with the consequence of the current dependence on imported natural gas and oil) and the strategy of the national companies, without a state company like the former YPF, will also be to maximize their income. The fact is that a large number of Argentine businessmen are "rentiers," their aversion to risk is very high, and many have grown up as contractors or concessionaires of the state in public services, a task that has a very low level of risk and a secure demand. This characteristic comes from the time of the colony, from the country of cattle raising and later agriculture, where the large estates enjoyed the quality income of the humid Pampa and the only risk was that it may not rain. Thus, what will be left for the country, only royalties, occupation of labor and income tax, as it happens with all extractive activity.

Another of Vaca Muerta's problems is related to the lack of infrastructure, essentially oil and gas pipelines (beyond the idle capacity of some of the existing ones) and heavy drilling equipment. The construction of oil and natural gas evacuation infrastructure requires time, as well as investment, and this clashes with the current government's idea that Vaca Muerta oil field will be the solution for the contribution of the enormous amount of foreign currency that the country requires almost immediately. On the other hand, the multinational companies want that the state builds much of the missing infrastructure, especially roads, railways, hotels, homes, hospitals, running water and other services. These "externalities" would then have to come out of the state coffers, and the big oil companies would be dedicated to extracting and transporting the oil and natural gas. How far from the role played by the State YPF in the past, creating and sustaining the villages around its operations!

Perhaps the solution to the Vaca Muerta case would be a controlled exploitation, with the consequent social license, due to the environmental effects of fracking, conserving the resources so that future generations should not have to import the hydrocarbons. Essentially industrialize primary production by converting it into petrochemical goods and fertilizers, using the surplus income to contribute to the change in the productive structure of the country through industrialization and technological development. But this would only be possible if Vaca Muerta's hydrocarbon resources were managed by the National State and the provinces through a federal company as YPF should have been. But as of May 2018, almost 80% of Vaca Muerta has been granted in concession for 35 years, so the companies that hold the concessions already have acquired rights and it would be very costly, except for breaching the contracts, for these areas to revert to the Provincial States.

What will be left for the country of Vaca Muerta oil field? Royalties, some taxes paid in the provinces, income tax, low impact on the local equipment industry (as duty-free imports are allowed), low-quality local inputs, sand mining and little else. Today, there are almost 70,000 poor people in Neuquén, and it does not seem that the Vaca Muerta project is meant for them.

Will Vaca Muerta be another lost opportunity?

2.3.5 Liquid Petroleum Gas (LPG)

(i) **Argentina is a net exporter of LPG**

With Approximately 33% of the production is exported, and therefore, there would be no problem to satisfy the internal consumption of LPG and less for the 2.5 million users that would constitute the universe of "poor subsidized" without access to natural gas by networks. This situation led to the implementation of the so-called Plan Hogar from 2015 during the government of Cristina Kirchner.

(ii) **Production**

Most of the LPG comes from natural gas fields and the rest from distilleries. In both cases, it is a by-product of the process.

(iii) **Concentration of sales**

Almost 70% of the gas sales in the domestic market are concentrated in the residential sector.

(iv) **Consumption**

It is assumed that, depending on the area and number of household members, these users would consume between 500 and 750 thousand tons of LPG annually, equivalent to 50 or 74% of total residential LPG consumption. It is estimated that there are 4.5 million residential LPG users, of which 2.5 million would be subsidized.

(i) **Subsidies**

The amount of the planned subsidy would have been 3 billion pesos in 2015, equivalent to 1.6% of the country's total economic subsidies in 2014 and 11.6% of social subsidies. In the first eight months of 2018, it would have reached 5.563 billion pesos, equivalent to 1.5% of the total allocated to the energy sector.

(vi) **The Household Plan**

In fact, the implementation of the Plan Hogar, which was an excellent idea, clashed with the scarcity of containers and with the virtual impossibility of accessing subsidized carafes. This situation worsened after the current government took office.

Periodically, the LPG prices foreseen for the Plan Hogar are updated and the price of the kilocalorie of this fuel is 2.6 times greater than the natural gas by networks for the non-subsidized users and 22% greater for the subsidized ones. So, fuel for the "poor" is still the most expensive.

(vii) **Users**

It is estimated that there are about 5.3 million residential LPG users and almost 50% should have access to the Plan Hogar, but in almost all the cities of the country the subsidized cylinders are very scarce and the users end up paying not 240 or 260 pesos per 10 kg container.

(viii) **The containers**

It would then be necessary to have much more 10–15 kg containers available and to ensure the supply at the users' homes, increasing the amount of subsidies since fuel is a necessary and indispensable good. Otherwise, the substitute in some regions is wood, which, if it cannot be appropriated for free from the forest, is the most expensive useful calorie for the poor.

(ix) **Caloric consumption**

Therefore, LPG is a fuel that essentially supplies the caloric consumption of the families who use it because they do not have access to natural gas by networks.

(x) The "Home" Program would work with much less inconvenience if the State Gas Company were recreated and assigned the role of executor of the plan throughout the country. It is very difficult to want to convert by means of standards and inspectors the current LPG market, whose sellers are governed by profit, into a public service as residential LPG consumption is in reality. The Gospel says that it is very difficult to "serve two masters" at the same time.

2.3.6 Compressed Natural Gas (CNG)

Argentina is one of the most advanced countries in terms of CNG use for automotive transportation. In addition, it has an industry developed for the manufacture, installation and maintenance of CNG equipment in vehicles and is even an exporter.

(i) **The CNG vehicle fleet**

The fleet converted to CNG has increased from 1,742,762 vehicles in 2015 to 2,035,301 at the end of 2017.

The 47% of the fleet is patented in the city of Buenos Aires and the province of Buenos Aires.

Around 52% of the present fleet is less than nine years old.

(ii) **Evolution of Consumption**

The relative increase in the price of natural gas with respect to naphthas explains the fall in consumption of CNG from 2105 to 2017.

The above situation has been reversed so far in 2018, and there has been an increase in conversions of gasoline vehicles to the use of CNG. It is that in the first eight months of the year 2018, the price of gasoline has increased by almost 50% while CNG has done so by 22%.

The cost of converting vehicles to CNG ranged in August 2018 between 17,000 and 34,000 Argentine pesos, with a lifetime of fifteen years. With a Premium naphtha price of 39.88 Argentine pesos per liter, to travel 15,000 km would cost 58,465 Argentine pesos, and with a cost of CNG m^3 of US $14, the same trip would be 21,000 pesos, with a saving of 37,465 pesos, amortizing the cost of fuel conversion in less than a year. Of course, the relative prices of naphthas/CNG will continue varying and with them the tendency to convert or not.

Conversion equipment

An element to consider is that the imported component of the conversion equipment reaches almost 55% of the total cost (even in middle 2018 inputs were lacking due to the increase in conversions), and this disadvantage should be compared with the lower environmental impact of CNG with respect to the naphthas and the development of a major industry of these teams in the country that also exports.

For these reasons, neither would be advisable, as it arises from some information, the idea to enable the import of vehicles converted to CNG use, and on the contrary, it would be necessary to facilitate (which does not imply to subsidize) this activity that creates local work and to foresee the substitution of the mentioned imported components, which are essentially tubes and electronic parts.

2.3.7 *Electrical Energy*

(i) **The Electrical System**

Until the early 1990s, practically the entire Argentine electricity system was in the hands of the State, with the companies "Agua y Energía" and "HIDRONOR" standing out. Subsequently, almost all the plants were privatized, leaving only the national government in charge: the binational Yacyretá (shared with Paraguay) and Salto Grande (shared with Uruguay), as well as the three nuclear plants (Embalse, Atucha I and Atucha II).

Transener is a mixed company that deals with the transmission of electricity at high voltage.

The distribution is divided between private companies such as Edenor, Edesur and Edelap that operate in the city of Buenos Aires and in the city of La Plata and provincial companies that distribute in territories of the provinces.

Cammesa is a mixed company that operates and dispatches the Interconnected Electric System. The Patagonian System has also been integrated into the National System.

The ENRE is a public agency in charge of regulating the national electricity system.

It is worth remembering that the main hydroelectric works built in the country were developed by the National State through the specialized agencies Agua y Energía Eléctrica and HIDRONOR following perfectly defined study stages as to their scope and content, such as inventory, technical-economic pre-feasibility, technical-economic feasibility, preliminary or basic project and finally, the executive project.

(ii) **The hydroelectric potential**

As of December 2017, this potential was 34,733 MW, with a capacity to generate 142,228 GWH. The hydroelectric potential was equivalent to 97% of the country's total installed power in 2017. Of the hydroelectric potential, 66% was in different degrees of progress for its use; 0.3% was under construction; 34% with a basic project; 7% with a feasibility study; 34% with a pre-feasibility study; and 24% with an inventory. The hydroelectric power installed at the end of 2017 was only 34% of the detected hydroelectric potential. The hydraulic plants in the pipeline by the current government for its construction would be about 2522 MW until 2025, but almost all of them are very delayed in some cases due to conflicts between provinces (for instance, Portezuelo del Viento), but essentially due to financing problems. The only one that seems to be advancing is the binational Aña Cuá. The country's budget for the year 2019 only registers the contribution of some funds for Chihuido I and two dams in Santa Cruz. The others are subject to the Public–Private Partnership Program (PPP), which will be discussed later.

In recent years, the execution of new hydroelectric works has been neglected, with the satisfaction of demand falling on the thermal power plants, with the consequent negative effect on the energy trade balance, since the weight fell on the natural gas that had to be imported in increasing values for this purpose.

The only hydroelectric use of certain installed capacity has been H.A. Punta Arena in the province of San Juan with 62 MW.

(iii) **Electrical power**

In 2017, the installed power was predominantly thermal (58%), and within it, although combined cycles predominated, the inefficient equipment of gas turbines, which did not always operate at the top of the load curve, was still important. The three nuclear power plants only contributed 2.3% of the available power and 4.9% of the installed power (due to the maintenance shutdown of the Embalse plant) and the wind turbines and photovoltaic panels were insignificant.

The power available in 2017 was approximately 18.2% higher than the maximum with a reserve margin of 4668 MW (in 2015, the reserve reached 1845 MW).

Between 2009 and 2011, the maximum power grew at a rate of 4.2%, while between 2014 and 2017 it grew at a cumulative annual rate of 2.2%, showing the effects of the recessive economic process.

Between 2015 and 2017, the maximum power grew by 1679 MW, while the available power grew by 4502 MW.

It should also be noted that the works that enter a period have been the result of decisions and executions made in the previous period.

As for the rate of installation of new power, to meet the demand of a developing country and not stagnate, it should not be less than 1000 MW per year, plus a power that ensures an adequate level of available reserve, with an appropriate mix of plants that supply the peak and base of the system.

According to the Secretariat of Energy and Mining, between 2017 and July 2018, 4100 MW (almost 95% in conventional thermal power plants) will be installed in 41 power plants.

(iv) **The demand for electricity**

Total electricity demand by destination is concentrated in the wholesale electricity market whose consumption grew between 2009 and 2014 at an annual cumulative rate of 3.9%, and between 2014 and 2017, it decreased to 1.6%.

Generation by type of energy source showed the growing predominance of fossil fuels, which each year displaced hydroelectric and nuclear power, evidencing the neglect of the abundant hydroelectric developments and the practical paralysis of nuclear power, despite the entry of the third nuclear power plant in 2014, after thirty years of the second one which was a reservoir.

Electricity imports were insignificant.

Renewable energies (except for hydroelectric power of more than 30 MW) contributed only 2.0% of the needs of the wholesale electricity market. The greatest contribution was made by small hydroelectric plants, and there was a growing supply of wind power among the renewable energies detected.

The turnover by type of user, in the year 2016, was concentrated in two sectors: the residential (37.7%) and the industrial (32.1%), increasing the participation of the residential one and diminishing that of the industrial one as it shows, perhaps, the process of deindustrialization. The spatial concentration of the population and of the country's productive activity was evidenced by the fact that the city of Buenos Aires, Greater Buenos Aires and the provinces of Santa Fe and Córdoba absorbed almost 67% of the electricity turnover.

Almost 99% of the population had electricity service.

In the first nine months of the year 2018, compared to 2017, the demands decreased sharply: in the residential sector 5.5%, in the commercial sector 6.9% and in the industrial sector 5.3%. These data are in line with the recessionary process, rate increases and fall in real wages, so far in 2018. Even the distributors mentioned the increase in the number of delinquencies and in electricity thefts.

(v) **Energy used for electricity generation**

The high participation of natural gas, the low relevance of wind and solar energy and the decline of hydroelectric power are evident.

(vi) **Generation of electricity from public services and self-production**

The self-production of electricity represented 10% of the total generation of the country. Surely with the implementation of the "distributed energy" policy, this percentage will increase in the future. This issue will be discussed later.

(vii) **The MONOMIC spot market price**

The gap between Cammesa's revenues and the MONOMIC price fell from 86% in 2015 to 51.3 in 2017 and 41.3 in August 2018, essentially as a result of the tariff increases.

The hourly Spot Market MONOMIC price, without transport charges, showed a very important difference with the amount collected by the distributors, which was covered by increasing subsidies. In 2014, they barely covered 17% of the MONOMIC price and the rate increases took it to 49% in 2017. Wholesale market prices were out of line with the marginal costs of electricity generation. The application of Resolution 240/03 of the Secretariat of Energy allowed prices to be sanctioned as if there were a full supply of national natural gas, without any restrictions, and also set a maximum price for energy in the wholesale electricity market of US $120/MWK, both of which caused a significant distortion in relation to real costs. This, in turn, led to a substantial increase in electricity subsidies provided by the State to keep the electricity system running. Thus, Cammesa accumulated a considerable debt to the generators.

The payments made by the consumers were not enough to level out the real costs that were covered by contributions from the treasury.

The wholesale cost of electricity would fall in 2018 to US $66/MWH, essentially because Cammesa, instead of paying for natural gas at US $5.20 per million BTUs, would pay it at US $3.40 per million BTUs and, to a lesser extent, because of the income from generation of renewable sources.

Despite this, the price paid by users will continue to increase, due to the strong devaluation of the peso as tariffs are dollarized.

From August to October 31, 2018, users will pay about pesos 1400/MWH at a cost of 3000 pesos/MWH.

So if the cost was US $66 /MWH (2450 Pesos/MWH), the subsidy, by the State, would be 40% of the cost and the user would pay the remaining 60%.

It is estimated that the amount to be subsidized in 2018 would be more than 65 billion pesos.

On the other hand, the 2019 budget establishes a subsidy to the electricity sector (not only Cammesa), of about 99,484 million current pesos.

(viii) **Transportation, transmission and system failures**

By 2017, all transportation systems were interconnected.

The electrical system showed faults in distribution and transformation that were recorded with the supply cuts, especially in the areas served by Edenor and Edesur despite the very large increases in rates received since 2016.

When the temperature exceeds 30 °C in summer, the cuts begin.

This situation led the ENRE to issue Resolutions 198 and 199 that modified the service quality standards.

For example, at Edenor, between March and August 2017, there were 63 faults in medium voltage feeders and between September 2017 and February 2018, 143.

Edesur also verified 63 and 143 in the same periods.

The ENRE warned that the number of service interruptions and their duration greatly exceeded the established guidelines.

As a result, fines of between 330 and 600 KWH per affected user were applied.

The ENRE also established that the above-mentioned companies should return about 500 million pesos to the affected users.

As an example, we cite the cut at the end of August 208 that affected 300,000 users of Edesur out of a total of 2.2 million and of Edenor 20,658 on the same date.

It would seem that it is cheaper for the companies to pay the fines than to make the necessary investments.

Edenor had planned to invest some US $5.211 million during 2018 and Edesur some 4.400.

(ix) **The case of TRANSENER**

The sale of the state's participation in TRANSENER to private shareholders, which is a consequence of Decree 882 of 2017, would not be convenient. Essentially, TRANSENER is profitable, so why privatize it?

The state's participation in the electricity transmission company (which is 26%) was worth between US $350 million and US $400 million until a few months ago, but in October 2018 it fell to approximately US $120 million. The dollar share lost 50% of its value in almost a year and 64% from its peak in February 2018. Nor did it make sense to leave the sale process of these shares in the hands of the consulting firm Price WATERHOUSE & Co Asesores de Empresas S.R.L (PWC Argentina), which won the bidding for $13.3 million.

It would seem that one of the main interested parties in the purchase would be PECOM of Pérez COMPANC.

The State had bought the shares at $7 in mid-2016 came to quote $65 and in October 2018 was at $50.

(x) **Plant privatizations**

It would seem that the current government intends to privatize the thermal power plants built by the previous government. This is Brigadier López and Ensenada.

But due to irregularities the sale could be complicated, and it is denounced that the price set for the sale is below the market.

The requirements for acquiring the plants would also have been reduced, opening the door to investment funds and not just to experienced operators.

In addition, the Appraisal Court would have set a value of US $305 million for Ensenada with a minimum cash offer of US $229 million and an official valuation of US $207 million and a minimum cash offer of US $155 million for Brigadier López.

These figures contrast with the construction cost of US $1 billion for both.

In addition, the balance sheets presented by ENARSA showed a result of 121, 2 for Ensenada and 62, always in millions of dollars, for Brigadier López, so investors would recover the disbursement in a very short time.

2.3.8 *Nuclear Energy*

(i) **The situation**

The country has three nuclear power plants.

All of them run on natural uranium and heavy water.

The fuel assemblies of the Atucha I and Embalse nuclear power plants were manufactured until 1995 using uranium of national origin.

In 1995, due to a sharp fall in uranium prices on the international market, activities at the Sierra Pintada site in the province of Mendoza (the only one in operation) were suspended and uranium concentrate was imported; the rest of the transformation processes being carried out in the country.

The country imports all the foodstuffs it consumes.

In terms of reserves, Argentina has the certified reserves between the Sierra Pintada and Cerro Solo deposits of 7000 tons of uranium, while the inferred and predicted reserves are an additional 3000 tons.

On the other hand, the current consumption of uranium is 215 tons per year counting the operation of Atucha II.

Consequently, if the fuel elements were to be produced again in the country, the duration of the 7000 certified tons of uranium reserves would be thirty-three years.

The share of nuclear energy in the installed power was 4.8% in 2017 and 5.5% in generation (in this case, it could be somewhat higher since the Embalse plant was shut down for repowering).

(ii) **Paralysis of the nuclear plan**

An article by Andrés J Kleine is transcribed, and the concepts of which are fully shared by the present author.

"Throughout almost 70 years, Argentine society has managed to generate a scientific-technological-industrial conglomerate that has placed our country in an important and respected position at the international level in the field of peaceful applications of nuclear and perinuclear technology. The main driving force has been the State through its investments and its great purchasing power, managing to organize first-rate Science and Technology institutions (CYT) such as the National Atomic Energy Commission (CNEA) but also a conglomerate of public companies, such as

Investigación Aplicada Sociedad del Estado (INVAP), owned by the State of Rionegrino; Empresa Neuquina de Servicios de Ingeniería Sociedad del Estado (ENSI), partially owned by the State of Neuquino and the CNEA, with its Heavy Water Industrial Plant (PIAP); DIOXITEK S.A. and Nucleoeléctrica Argentina (NASA). In addition, it has been able to convene national private companies to form joint ventures such as Combustibles Nucleares Argentinos (CONUAR) and many other large, medium and small companies that have bet on the development of this technology as qualified suppliers.

It is worth mentioning that Argentina opted, in the first half of the 1970s, for the line of natural uranium and heavy water reactors in order to become independent from the expensive and proliferating technology of uranium enrichment, which we still do not possess at the industrial level. Our three power plants work safely with this technology, which today we master after many years of work and investment and which also has a great potential for innovation. The government has just dismissed 250 workers from NASA's permanent plant as a first consequence of the cancellation of the project of the fourth heavy water plant, with very convenient financing assured, of enormous interest for our country, and talks about a turnkey purchase of a Chinese reactor in 2022. This is the worst possible combination. The fourth plant was to be financed 85% by China and our country's share was to be dominant (70%) with a perspective of continuity for many small, medium and large companies that had bet and invested in technology and installations. We want to avoid the cancellation of the fourth power plant project because of the devastating consequences it will have and is already having. The destruction of NASA's management unit, where there are highly trained personnel who participated in the very meritorious construction and start-up of ATUCHA II and who are participating in the extension of the life of the Embalse plant, is not only an unjustifiable squandering of very valuable human capital, but is also a risk for the operation of existing nuclear power plants that require local knowledge and design capacity. This human capital will be taken advantage of in other latitudes where many nuclear power plants are being built, forming a new chapter in the brain drain of our country. We should also remember that the cancellation contravenes Law 26,566, which explicitly declares the construction of the fourth plant to be in the national interest.

In addition to this desolate situation, Dioxitek, the company in charge of manufacturing the base compound for nuclear fuels, has been dismissed.

These decisions will destroy thousands of highly qualified jobs and seriously jeopardize the continuity of the sector, squandering the capital accumulated during almost 70 years of continuous and fruitful work that has positioned our country among the most advanced in nuclear technology."

Andrés J. Kleiner, Senior Researcher at CNEA-Conicet. Professor of the UNSAM (Página 12, newspaper, Buenos Aires, 20 August 2018)

(iii) **Paralysis of the heavy water manufacturing plant (HWMP)**

One of the biggest and most successful investments to sustain this line was the construction of the PIAP, a real technological jewel that we need to provide very high-quality heavy water to our present and future plants and to the research reactors that INVAP manufactures and exports. But it would seem that the funds to pay the salaries of the PIAP workers have run out (taken from André J. Kleiner opinion quoted above).

"PIAP" is controlled by the Neuquén Engineering Services Company (ENSI), a partnership between the province of Neuquén (51%) and the CNEA (49%). This is where the heavy water that moderates and cools the nuclear reaction in atomic power plants using natural uranium is produced. The company halted production in May last year due to a technical shutdown, and had initially promised to restart in August, but that never happened.

PIAP produced last year a stock of 23 tons of heavy water that could be sold to the state-owned NASA, the firm in charge of operating the three nuclear power plants, and thus get out of this delicate situation at least for a few months. First it was argued that NASA did not have enough resources to pay for the production because they also cut its funds and had their tariffs frozen. At the end of February, the government raised, through Resolution 73/18, the remuneration that the company receives for the sale of energy, but NASA did not buy the heavy water anyway (fkrakowiak@pagina12.com.ar)".

One of the alternatives for the PIAP to continue working is to change its objective, and instead of heavy water, it would produce fertilizers.

2.3.9 Coal

Unlike what happens in Europe, in Latin America, with the exception of Colombia and certain regions in other countries, coal has not been used by families and barely by industries to generate steam, its use being limited essentially to the generation of electricity and steel.

(i) **Reserves**

In 2016, the Measured Coal Reserves reached 477.9 million tons, which added to the indicated reserves of 206.5 and inferred reserves of 67.8 gave a total of 752.3 million tons.

(ii) **Production**

The current production of Río Turbio, the only field in operation, in 2017, reached 20,340 tons. The maximum production peak was reached in 1979 with 1,326,254 tons, and since then, the drop has been permanent, after the privatization of Yacimientos Carboníferos Fiscales (YCF) and despite the investments made by the previous government in the Río Turbio mine.

In 2016, the ratio of reserves to production reached 23,500 years, showing the waste of this resource.

(iii) **The markets**

As Río Turbio's coal is not suitable for economic coking, the almost only consumer market is the thermal power plants. As practically the only generating plant was San Nicolás, this explains the drastic decrease in production. Most of the coal consumed is imported, of the coking type, and destined essentially for the iron and steel industry to feed the coking plants. The rest of the coal is used by the San Nicolás thermal power station. The permanent decrease in consumption is essentially due to the crisis in the country's steel industry.

(iv) **The problem of the Boca de Mina plant**

In 2018, a 240 MW capacity power plant was being completed at the Boca da Mina in Río Turbio, which will consume 1.2 million tons of raw coal per year. But until the mine's production, which did not exceed 20,300 TN/year, is reactivated, the plant should operate with natural gas or imported coal. Important investments have been made in the mine that could, with an additional effort, reach the production required by the plant, while the plant is practically finished. In the Río Turbio area, the population considers that closing the mine is condemning the population to disappear, while environmental groups oppose the use of coal to generate electricity for environmental reasons. The issue is complex, and the worst way out is to do nothing.

2.3.10 Renewable Energies

(i) **The scope**

Here, we will analyze biomass energies (essentially biodiesel and bioethanol with some mention of others), wind and solar energies.

A large part of the consumption of these sources is destined for the generation of electricity.

(ii) **Large hydroelectric plants (more than 30 or 50 MW), which were also included with electricity, will not be dealt with here**

According to the law, the contribution of renewable sources should reach 8% of the total electricity generation in 2017, but in 2017 it only reached 2%.

The resources of these energy sources are really important in Argentina, and only, bioethanol and biodiesel are used in relevant quantities.

The plans formulated to develop renewable sources have been the RENOVAR, PROBIOMASA and MATER. From many years ago comes the PERMER.

(iii) **Origin of the promotion of renewable energies**

The idea of promoting the development of renewable sources came from the previous government, but in fact it did very little about it. Perhaps the achievements of the PERMER should be highlighted in some provinces.

The current government, on the contrary, has given a great impulse to the subject, and the results in terms of power in construction and contracts signed show it.

(iv) **National participation**

As the program did not contemplate the forward and backward effects of a massive income, it has generated the opening, almost unrestricted, of the market to foreign companies essentially from Europe and China that have developed a capacity to produce very large equipment and to deliver it in short periods of time. This has dislocated national metalworking companies that cannot compete in terms of quantity and deadlines. This is beyond some weak clause that encourages them; the law allows practically the import of all the metal mechanical parts of the equipment and exempts them from any import duty.

(v) **The implementation alternative**

Consequently, Argentina is not repeating the successful experience of nuclear development that positioned it at a global level. In this case, the effort would have been much less, since we are not dealing with state-of-the-art technologies.

The alternative would have been to equip at a slower pace that would have allowed the massive participation of the national industry, which could have even generated the possibility of exporting to neighboring countries.

It seems another opportunity lost.

(vi) **Prices**

As far as prices are concerned, they started out very high, but by October 2018, the average equipment was below the money price paid by CAMMESA (54.72 for renewable sources compared to 61.73 for CAMMESA's contracts). This is due to the predominance of wind and solar power, which contributed 55% and 38%, respectively, of the power in operation and whose prices were below the aforementioned average.

(vii) **Biodiesel and Bioethanol**

As for biodiesel and bioethanol, there is still a "conflict" between oil trading companies and sugar, corn and oil companies over the prices established by the state for these biofuels and the percentage of participation of both in the respective mixtures with gasoline and diesel. In this last aspect, biofuel producers aspire to increase their participation in said mixtures, even in some cases up to 100% in the case of biodiesel.

(viii) **Exports of Biodiesel**

Another not insignificant aspect is the magnitude of biodiesel exports, which by 2017 were reaching almost 58% of production. The fact is that Argentina has one of the most important oil complexes in the world and these exports provide added value as do not the beans and residues of this industry which represent one of the largest items in the country's export income.

The USA has applied strong tariffs to imports from Argentina, and the European Union has also opposed allowing them.

(ix) **PROBIOMASS**

It is a good initiative where the participation and contribution of institutions such as INTA and the National Institute of Agricultural Technology, which has many years of experience in the field, would be essential.

(x) **The MATER**

It introduces a new complexity in the electricity sector and makes us miss the simplicity of the system when it was managed by State companies, and they actively participated in the unified dispatch of charges and in the construction of energy works with the financing of resources from energy consumption taxes (essentially liquid fuels) with the contribution, always a minority, of the International Development Banks. This scheme was destroyed in the 1990s, and the electricity system is no longer functioning.

(xi) **Results of the RENOVAR and MATER Plans**

As of October 2018, there were 16 facilities of these plans in operation, providing a power of 348 MW.

The first development began operating in mid-2017, and the last was connected in October 2018 by YPF when the Manantiales BEHR plant was completed.

The 16 developments in operation range from small-scale hydroelectric developments to biogas and biomass plants, solar and wind farms. In addition, 86 of the 157 projects already awarded are under construction, while the installation of large hydroelectric and nuclear plants is delayed, affecting the Argentine nuclear development.

2.3.11 Public–Private Partnerships (PPP)

(i) **What they are and where they originated**

They arise from the conception of the current government that does not favor state investment in infrastructure works such as energy and more specifically those in the electricity sector.

PPPs are contracts between the public sector and the private sector to carry out works or supply goods. In these contracts, the (private) contractor assumes the responsibility to obtain a substantial part of the financing of the work.

They are projects aimed at infrastructure, housing, public services, productive investments, applied research and technological innovation.

But there is no room for those destined only for the provision of labor, supply and provision of goods and those financed essentially with National Treasury Funds.

They are obtained by public tender or competition.

Payments can be made by directly affecting resources or transfers or through duly regulated trusts.

The Trust Agent is the Bank for Investment and Foreign Trade (BICE), and the trustees are the ministers or secretaries under whose jurisdiction the work is carried out.

The duration is for thirty years.

This figure originated in the United Kingdom in the 1970s and was developed mainly in the transport sector with mixed results, to be benevolent.

(ii) **The Legal Part**

The law and the regulatory decrees are very interesting for the benefit of private contractors, since they eliminate the obligation to pay VAT for the funds they participate in, exempting them from income tax and eventual property tax.

If the contractor is a foreigner and there are disputes, courts of countries with headquarters in member states of the convention on the recognition and enforcement of foreign arbitral awards act.

(iii) **Control**

This control is carried out by a Bicameral Commission for the monitoring of contracts, presided over by a member of the opposition and vice-presided over by a member of the government. It is composed of seven deputies and seven senators.

(iv) **Advantages, disadvantages and criticism**

A study by the CECE Foundation, composed mostly of professors from the University of Buenos Aires (UBA), adds that this system, developed in the early 1970s in the United Kingdom, generally involves "the direct provision of a service to the public sector by a private company, through a contract that includes the design, construction and then operation and maintenance of an infrastructure."

According to the CECE Foundation's report, written by economist Alejandro Einstoss, "the greatest advantage for the public sector is that the works are financed by the private sector, thus avoiding budgetary restrictions, without generating public debt and taking advantage of the private sector's management capacity." On the side of the private sector, the advantages "come from being able to participate in and structure projects that would be out of scale without the participation of the public sector," where the state assumes certain risks or reinsured an amount of income.

However, PPPs can present different disadvantages in relation to traditional public works. Among them are possible failures in the preparation of contracts that result in cost overruns for users, errors in the choice of projects and in the distribution of risks, high financial costs and recurrent renegotiations.

"The origin of the PPP is Margaret Thatcher's United Kingdom, and so this type of public-private partnership scheme was extended to other countries (Spain, Chile, Brazil, Panama, Colombia, Peru, France).

However, the British National Audit Office (NAO) denounced the poor results of the system because the over costs were up to 40$ in the works.

Furthermore, in Spain, the bankruptcy of the construction companies that participated in this system pushed the Spanish administration to nationalize four highways built through PPPs, resulting in a loss of 2.5 billion dollars to the Spanish government (Taken from Conclusion, interview with Pino Solanas, April 29, 2018)."

2.3.12 Distributed Energy

Some concerns

Because of economies of scale, electricity generated by large hydro and nuclear power plants has to be cheaper than electricity produced by small (less than 300 KW) residential or commercial generators or "large" generators (more than 300 KW) that are generally industrial (except for very special cases of cogeneration), and it does not seem that economies of series prevail over economies of scale.

If this were to become widespread and not a small portion of transactions, it would atomize the electricity market and could raise generation costs and make the country less competitive.

On the other hand, if one of the advantages of this alternative were to eliminate the needs of the distribution networks, transactions could not be made from the generator to the system when there was a surplus of energy, nor could they be bought from the system otherwise.

In short, it seems interesting if it represents a small part of the transactions, for example, up to 10%.

In any case, foreign sellers of renewable source equipment are happy.

2.3.13 The Rational and Efficient Use of Energy (UREE)

(i) **Variables to be taken into account: prices**

One of the variables usually taken into account in UREE's analyses and proposals is that of the absolute and relative prices of the various energy products.

Given that these are generally goods with low or very low price-demand elasticity, tariff incentives or disincentives will not necessarily serve to make UREE policies effective.

(ii) **Variables to be taken into account: type of activity**

Another aspect, and this is more related to the industrial sector, when it does not involve electro-intensive activities, is that the incidence of energy costs on production costs is usually very low, especially when compared to salary and financial costs. So on the side of the tariffs, this type of industrialists will not have too many incentives to apply UREE in their processes. Then, we must look for other types of advantages such as tax relief. Then, it will be the State that will have to bear the cost of the programs, under the assumption that a saving in the energy supply will be economically and environmentally favorable to the interests of the country.

Although the low price-demand elasticity of energy products does not help to generate UREE policies in the residential sector, considering also that these are necessary goods and in several cases not replaceable, very low relative prices, as occurred in much of the previous government, make UREE policies unviable.

But at the other extreme, excessive tariff increases such as those applied by the current government cause, when not accompanied by increases in wage income, an artificial incentive to UREE policies.

The situation is not sustainable over time, and there is a risk that users will not be able to use the services, especially natural gas, LPG and electricity. In any case, the measures to promote the labeling of household appliances are welcome, but they are only valid for new equipment and it is unfeasible for the majority of the population to recommend the change of appliances due to the deterioration of the level of income.

Even more so when the so-called social tariffs practically disappear and the discounts for lower consumption compared to similar periods of time in the past are eliminated.

In these cases, UREE policies clash with the satisfaction of basic needs.

In the transport sector, consumption of petrol and diesel, traffic congestion in cities, GHG pollution, the absence of public transport for people and the virtual elimination of rail freight make this the most important sector for implementing UREE policies.

Measures to replace lamps in public lighting have also been very effective, as well as projects to label homes as energy savers. However, this must go hand in hand with plans to build housing adapted to the country's climatic realities.

(iii) **The work of the current Undersecretariat for Energy Saving and Efficiency**

In summary, the work of the current Undersecretariat for Energy Saving and Efficiency would appear to be very active and positive.

2.3.14 Energy Prices and Tariffs

2.3.14.1 Natural Gas Tariffs

The production of natural gas at wellheads is subsidized, since at least several producers are recognized for a price of US $7.5 per million BTUs, which then falls by US $0.5 per year to US $6 and then is guided by the "market" (estimated at no less than US $4) above the development cost of US $1.7 and the Henry HUB of US $3.

Thus, for producers, the subsidy is not a "bad word."

The transport and distribution of natural gas are monopolistic activities, unlike production, and the consumer is captive to the company from which he buys the natural gas; that is, he cannot change companies.

The fact that the current rates are very high is proven by the enormous profits of the distributors.

There were public hearings, and in them, ENARGAS presented a report justifying the increases.

But who audited the costs of the transport and distribution companies?

Who verified and is controlling their future investment plans to support the new tariff levels?

The legislation of the 1990s that culminated in the privatization of Gas del Estado established a periodic tariff review with the intervention of the regulatory body (ENARGAS).

The enormous increase in prices after 2015 was carried out in order to reduce the large subsidies, reduce the fiscal deficit and favor the profits of the companies that, as was said, did not make the investments that these price increases facilitated and instead increased the payment of dividends to the shareholders.

2.3.14.2 Electricity Rates

As with natural gas, users are captive to the distributor who sells them the electricity. That is to say, I could not choose another one either. It is a monopolistic activity.

As in the case of natural gas, non-binding public hearings were held to explain the tariff increases.

Here too, who audited the costs mentioned by the companies? Who is controlling the realization of the investments, supposedly committed, that are the basis for the increases?

What can be verified is that in the city of Buenos Aires and in the suburbs the cuts have continued, essentially due to distribution and transformation problems (responsibility of the distributors).

For example, the Central Port, the main private generator, has earned US $310 million in the first quarter of 2018.

2.3.14.3 Some Effects of Rates

Other effects of the TARIFFs are given by the analysis of the balance sheets that are eloquent with respect to the brutal change of relative prices and the transfer of income that, via tariffs, is made in favor of the companies that make up the privatized energy sector, to the detriment of the productive sector and the households. Thus, it can be seen that within 53 companies eight firms stand out that have a direct relationship with the policy of tariffs of the national government: Edesur, Edenor, Metrogas, Camuzzi Gas Pampeana, TGS, Pampa Energía, Endesa Costanera and Central Puerto.

For these companies, the growth of their turnover was 99%, their gross profits expanded by 209% and their final net profits by an even greater percentage, 772%. Compared to 2016, a company like EDENOR expanded its gross profits by 2273.3%.

2.3.14.4 The Fuel Tax

(i) **Before and after deregulation**

Prior to the deregulation of the 1990s, there were four types of specific taxes levied on activity at different points in the oil chain. These were:

The tax on the processing of crude oil (equivalent to 10% of the FOB value of the crude oil),
The one applied to the transfer of liquid fuels,
The one applied to the social security funds, and
VAT.

This tax mechanism was in place until the end of the 1990s and was modified when oil deregulation was implemented.

National Law 23,996 established a wide range of reforms in line with the Convertibility Plan of late 1991.

(ii) **The destination of fuel taxes from convertibility**

After convertibility, the National Energy Fund, the fund for Chocón and large electrical works and the tax on crude oil ceased to exist.

The fact is that, with the privatization of practically the entire energy sector, the private companies would be in charge of making the investments, and consequently, these funds would not be necessary.

We only have to remember that, thanks to the energy funds, it was possible to build the extensive network of gas pipelines, hydroelectric and nuclear power plants in Argentina.

(iii) **Immediately before and after the March 2018 amendment**

The fuel tax was modified in March 2018 by introducing changes to title IV of National Law 23,966. The two most relevant aspects introduced are that the taxes were changed from a variable amount to a fixed amount and a new tax was added: that of carbon dioxide emissions (discounting biofuels (biodiesel and bioethanol) added to diesel and gasoline).

The fixed amount will be updated quarterly according to the variation of the Consumer Price Index published by the INDEC.

As of March 2018, the share of fuel taxes and carbon dioxide tax represented 22 and 20% of the retail prices of super gasoline and common diesel oil, respectively. This is almost 40% of what they represented historically for gasoline and almost 30% of what they represented for diesel.

This change of modality favors the oil companies that distribute the fuels because now the increases in the prices of these products will go entirely to them, the State will lose and the consumers will assume the totality of the increases.

There are two reasons for this: The prices of oil derivatives (essentially naphtha and diesel) increase much more than the Consumer Price Index (e.g., in the first nine months of 2018, the prices of derivatives increased by nearly 65% and the index by 35%), and the relative delay is one quarter.

As for the destination of the taxes, nothing goes to the energy sector (for the same reasons mentioned when analyzing the change caused in the nineties by the oil deregulation) and an important part will go to the trust funds that will be derived from the Public–Private Participation Programs already mentioned.

2.3.14.5 Prices of Crude Oil and Its Derivatives

(i) Price release

In a letter, the former Minister of Energy and Mining, Mr. Juan José Aranguren, informed all the actors in the hydrocarbon market that "in relation to the agreement for the transition to international prices of the Argentine hydrocarbon industry," and depending on the dynamics of external quotations, it had been decided to free the price of fuels.

From then on, the oil companies would be authorized to modify the sales price of their fuels for consumption in the automobile market. In other words, the internal prices of oil derivatives would be freed up.

The winners of this measure were the refineries—those that sell naphtha, such as YPF, Shell, AXION, OIL, REFINOR and Petrobras—that set prices according to what they pay for the crude oil they buy from local producers (YPF, Pan American Energy, PLUSPETROL, Petrobras, mainly).

This will not favor the companies that extract the crude in the country as long as they receive a price for the crude that is lower than the international market price. In June 2018, this difference was 11%, as the international reference crude oil that had always been the WTI had been exchanged for Brent, which was approximately 10–15% more expensive.

In short, the oil companies in Argentina never lose.

From this decision of the authorities, the prices of oil derivatives in Argentina are freed, they are dollarized, and it will be necessary to take into account the international quotation of Brent crude (with its equivalent in pesos), the value of the dollar, the price of biofuels (since they are mixed in 10% with diesel and 12% with naphthas) and the quarterly variation in the INDEC retail price index.

(ii) This change in rules put an end to a period of more than fifteen years in which local prices were regulated and independent of the international price of oil.

When crude oil reached more than 140 dollars a barrel, fuels had low prices due to progressive export taxes. This situation began to be corrected, through small price adjustments.

From mid-2014, the price of crude fell to less than US $44 in nine months and forced the government to agree with the oil companies on a much higher sustained price ("criollo barrel" in pesos), of US $73 on average. But with oil on the ground, fuel prices became the highest in the region behind Uruguay. That pattern was maintained by the current government, whose assumption coincided with the lowest Brent (US $36.2 per barrel) since 2009. The difference is that withholding taxes were eliminated, and then, a decreasing path was set for the "criollo barrel," which was actually a consumer subsidy to the oil companies, as opposed to the years of expensive oil and low prices.

(iii) As of November 2018, gasoline and diesel fuel prices have already risen by an average of almost 70%, with the impact on freight and transport costs adding to inflationary pressures. Despite the fact that during this period the Brent fell, the dollar rose by almost 100%. In short, with free prices, fuels can rise when oil does not, but the dollar does, or vice versa. And they should fall if crude falls, and the exchange rate will stabilize with the current volatility in both markets; the unknown is the frequency of changes.

(iv) In the first days of November 2018, fuel prices increased again.

On the other hand, the price of Brent crude oil (a marker for the Argentine market and this was changed by the former Minister of Energy, since it was previously the WTI, which in November 2018 was 14% cheaper than Brent, which favors the oil companies and harms consumers) in the same period fell by 5%.

In other words, it is verified that in the Argentine market the prices of oil derivatives are inelastic to the fall of the international price and elastic to the rise.

(v) Another aspect that deserves to be taken into account is that with the current price policy the value of, for example, naphtha and diesel oil is much more expensive in some provinces than in others. In other words, prices are not fixed and homogeneous as they were before the 1990s.

2.3.14.6 LPG Prices

Since the previous government, the so-called Plan Hogar has been in force with the aim of making the carafes more accessible to the lower-income sectors. It is that the price of the calorie that pays a user of LPG in carafes of 10 kg was in June of 2018 1.5 times more expensive than the one paid by a consumer of the category R 23 of natural gas.

2.3.14.7 Biofuel Prices

These prices are updated periodically by the Secretariat of Energy and Mines (called reference values) and are different for small and large producers. In the case of bioethanol, the price is different if it is obtained from corn or sugarcane.

There is a difference between biofuel producers and oil refiners, who argue that this mixture makes their costs more expensive, but in this and other differences it is always necessary to consider the positive and negative impacts on the productive fabric.

For example, there are experiences that increase the addition of biofuels over the current limits, reaching, in the case of biodiesel, the operation of buses with 100% of this fuel and the case of Brazil that generated a fleet of special cars that used totally bioethanol. In other words, this is an issue that cannot be left to the producers of biofuels and oil derivatives alone.

Environmental impacts cannot be ignored, as is the case of biodiesel obtained from palm oil in Colombia.

2.3.15 The Grants

During the previous government:

In the last years of the previous government, there was an irrational tariff structure that led to a policy of indiscriminate subsidies (not for all provinces) which, together with the delegation of a large part of the investments to the private sector (it should be remembered that since the 1990s almost all the state-owned companies had disappeared), led to problems of shortage of electricity and natural gas and the need to import fuels (the energy export policy of the 1990s and early 2000s and the late reaction, only in 2012, with the particular form of nationalization of YPF, are responsible for this).

But the distorted tariff structure of the current government, especially in the AMBA, was the product of a correction, almost overnight, that should have been carried out very gradually and elaborated a new one far from the neoliberal principle that says "that everyone should pay according to their costs" because this led to a regressive tariff structure that seriously affected fixed-income family economies, such as salaried and retired workers, SMEs and neighborhood clubs, for example.

The government that emerged from the December 2015 elections with its intention to cancel the primary fiscal deficit (increasing the secondary deficit with the increase in interest on the contracted foreign debt) has implemented a strong policy of decreasing subsidies, in this case energy subsidies, with the objective that as soon as possible each user ends up paying full rates for electricity, natural gas and very high rates for LPG.

During the period that covers the years 2004 and 2014, the subsidies destined for the energy sector added up to almost $342,000 million current pesos (equivalent to 2,695,658 million pesos of the year 2018).

The energy subsidies were channeled mainly through two large companies, CAMMESA (55.8%) and ENARSA (30.7%), which among other purposes are responsible, respectively, for subsidizing the generation of electricity (either through compensation to the generators or through the purchase of fuel) and the purchase of imported gas to supply the domestic market.

With a much lower incidence, there were also transfers to gas-producing companies resulting from the program for injecting surplus natural gas (known as the "Gas

Plan") approved at the beginning of 2013, aimed at increasing production of this resource by granting subsidies to companies (5%).

Yacimientos Carboníferos Río Turbio (2.2%) for operating expenses, essentially salaries.

Trust fund for residential LPG consumption (1.5%) for the Plan Hogar for the so-called social container.

Yacyretá (1.4%) for payment of electricity to Paraguay.

Trust fund for residential natural gas consumption (0.2%) for tariff compensation for consumers in Patagonia and other areas of the country.

Other destinations (3.1%).

During the current government:

In relation to 2016, subsidies would have decreased by almost 16%, with strong falls in the Gas Plan and important increases in those destined for the former ENARSA and LPG consumers.

But the enormous magnitude of tariff increases (NG and EE) has not been enough to significantly reduce subsidies.

So the government will continue to increase rates in real terms?

If this happens, a considerable number of families from the lower and middle classes will lose their right to consume energy and hundreds of SMEs and public good organizations will tend to disappear.

The solution would seem to be different.

2.3.16 The Energy Matrix

Some Comments

First, it should be clarified that intermediate consumption, for example, those destined to produce electricity, is not included in the total consumption.

Total energy consumption has hardly grown between 2014 and 2017.

This is not as a result of energy-saving measures but as a consequence of a drop in the country's economic activity and the income of its inhabitants.

There are few changes in the share of sources: For example, the fall in natural gas and derivatives would be due to strong tariff increases and the aforementioned lower activity in the productive sectors, and on the other hand, the advance of oil and derivatives to the higher consumption of gasoline and diesel, in spite of the increases of prices of the same ones.

As for BIOENERGETICS, the relative gain would have been due to that of naphtha and diesel since bioethane and biodiesel are mixed with them.

As for electricity consumption, it would have practically stagnated, and the commentary included in this section is valid.

In the energy consumption to generate electricity, the impact of the income from Atucha II was noted, as the use of natural gas for conventional thermal generation has continued and the effects of the renewal plans in terms of the relative weight of the wind turbines and solar plants are not yet observed.

There would be a slight drop in energy efficiency. That is, system losses would have increased (primary, secondary and transformer station losses); the level of self-supply would also have fallen (obviously due to lower oil and natural gas production and higher liquid natural gas imports); and it would be noted that less energy was needed to generate one unit of GDP in constant currency.

2.3.17 Guidelines for a Different Energy Policy

A previous budget.

Without an integral and sustainable development plan that gives an idea of the type of country that Argentines want in the medium and long term, all energy policy will cease to have any support.

As energy, at least in Argentina, is a satisfactory need of the socioeconomic system, the quality, quantity and diversity of it will respond to the model of country that is decided.

Otherwise, the market or, better yet, the banks will recreate energy planning as an indicator of the most appropriate allocation of energy resources, with the main objective of ensuring the supply of energy in the quantity and quality required by the socioeconomic system not only for present generations but also for future ones. These plans must be independent of the periods of government and must be updated annually or biannually.

Make the Secretariat of Energy responsible for implementing the policies of the energy plan. Recreate the energy information systems by recovering, for example, the annual fuel and electricity bulletins, which of course have been improved and made electronic and which greatly facilitate the tasks of energy scholars.

Drafting an energy code and a new hydrocarbons law, which will govern activity throughout the country. The fact is that the constitutional reform transferred natural resources to the exclusive domain of the provinces and this is generating, to be mild, imbalances and asymmetries in the management of these resources and makes it very difficult to design and implement a coherent energy policy. This code should be elaborated by the National Congress.

A strong role for energy regulators. As long as private energy companies exist, their annual and medium-term plans should be monitored for compliance by these entities. Ensuring that they are compatible with energy plans and particularly that they comply with investment plans in a timely manner. In this way, the supply of all energies would be assured in terms of quantity, quality and time. Public energy companies should also be subject to the rules established by the entities.

In general, massive energy exports will not be encouraged, except in the case of interconnections and agreements with Latin American and Caribbean countries or exceptional discoveries. In the latter case, a policy similar to that of Norway and not Nigeria should be followed, and surplus resources should be used to improve the quality of life of the population, ensure the supply of future generations and industrialize the country. In other words, break with the policy of exporting raw materials and importing industrialized goods.

A necessary condition for the implementation of an energy policy that takes into account the present and future interests of the country and its inhabitants is the recreation of the state energy companies, each with its own sphere of influence and coordinated through the Secretariat of Energy.

But these companies must have management autonomy, their management personnel must be selected by competition and not by political cronyism, they must present annually their plans inserted in the National Energy Plan, and they must account for their actions in terms of compliance with the goals contained in the plan, not only those of production and be federal; that is, the provinces must join their boards and participate not only in decisions but also in the "benefits" that the companies produce.

Perhaps the model, adapted to our reality, is that of companies such as Electricidad de Francia and its plan contracts.

The provinces should be integrated into the boards of companies so that they are not just recipients of royalties.

In this way, the natural resources existing in provincial territories will now belong to the provinces, and the nation and the same should happen with the existing energy resources in marine waters.

As today there are no national energy companies, and in the case of hydrocarbons, private companies exercise de facto control over the resources in their concessions, it seems important to elaborate a new hydrocarbons law that replaces the inadequate hydrocarbons National Law 17,319 in force, reformed in 2007, for the purposes proposed here.

In this way, it would be necessary to suspend the extension of the concession contracts that the provinces are making, review those granted and convert the exploration permits and those exploitation concessions into service or association contracts with the majority of the state YPF and without free availability of foreign currency or oil and natural gas.

The two points that follow must be analyzed in the following context that will give an idea of the magnitude of the "business" of hydrocarbons in Argentina.

The sale of oil derivatives plus exports and minus imports would generate no less than 33 billion dollars a year from 2014 at the official exchange rate.

The economic value of the oil and natural gas reserves, at international prices, would reach 214 billion dollars without counting the resources of Vaca Muerta and the rest of the non-conventional deposits.

But how will the implementation of the energy works be financed?

Years ago, a large part of the financing of the energy works came from the taxes on fuels that were assigned to build hydroelectric, nuclear and conventional plants and to construct the country's gas pipeline networks.

During the 1990s, everything was privatized and the state delegated the investments to private companies because, it was said, "the market was a good allocator of resources."

This policy continued with subsequent governments.

But the private entrepreneur must respond to the interests of his shareholders and these do not necessarily coincide with what the country needs in terms of the type and opportunity of the works.

The objective of making the market more transparent and more competitive was practically not met and the consequences are clear.

In addition to good management and resources from fuel taxes, via the energy fund, resources can be obtained from equipment suppliers' credits, from multilateral banks, from a regional bank like Banco Sur, from the recreation of the National Development Bank (i.e., redirecting the current banking system) and associated through bond issues to Argentineans.

But what role does the private entrepreneur have in this scheme?

In a very general way, it can be said that the equipment of the energy sector in Argentina requires between 5 and 8% of the GDP, depending on the delay of works and the growth of the demand.

Therefore, the private sector that wants to be linked to energy must reconvert, leave the operation of the fields and power plants, gas pipeline, etc., to state-owned companies and devote itself to the manufacture and installation of the equipment required by the activity.

To do this, it will have the purchasing power of state enterprises that can range from 15 to 20 billion dollars a year (this will emerge from the energy plans that will be drawn up).

It seems that this role of entrepreneurs may be more useful to the country than that of the current Argentine "entrepreneurs" participating in YPF S.A.

This "Argentinization" of YPF is quite similar to that of YPF's old contractors, when some created their empire with virtually no entrepreneurial risk.

Energy price and tariff policy and subsidies

The most genuine way for a company to finance the capacity of its productive structure is through the income it receives from the sale of its products.

In the case of the energy sector, it is not usual for a public or private company to self-finance all of its investments. The fact is that often large investments have to be made that mature several years later.

For this reason, bank and supplier credit and sometimes company capitalization are often used.

In any case, it is not desirable that one's own income is not enough to cover a more or less important part of the investments.

The country is applying an erroneous pricing and tariff policy for energy products, especially for electricity and natural gas.

The case of oil derivatives is different, and the current prices, measured in constant dollars per liter, for diesel and super naphthas are today above those in force during convertibility.

Trust funds have been another instrument for private companies to take ownership of power plants.

In other words, tariff structures need to be revised. This should not be a policy for the economy, but should be the country's tariff policy.

Of course, if the operation of the sector were in the hands of state-owned companies, implementation would be simpler, because the values should cover the costs and an adequate expansion of the productive capacity and not allocate amounts to the shareholders who own the private companies.

Subsidies such as taxes are economic policy tools and their goodness or opportunity will depend on how they are used.

Families that consume energy (e.g., electricity, natural gas and LPG) do not have the same level of income and consume in general in relation to that level of income.

A family living in a 20–40 m^2 apartment will consume less than a family living in a 200 m^2 house.

Therefore, not only the average rate should be taken into account, but also a rate structure should be implemented. This structure can be progressive or regressive.

It is progressive when the price of the unit consumed (KWH or m^3 or kg) is lower while consumption is lower and reciprocally. In other words, those who consume less will have a lower unit rate than those who consume more. This generates what in economics is called a cross-subsidy, where those who consume more subsidize those who consume less.

The important thing is that the average rate covers all the operating costs and a part of the investment costs (not all of them, but the rate would be very high and the present generations would be financing future generations).

If the average tariffs were very low, future generations would either have no energy or they would have to finance companies with very high average tariffs if they want to have energy and they would be financing the past generations that had very low average tariffs.

The same should happen with the service and productive sectors. In this case, the final consumption sectors (commerce) should have higher unit rates than the intermediate consumption sectors (industry, especially SMEs) and public service providers (transport).

This should lead to the design of a tariff structure that has different unit rates according to whether the families consume little or a lot; whether the sectors are of final consumption or intermediate consumption; and whether the temperature where they are located is more or less rigorous.

If the problem is analyzed as a whole, i.e., medium- and long-term plans are made, tariff structures will emerge that are equitable, socially, economically and in terms of generation.

This should lead to the stratification of household consumption and even determine a reasonable minimum consumption (which does not imply having a 40 W light bulb per room or a consumption of gas only for cooking) that will have the minimum tariff.

Of course, everything would be solved if all the inhabitants of the country had a stable, dignified and fairly paid job. Then, the tariff structure could be different.

In summary, in recent years, in Argentina, there was an irrational tariff structure that led to generate a policy of indiscriminate subsidies (not for all provinces) that along with the delegation of investment in the private energy sector (had disappeared the state enterprises) led to supply problem in electricity and natural gas and the need to import fuels (this is essentially responsible for the policy of energy exports in the 1990s and early 2000s and the late reaction, only in 2012, with the particular form of nationalization of YPF).

But on the other hand, the current distorted magnitude and tariff structure, especially in AMBA, should not have been corrected in such a short term (and it is not yet finished) but gradually and elaborating a new tariff structure away from the neoliberal principle that says "everyone should pay according to their costs," because this would inevitably lead to a regressive tariff structure that would make it impossible in many cases to pay a large part of the population, especially those with fixed incomes that do not keep up with the rates.

The issue of energy saving or rational energy use can and should, if well managed, influence tariff structures, but far beyond a system of rewards and penalties. It seems that even the deductions from energy service payments for lower relative consumption in comparison with similar periods in previous years would have disappeared.

The application of the so-called social tariff would also have been greatly restricted.

But why did these extreme imbalances not occur before the 1990s?

Essentially because of the existence of the energy companies among them YPF, Gas del Estado and Water and Energy that were privatized in that decade.

The financing of energy investments came from fuel taxes (essentially gas and diesel) which together with contributions from national banks such as the Development Bank and the Banco Nación (the most important public bank of Argentina) contributed almost 70 or 80% of the cost of the works, and the rest came from multilateral banks such as the IDB and from equipment suppliers.

The public energy companies set their tariffs and the political authorities sometimes used them to avoid increasing inflationary rates, which delayed the work.

Another aspect was that the public energy companies did not need the profit item, but the one that contemplated the expansion of the service, and this meant that the tariffs did not owe so high, because the "shareholder," to whom they had to be accountable, was the State, that is, the population.

Furthermore, energy was considered a public service and energy consumption a right.

The fact is that energy meets the needs of the productive sector and the population.

On the other hand, neoliberalism considers the energy sector as just another sector producing goods and services.

Therefore, for this conception, energy activities are self-contained and isolated business units and each of them must be self-sustaining.

Consequently, they do not accept subsidies, and ultimately, if a business unit gives "losses," it must import what it produces.

There is no connection between the energy sector and the rest of the sectors: in political terms, between the Ministry of Energy and the Ministries of Finance, Production or the Central Bank.

So autonomous decisions taken in isolation, for example, by the energy authority, with an exorbitant tariff increase, do not take into account the repercussions on the rest of the system (e.g., inflation, fall in consumption, standard of living of the most vulnerable population).

The then energy minister was given the directive to increase tariffs to end subsidies and reduce the deficit, considered to be the sole cause of inflation. He fulfilled his task, did not consider the repercussions to be his responsibility and only agreed to take into account the so-called social tariffs.

In other words, there is a problem of "ideological" conception underlying the decisions.

That is why the problems of tariffs and subsidies are a political issue that must be addressed, of course, technically.

(x) **In summary**

A comprehensive national plan for the country's sustainable development.

Energy planning.

Re-creation of the decision-making power of the Secretariat of Energy.

Preparation of an Energy Code and a new Hydrocarbons Law.

Re-creation of national and federal state energy companies.

Re-creation of the energy fund from the taxes on oil and natural gas derivatives (which are non-renewable energies and are putting a lot of pressure on the energy matrix structure) to finance part of the expansion of productive capacity, including the mining risk involved in exploration.

Assign the private sector the role of supplier of the equipment and inputs required by the companies' investments.

Re-create the National Development Bank as a genuine source of financing for part of the sector's investments.

To reformulate another structure of prices and tariffs to leave the current one and also to serve as an instrument to modify the energy and electricity generation matrix.

Finally, it will not be possible to implement an energy policy at the service of present and future generations in the country if state energy companies are not recreated.

2.4 Contents

This document attempts to analyze the energy policy applied in Argentina from December 2015 to the end of 2018.

The aim is to observe the most prominent aspects of this policy.

Of course, the period studied is very short compared to the period during the Kirchner's government, which lasted about twelve years.

The energy policy of the "Kirchnerist" period was examined in two works by the author and in another by Roberto Kozulj.

Bravo, Víctor, 2015. "Análisis de la ley 27007, llamada de Hidrocarburos, y de la política hidrocarburífera del período 2002 a 2014," Working Documents of Fundación Bariloche, January 2015. In Spanish. In English in this volume.

Bravo, Víctor, 2015. "Panorama de la Energía en Argentina," Fundación Bariloche Working Documents, August 2015.

Kozulj, R., 2005. "La crisis de la Industria del Gas Natural en Argentina," ECLAC, March 2005.

Both the three works mentioned and the current one are critical, but with a constructive intention. That is to say, they particularly point out the measures that, in the authors' opinion, were not convenient for the country, without forgetting to mention the others.

This paper attempts to contribute to what should have been done or should be done, always in terms of energy policy of the current government.

That is, it seeks to give an opinion on the most serious energy problems, or not so much, that Argentina has in mid-2018.

That is:

Is there coordination between the measures taken by the Ministry of Energy and the other Ministries?

Is there energy planning, and if so, what kind?

What is the role of the National State, the Provincial States and the Congress in the energy sector?

Why is oil and natural gas production falling?

What is happening with Vaca Muerta oil field and with conventional hydrocarbon deposits?

How is the social license ensured and the environmental impacts of fracking considered?

Why is the import of oil and derivatives increasing?

Why do we want to export natural gas?

What about the refineries?

What is the role of YPF?

Are investments in transmission and distribution generation contemplated to ensure the supply of electricity?

What will be the role of the hydroelectric plants and why is the construction of several of them delayed?

What will be the nuclear plan and the local participation in its expansion?

Why does TRANSENER want to be privatized?

How is the participation of local industrialists in the development of renewable energy sources ensured, specifically wind and solar energy? What are the costs of generating electricity with renewable sources? How are they dispatched?

What will be the role of ethyl alcohol and biodiesel and their prices in the mixtures with gasoline and diesel?

Why the big tariff increases for natural gas and electricity? And those of the petroleum derivatives? What happens with the tariffs in the interior of the country?
What is the role of rational energy use?
What about the social carafes?
What will happen with the Río Turbio coalfield?
And many other questions.

First, we will mention the aspects related to energy planning, and then, we will go into the aspects related to each one of the energy sources (oil and derivatives; natural gas and liquid petroleum gas; electricity; nuclear energy; biofuels, wind energy; solar energy; and mineral coal).

Special treatment will be given to prices and tariffs, including fuel taxes, and subsidies.

The energy matrix will be analyzed.

It is intended to present the topics in such a way that they can be read not only together but also individually.

Although the policies for each energy source are, to a lesser or greater extent, interdependent, the policy of one source in particular may be of special interest.

2.5 Something, Very Brief, About "Ideology"

In the area of energy policy in Argentina, there have always been two main concepts regarding the role of the State and the private sector.

One considers the energy sector as similar to any productive sector of goods and services and therefore subject to the same rules. Its profits must pay taxes and when its costs, for internationally tradable goods (this includes oil, its derivatives, coal, liquid natural gas and nuclear fuels), are higher than the international ones, those goods must be imported, and if they are lower, they can be exported.

The idea is based on an unrestricted opening of international trade in a market where these goods abound and there are numerous suppliers and buyers.

The internal market must be freed up and let the various private players compete, and this will ensure the best service to consumers. Then, the price for Argentine consumers will be adjusted to the international price and the dollar–peso value. Then, the presence of the business state is unnecessary and should be limited to ensuring competition, avoiding the domination of the market by monopolies.

This is "what the private sector can do should not be done by the State."

In the case of goods, in principle, that are not internationally tradable, such as electricity or natural gas, the State should only have a regulatory role, but never an entrepreneurial one.

Another political conception considers energy activity as a public service and producer of non-replaceable goods that are essential for the daily functioning of the socioeconomic system. For people, it considers it a right to have energy.

It also considers the markets of these goods as monopolistic or oligopolistic, which, in addition, as in the cases of natural gas distributed by networks and electricity, are captive of the supplying company.

These markets, by nature, cannot be competitive (there are a much smaller number of suppliers than the number of buyers) and are generally either oligopolistic, as in the case of oil and its derivatives, or monopolistic as in the case of natural gas distributed over networks and the transport and distribution of electricity. For this policy, state-owned energy companies are essential. In this conception, the role of the private sector is essential as a supplier of the equipment required by all energy works and which constitutes the majority of the large investments. It does not accept the role of private concessionaire operator of the sector's installations and reserves it for State companies.

In Argentina, throughout history, one or other of the above-mentioned concepts has been in power.

In Argentina, there was a period when almost the entire energy sector was in the hands of state companies.

YPF, created in 1922, was the first state-owned oil company in the so-called Western Hemisphere and survived until 1992, when it was privatized, with private national and international companies, and then, it was bought by the Spanish Group Repsol in 1999 and "re-stratified" in 2012.

Agua y Energía Eléctrica was created in 1947 and closed in 1992.

Gas del Estado was created in 1946 and privatized in 1992.

Yacimientos Carboníferos Fiscales, created as a management company in 1958, was privatized in 1994 and renationalized in 2002.

HIDRONOR was created in 1967 and privatized in 1992.

Atucha I began operating in 1974, Embalse in 1984 and Atucha II in 2014, all three of which continue to be owned by the National Atomic Energy Commission of Argentina.

The Uruguayan-Argentinean Binational Entity, Salto Grande, a hydroelectrical dam, was created in 1946 and has been operating since 1974.

The Paraguayan-Argentinean Binational Entity, Yacyretá, also a hydroelectric dam, was created in 1973 and has been operating since 1994.

Both binational entities depend, at present, on the recently created company IEA S.A. (Integración Energética Argentina).

In addition, there have been and are provincial companies and cooperatives, generally distributors of electricity and even in some cases of natural gas, liquefied petroleum gas and oil.

It is not intended here to analyze the history of the Argentine energy sector.

The period of the "Menemism" (when Mr. Carlos Menem was President) and previous ones was examined, among many others, in the following publications:

"The Policy of Argentine Oil Deregulation, Background and Impacts."

R. Kozulj and Victor Bravo Centro Editor Latin America, 1993.

"The New Regulatory Framework and the Privatization of State Gas" R. Kozulj Energy and Development Magazine of the Institute of Energy Economics of Fundación Bariloche, Volume 4, October 1993.

"The results of the Argentine electricity reform" Graciela Díaz de Hasson, Energy and Development Magazine of the Institute for Energy Economics of Fundación Bariloche, Volume IV No. 8, October 1995.

The "Kirchnerist" period (when Mr. Néstor Kirchner and his wife Cristina were presidents), was studied in other publications already mentioned.

Now, it is time to examine the "Macrist" period.

2.6 Energy Planning

2.6.1 Some Concepts

This is not the time to justify the need to plan the development of a country's energy sector, but there is a political conception that considers that the market is giving the signals for the future and that planning is out of place.

It is the private sector, in this conception, that is in charge of everything and the State should only ensure clear rules of the game in order to avoid monopolies.

The author has a "CEPAL" (the Commission for Latin American Development) background from the 1960s and participated in the elaboration of the national development plan during the government of Dr Arturo Illia and the "Triennial" energy plan of the third presidency of Juan D, Perón.

He always saw with sympathy the Five-Year Plans of the first two presidencies of Perón.

Briefly some concepts.

Energy is considered to be a satisfactory input for the socioeconomic system of a country.

Then, depending on the future image of a country, it will be the energy sector that country needs.

This implies the need for socioeconomic planning, which serves as a framework for energy planning, not as an immovable corset but as a guide that indicates the direction.

Since energy sources are used specifically, or in some cases, substitutable for one another, they need to be considered together.

The relationships between oil and natural gas and both with electricity are examples of this.

Not to mention the relations between the tariff structures of the energetic ones.

Furthermore, investments to ensure supply have a long period of maturity. This means that today we must "know" the energy demand for the next five or ten years, in order to know when to start investing.

If this is left to private investors, there is a risk of energy shortages, generally due to underinvestment, or of resources being depleted by over-exploitation if market conditions warrant strong exports

2.6.2 Does the Current Government Have an Energy Plan?

The Secretariat for Energy reports to the Ministry of Finance, together with the General Coordination Unit, the Secretariat for Energy Planning Coordination and the Secretariat for Energy Policy Coordination.

There are seven subsecretariats: the Legal Subsecretariat, the Hydrocarbons Subsecretariat, the Electrical Energy Subsecretariat, the Renewable Energy Subsecretariat, the Nuclear Energy Subsecretariat, the Energy Saving and Efficiency Subsecretariat and the Administrative Conduct Subsecretariat.

Energy planning is concentrated in the Secretariat for Energy Planning Coordination.

What is important for defining and implementing an energy policy is not the administrative structure but the ideas generated and executed by the actors involved.

Thus, this structure will not be commented on or criticized. The objective of this paragraph is different.

The Secretariat for Energy Planning Coordination has prepared two energy scenarios: one to 2025 and the other to 2030.

Of course, they have been very careful to call them plans.

The one to 2030 presents four scenarios, with different hypotheses of occurrence.

The objective is to provide inputs to all sector actors.

In other words, it does not intend to be a guide for decision-making but, given its exploratory and non-predictive nature, shows what could happen if the different hypotheses used to develop each scenario were to be fulfilled.

Furthermore, it rightly clarifies that the results achieved will not remain static, but will change as different realities affect the hypotheses of the starting point.

In other words, "Planning" is seen as a dynamic tool that must be updated, for example, annually.

First, the energy plan to 2030 will be described based on a PowerPoint presented by the Secretariat of Energy and Mines in October 2018, and then, some comments will be made.

It is not intended here to describe the tools or the results achieved.

The analysis that will be made will then be available, and since the basic documentation that allowed the Plan to be prepared is not available, the comments will have this relativity.

The guidelines of the plan and the main objectives would be the following:

To increase the local supply of crude oil;
Increase natural gas production;
Enhancing the use of renewable energy resources;
Achieving a significant trade surplus in hydrocarbons.

To achieve this, the fundamental action would be to accelerate the development of non-conventional hydrocarbons from the Vaca Muerta oil field, almost as a fundamental objective of the plan, including also the exploration tasks in the Argentine Sea.

Non-conventional hydrocarbon resources and conventional reserves are mentioned.

Natural gas exports are quantified, even indicating the markets (Chile and Brazil for the gaseous one and the entire world for liquid natural gas, in huge volumes in this case).

For electrical energy, value is given to the incorporation of power until 2030, concentrating the effort on renewable energy (essentially wind and solar, accompanied by bioenergy), completing with conventional thermal, some hydroelectric and some nuclear.

The medium and high voltage transmission networks would be left in the hands of the Public–Private Partnership (PPP) modality discussed later in this text.

It is interesting to note the contribution to the country's exports from Vaca Muerta oil field and to compare them with those of the agricultural sector, especially those of soybeans.

It could be said that, just as the previous government was strongly linked to soybean, it would seem that this government wants to be linked to Vaca Muerta oil field.

Always based on extractive activities with relatively little generation of industrial added value typical of the enclave countries.

There is no mention of the evolution of the internal demand for energy. It is assumed that they have considered it because otherwise there would be no way to obtain the exportable balances.

Although it must have included socioeconomic inputs provided by the INDEC and the Ministry of Production, these inputs are not the product of socioeconomic plans or scenarios that would provide the possible long-term images of the country. In this way, the energy sector is isolated from the context and only reflects the impact of some of these variables on energy. Because there is no socioeconomic reference framework for the country (a National Development Plan), it is not known what type of future country the plan is referring to. It seems that it would be a "primarized" country, without a boost to the light and heavy industry, which is the one that generates more labor occupation and at the same time drives and is driven strongly by science and technology.

The presentation seems to be aimed at offering foreign investors' business opportunities as if the country were on the verge of collapse.

The role of the provinces in developing the scenarios is also not clear.

Despite the fact that on October 26, 2017, Decree 854/2017 created the Federal Energy Council, within the scope of the Ministry of Energy and Mines, with the participation of the Nation, the city of Buenos Aires and all the provinces (with the exception of the provinces of La Pampa, Tucumán and San Luis) and with the agreement of the presidents of the Energy Commissions of the Senate and the House of Representatives, the Federal Energy Council, never came into operation.

The objective of this council would be to consider the planning and development of the energy sector in the medium and long term; to advise on the modifications required by the legislation on energy, including the regulatory frameworks of each public service; to act as an Advisory Council to the National Executive Branch and the Provinces that require it in all matters concerning the energy sector, including the development of specific programs to promote alternative energies and energy efficiency, tariff harmonization, investment projects of national or provincial interest, among others.

It is not known what the participation of this council has been in the development of the energy scenarios prepared by the Secretariat of Energy and Mines and in the provision of the socioeconomic variables used.

It is that a socioeconomic planning would seem to go against the ideological conception of the current government for which the market would be in charge of setting the course and private actors would act accordingly.

In summary, answering the question posed above, yes, there are energy scenarios, but there does not seem to be energy planning.

2.7 Hydrocarbons

2.7.1 Oil and Derivatives

2.7.1.1 Some Data

Physical data and monetary data will be included. The latter refer to foreign trade (Tables 2.1, 2.2 and 2.3).

As of July 2018, almost US $871 million of oil and oil products had been exported, quadrupling external sales for the same period in 2107. More than the increase in crude oil production, this was due to the drop in crude oil processed by the distilleries. This is not a good sign since it implies a preference for exporting raw material (crude) and importing derivatives due to policies that favor this modality, as evidenced by the appearance of TRAFIGURA.

The Chinese and US markets absorbed almost all of these exports.

The crude oil exported was from the Cerro Dragón fields of the Pan American Energy Company.

Despite this, the country's trade balance was in deficit by US $789 million between January and July 2018.

2.7.1.2 Some Thoughts

(i) Reserves

Proven reserves fall from 380.7 million cubic meters in 2015 to 318.5 million cubic meters in 2017. This is due to the fall in exploratory wells from an average of 80 between 2012 and 2015 inclusive (or an average of 70 between 2014 and 2015) to 63 between 2016 and 2017 inclusive. This shows that the search for "conventional"

Table 2.1 Some EP and DP physical data

Year	Checked reservations (+) (*)(*)(*)	Total production	Non-conventional raw materials production	Production raw conventional	Raw processed	Exploring wells	Total wells	Motorcycle sales	Gas oil sales
Unit	10^6 m^3	10^3 m^3	10^3 m^3	10^3 m^3	10^3 m^3	Nº	Nº	10^3 m^3	10^3 m^3
2015	380.7	31,970	1523	30,447	31,036	63	1440	8538	12,883
2016	343.7	30,763	1981	28,782	29,698	59	938	8652	12,606
2017	318.5 (*)(*)	27,826	2575	25,251	28,963	67	876	9249	13,023
2018 (*)		6917							

Source Ministry of Energy and Mines of Argentina

Notes

(*) January to March

(*)(*) Estimate

(*)(*)(*) Proven Reserves do not include Non-Conventional Oil. According to the International Energy Agency report for 2013, the resources, not the reserves, of this type of oil would be equivalent to eleven times the proven reserves of Argentina in 2012, that is, 2933 million m^3

Table 2.2 Foreign trade of oil and its derivatives (10^6 m^3)

Year	M	M	M	X	X	X	X	X	X
	Crude	Oil	Naphtha	Crude	Diesel oil natural gasoline	Virgin naphtha	Other naphthas	Diesel oil	Fuel oil
2015	986.2	50.5	3873	2340.3	338.3	363	498		
2016	920.5	331	4167	2914.7	349	214.1	400	5.3	11.2
2017	1255.4	540.6	3563.7	1652.8	366	374.6	246.8	4.2	317.7
2018 (*)	258.6	163.8	301.8	538.6	51.7	34.2	65.6	23.5	164

Source Ministry of Energy and Mines of Argentina
Notes
M: Imports. *X*: Exports
(*) January and February

Table 2.3 Foreign trade in oil and oil products (10^6 USD)

Year	M	M	M	M	X	X	X	X	X
	Crude	Oil	Gas oil	Total	Crude	Natural naphtha	Virgin naphtha	Other diesel	Gas oil
2015	355.8	29.3	1857	2242.1	670.6	92.8	103.3	147.8	
2016	269.3	141.1	1536.8	1947.2	726.9	84.6	59.8	113.7	1.9
2017	453.8	266.9	1501.8	2225.5	566	114.6	127.3	92.2	1.7
2018 (*)	114.3	864	161.6	362.3	219.8	19	14	28.1	10.5

Year	X	X	Commercial balance
	Fuel oil	Total	
2015		1014.5	−1227.6
2016	3.3	990.2	−957
2017	107.4	1009.2	−1213.3
2018 (*)	59	350.4	−11.9

Source Ministry of Energy and Mines of Argentina
Notes
M: Imports
X: Exports
(*) January and February

oil has almost been abandoned, while the exploitation of "non-conventional" oil (essentially from Vaca Muerta oil field) has not yet kept pace with expectations. In addition, the almost 24 million m^3 of annual reserves that are falling only due to the natural decline of the oil fields have not been replaced.

(ii) Investments

Investments in production and exploration fell from US $10,775 billion in 2015 (US $873 million in exploration) to US $6724 billion in 2017 (US $215 million in exploration).

(iii) Production

In the period from 2015 to 2017, 84.3 million m^3 of "conventional" crude oil would have been produced, so that only 19.5 m^3 of reserves would have been incorporated (i.e., "discovered"), in the same period, which is reflected in the 318.5 million verified reserves as of December 2017. This is serious, in particular, because it seems to depend almost exclusively on Vaca Muerta oil field to increase production.

It should be borne in mind that Pan American Energy continues to maintain and increase conventional production from the old Cerro Dragón field (one of the most important oil producers in the country). Also, recently, in Tierra del Fuego, also in conventional structures, ROCH discovered a field through a well with a very high productivity (almost 330 m^3/day).

The price of Brent oil oscillating between US $70 and 80 per barrel between June and August 2018 is an incentive to continue searching for conventional oil, as is also shown by the interest in exploring the Chubut plateau, since the development of this type of field requires lower investment costs than, for example, the Vaca Muerta oil field.

Although the costs depend on several factors (e.g., depth, type of rock for the conventional ones and for the non-conventional ones the number of fractures), some values will be indicated that may not be representative, especially for the non-conventional wells.

Thus, the oil reserve of a conventional well may vary between 40,000 and 550,000 m^3 and that of a Vaca Muerta well between 87,000 and 161,000 m^3. YPF announced in Loma Campana, in the Vaca Muerta area, a well with a reserve of 1.5 million barrels per day.

A conventional well costs between US $3 and 6 million and one in Vaca Muerta oil field between US $10 and 14 million.

YPF S.A., in its Five-Year Plan 2018–2022, plans to venture into conventional fields, including activating secondary and tertiary recovery.

In addition, the structure with the highest initial productivity found in Vaca Muerta does not reach 200 m^3/day-well, with the disadvantage that it declines much faster than conventional wells.

(iv) Reactivation of production

In order to reactivate the production of oil, two actions were implemented, in addition to the Vaca Muerta district.

One would involve granting "mature" fields with very low production (where almost 80% of water and only 20% of oil are extracted) to SMEs to exploit them, almost as if there were "one well for each SME." Pablo Bizzotto, executive vice president of Upstream at YPF, said the company will launch an aggressive plan to maintain production from mature fields and that the goal is to maintain leadership in Vaca Muerta oil field. "We have two pillars: optimization and minimization of decline of conventional mature fields based on an improvement of the secondary recovery process and an aggressive deployment of tertiary recovery with polymer injection," he said.

This policy has already been implemented in the past with very poor results.

The other, which emerged from the Secretariat of Energy and Mines, would involve exploring the continental platform.

(v) Exploration on the Continental Submarine Shelf

To collect data on hydrocarbons at sea, the Ministry of Energy and Mines, by Resolution No. 197 of May 16, 2018, extended the period of the surface survey permits from two to eight years, authorized the concurrence of more than one company for this activity in the same area and granted a period of ten years to market the information obtained. This norm modified Resolution 131 of 1970, which had established that the period of the superficial recognition permit could not exceed twelve months, extendable to twenty-four. The resolution clarifies that the ownership of the information shall be of the permit holder and the enforcement authority, but the latter may not disclose the information to third parties until the ten years have expired—in other words, a freeze on the availability of information, and possibly on the area, for a clearly excessive period of time.

The Ministry of Energy and Mines took the first step toward the next call for an international public competition aimed at granting offshore hydrocarbon exploration permits by inviting interested companies to submit proposals for the nomination of specific areas (there would be about fifty invited).

The companies had until June 7, 2018, to determine the areas that could be of interest for their inclusion in the future call, according to provision 90 of the Undersecretariat of Hydrocarbon Resources.

It includes 14 blocks in the Northern Argentina Basin (142,000 km^2), 6 in the Marine Austral Basin (5000 km^2) and 18 in the Malvinas Basin (90,000 km^2).

The water depth would be between 400 and 4000 m.

Each company will be able to nominate up to eight areas of its interest, without this step generating any special or preferential rights to be awarded, or any obligation to submit bids.

The companies seem to be more interested in oil than in natural gas, despite the fact that the Total Austral Company is producing some 20 million m^3/day of natural gas in the Southern Marine Basin adjacent to the bidding areas.

The State, in turn, will "enjoy total freedom" to decide which areas will be included in the public competition or in future calls. The presentations made by the companies will be supplied to the local subsidiary of the American company Schlumberger, a firm contracted for US $14.9 million to "provide services related to the public tender," which will prepare a report for the Ministry indicating the nominations received in respect of each area.

In the grounds of the provision, the invitation is justified "for reasons of efficiency, in order to achieve a better definition of the guidelines" of the next call, and with "the purpose of attracting greater interest from the hydrocarbon industry."

By Decree 872 of October 2, 2018, it was called International Public Bidding for the exploration of Argentine marine waters and was ratified by Resolution No. 65 of November 6, 2018, which approved the terms and conditions.

The specifications were acquired by twelve companies but there were about twenty interested parties.

> The Secretariat of Energy published in the Official Gazette of November 7, 2018, the call for International Offshore Public Bidding No. 1 for the award of exploration permits for the search of offshore oil and gas, through Resolution 65/2018.

There, it also submitted as an annex, the tender document, which was included in Decree 872/2018, published on October 2, 2018, in the Official Gazette.

With this measure, the government seeks to advance the exploration of oil and gas in the Argentine Sea—an area that is presumably rich in hydrocarbons, in addition to fishing resources—and, in this way, expand the reserves (resources that are technically recoverable at economic costs with the current technology available).

There would be about twenty oil companies interested in the bidding, among them the American giants ExxonMobil and Anadarko Petroleum, the British BP (ex

British Petroleum), the Anglo-Dutch Shell, the French Total, the Norwegian state-owned Equinor (ex Statoil), the Chinese CNOOC, the Malaysian Petronas, the Italian ENI and YPF itself.

Those interested in participating will have to pay US $50,000 to acquire access keys to the Digital Database of the areas and must be registered at least 30 days before the opening of the offers. They will be divided into four categories.

Operator A: may operate all the areas and will be the one whose net worth of the last fiscal year exceeds US $250 million and its average of the capital investments in the last three closed fiscal years exceeds US $700 million in economic terms, while in technical terms its production must be greater than 20,000 barrels of petroleum equivalents (BEP) per day on average during the three years prior to the International Public Bidding, or demonstrate having operated at least three offshore exploratory wells in more than 500 meters of water depth in the last 15 years.

Operator B: may only operate in shallow and deep areas and will be the one whose net worth of the last fiscal year exceeds US $100 million and its average of the capital investments in the last three closed fiscal years exceeds US $300 million in economic terms, while in technical terms its production must be greater than 10,000 BEP per day on average during the three years prior to the International Public Bidding, or demonstrate having operated at least three offshore exploratory wells in more than 100 meters of water depth in the last 15 years.

Operator C: may only operate in shallow areas and will be the one whose net worth of the last fiscal year exceeds US $30 million and its average of the capital investments in the last three closed fiscal years exceeds US $100 million in economic terms, while in technical terms its production must be greater than 5000 BEP per day on average during the three years prior to the International Public Bidding, or demonstrate having operated at least three offshore exploratory wells in the last 15 years.

Non-operator A, B or C: will participate only as an investment partner and will be the one whose net worth of the last fiscal year exceeds at least 50% of the amount required from each operator, according to the Type of Qualification or indistinctly, and its average of the capital investments in the last three closed fiscal years exceeds at least 50% of the amount required from each operator, according to the Type of Qualification (taken from Mr S. Spaltro, from the article "El Gobierno publicó los pliegos para explorar petróleo y gas en el Mar Argentino," published in the newspaper El Cronista, Buenos Aires, on November 7, 2018).

As it can be seen, there is no provision for reserving areas for YPF S.A., nor is the incorporation of Argentine technicians or local companies to assimilate a technology that the country must be aware of, since the continental platform is very wide and can be a depository of vast hydrocarbon resources.

Below is some additional information, which does not include the work of the Total Company that is producing natural gas in the Southern Marine Basin.

On October 2, the national government launched a bidding round to explore 38 blocks in the Argentine Sea and YPF's interest in the Northern Argentine Basin, 300 meters from the local coast, is known.

Bids would be received until February 2019 and a second round was expected.

This is the largest offshore exploration drive in recent decades.

It includes 14 blocks in the Northern Argentina Basin (142,000 km^2), 6 in the Marine Austral Basin (5000 km^2) and 18 in the Malvinas Basin (90,000 km^2).

The northern zone, on the continental slope, has very deep waters (up to 4000 meters) and is very risky but it is still one of YPF's central objectives.

In addition to YPF, the Anglo-Dutch company Shell, the French company Total, the German company Wintershall, the Chilean ENAP and also US companies would be interested.

This new bidding is framed in National Law 17,319, which has maximum exploration periods of 4 years each in the two "Basic Periods," which are added to the 5 years for the extension, thus reaching the maximum allowed of 13 years.

Between the 1960s and 1970s, 17 offshore wells were drilled, mainly in the western sector of the Colorado Basin, between Bahía Blanca and Viedma.

Only three were drilled during 1994 and 1995, in the eastern sector, by the Union Texas Petroleum.

The Cruz del Sur X-1 was the first in the basin that "tested oil, although not in commercial quantities," according to a Shell report. The other two, however, were "dry wells."

The drillings were carried out, without incident, in the Las Isletas field, about 40 km from the town of Pedro Luro.

The investigation of the subsoil that led to the discovery and definition of the Colorado Basin began in 1957, through a cooperation program between the Lamont-Doherty Geological Observatory and the Argentine Naval Hydrographic Service. The first results were published in 1963.

It is considered that it could have been an arm of the main rift that led to the opening of the South Atlantic Ocean.

At that time, a contract was agreed between YPF and Shell Production, to search for oil in an area of 30 thousand square kilometers. The area covered the districts of VILLARINO and Patagones, part of Río Negro Province and the sea coast, up to the town of San Antonio Oeste.

Between 1969 and 1997, 18 wells were explored, being, as mentioned, the Cruz del Sur X-1 the only one with positive results.

All the wells were located on the flanks of the basin and on the platform.

The continental slope has not yet been drilled.

In 2015, there was an intervention by YPF.

The company launched a 3D seismic campaign in at least some blocks, with the priority being to sweep a seismic line through the Colorado Marina area, located off the coast of Bahía Blanca.

The offshore exploration program developed between 2006 and 2011 required the drilling of eight wells, of which four were in the Gulf of San Jorge between YPF and Petrobras, three in the Southern Basin between YPF and SIPETROL and one in the Malvinas Basin carried out by YPF, PAE and Petrobras.

In 2017, the Ministry of Energy granted a 24-month reconnaissance permit over a large area of the sea.

The Norwegian firm Spectrum was also contracted to carry out 2D studies over 45,000 sq. kilometers in the waters off the provinces of Buenos Aires, Chubut, Río Negro and Santa Cruz.

However, it also studied the Colorado Basin in waters bordering Uruguay.

In addition, it carried out 2D studies with the Norwegian oil company Equinor (ex Statoil).

It is known that the sedimentary basins in our country total about 2 million square kilometers, of which 1/3 corresponds to marine areas in the continental platform which, including the slope, has a surface of about 3 million square kilometers. Unlike the land basins, where tens of thousands of wells have been drilled, only a little less than 200 were drilled in the sea.

This shows that almost everything remains to be done, especially because of the higher costs of marine exploitation, although there has also been significant progress.

In his work "History of the last 50 years of drilling," Petroleum Engineer Luis Rabanaque provided a complete report on what has been done and recognized four different stages in the process.

The first stage is prior to the last five decades and is centered on the coast of the city of Comodoro Rivadavia, on land flooded by the sea.

The other three stages correspond to the end of the 1960s—the first-; to the 1970s and 1980s—the second; and to the 1990s to date—the third.

"During the second half of the 1960s, due to a different focus in oil policies, concessions were granted for offshore areas, and exploratory drilling was carried out in the Salado, Colorado and San Jorge Gulf Basins. In the latter, the presence of hydrocarbons was found, but in non-commercial quantities," said Rabanaque.

As for the third stage mentioned by the specialist, this was driven by a new change in oil policy since it included a plan generated in the state firm called "YPF in the sea."

First a small self-raising platform was bought in the USA., the Ranger II, which had been working in the Gulf of Mexico. Rebaptized "Liberación," it did not reach its destination because it sank in the Caribbean Sea while being towed to Argentina. According to the overseas captain Omar Deus, that end is as dubious as it is inexplicable.

Despite the disaster, YPF redoubled its efforts and in 1975 acquired the Semi.submersile platform, built by France and a pentagon model, which arrived in the country in 1977.

Following the guidelines of the YPF at sea plan, the company drilled in the Colorado, San Jorge Gulf and Austral Basins.

Then, it followed the ESSO using the Mosconi platform. The last well was in the Austral Basin (Cyclone x-1). Then, the platform was rented by ESSO, which completed the planned program (about a dozen wells) in the two areas called Malvinas 1 and 2. It discovered two accumulations of oil considered at that time, not economic and, like Shell in Magallanes, stopped operations because of the Malvinas War.

"The platform was taken to the city of Puerto Madryn, in Chubut. This interval of inactivity was interrupted by an intermediate period during which it was rented to the company OXY to operate in explorations of that company in the area of the

Argentine coast. Two of the drills were successful, but unfortunately there was no commercial interest in the possible production versus the cost of exploitation," said Mr. Rabanaque.

Several years later, the Mosconi oil platform was sold to Breoker and moved out of the country to be used as an early production platform.

The fourth stage reaches our days and includes exploration drilling and also development drilling in various areas under concession to private operating companies and also in charge of YPF (taken from Adrián Luciradi; "Oil in Bahía Blanca? Next year the mystery could begin to be elucidated," La Nueva Provincia de Bahía Blanca Newspaper, October 21, 2018).

Comments

It is clearly inconvenient to hire a foreign firm to analyze the report on the presentations received by the State from the invited companies. This task should be carried out by personnel from the Secretariat of Energy, which does not need to be highly specialized in the offshore hydrocarbons industry for this task. If there is no trained personnel in the Ministry to carry out this preliminary report, there will be less to grant exploration permits to the bidders.

This recalls the Houston Plan and the Argentina Plan of other times and brings to mind the annulment of the marine exploration actions that were intended to be initiated in 1973 with the state YPF in charge of them. That is to say, forty-five years have been lost and now not even the training of Argentine technicians in the knowledge of an essential activity for the country's interest is assured, and the entire Argentine continental platform is left in foreign hands, since there is not even a reserve of areas for the "state-owned" company YPF S.A. to secure a part of the tasks. This is very serious.

(vi) The fall in production

Total oil production has fallen from 87,589 m^3/day in 2015 to 76,855 as of March 2018, while "conventional" oil production has fallen considerably more. Most regrettably, the rate of decline has accelerated from 4.4% between 2012 and 2015 to almost 12% between 2015 and March 2018. Since April, a slight recovery was noted, bringing the country's total production in May to 77,084 m^3/day.

In August 2018, national crude oil production reached 79,200 m^3/day, with a cumulative growth rate of 2% for the first eight months of 2018, while non-conventional crude oil production rose by 37.9 to 14.1% of the total.

The situation would have been worse if the crude oil processed in the refineries had not fallen in turn, despite the slight increase in imported crude oil processed.

(vii) The oil trade balance

The analysis in the previous paragraphs has repercussions on the oil trade balance, where, as shown in Table 2.2, the greater volumes imported (see Table 2.3) are compensated by the fall in import prices, a fall that is being reversed in May 2018 and this will further aggravate the trade deficit.

(viii) Production in the first quarter of 2018

A deceleration in the fall of oil production is noticed as indicated in Table 2.4 and that reaches 1.24%, compared to the same quarter of 2017.

(ix) Total production in July 2018

It reached 76,746 m^3/day almost 2% more than in July 2017. In other words, it was recovering, essentially because of non-conventional oil, which represented 13% of the country's total in July 2018.

(x) Imported Gasoline

The increase in the volume of imported naphtha due to the new free market policies and lower tax costs is worrying. Thus, companies such as Trafigura are appearing that prefer to import naphtha and sell it in the domestic market rather than build a refinery, process crude oil and produce naphtha. Shell, AXION (ex Exxon) and YPF have followed the same path.

(xi) Refining

As far as refining is concerned, except for YPF, which has announced investments to modernize and expand the Luján de Cuyo and Plaza Huincul refineries, there are no signs either of modernization (although AXION has hinted at something in the Campana distillery) or of the construction of new distilleries. This is consistent with the liberalization of the oil derivatives market that will make the import of naphtha and diesel oil more attractive than their production in the country (which in the short term implies less capital investment and higher profits). It should be remembered that the integrated companies in Argentina (i.e., those that produce oil and refine it in

Table 2.4 Oil production evolution by operator—In m^3—Period January–March

Companies	2017	2018	Dif.%
YPF S.A.	3181.538	3154.837	−0.83
Pan American ENERGY	1394.165	1442.917	3.49
PLUSPETRL S.A.	423.217	402.631	−4.86
SINOPEC	377.015	331.389	−12.10
Tecpetrol S.A.	193.168	171.150	−11.39
Associated Oil Companies	170.671	168.520	−1.26
Petrochemicals Comodoro Rivadavia	103.690	133.647	28.89
Total Austral S.A.	121.168	115.939	−4.31
Petrolera Entre Lomas S.A.	135.447	107.833	−20.38
ENAP SIPETRL Argentina S.A.	111.767	101.100	−9.54
Others	792.653	787.806	0.61
Total	7004.499	6917.770	1.24

Source Secretariat of Energy of Argentina

the country) are YPF S.A. and AXION which process almost 70% of the crude, and this allows them to process local crude which is cheaper than the price of imported crude. The companies that manage the South Dock and Bahía Blanca refineries are then left out of the picture, since the San Lorenzo refinery, which belonged to OIL, has an uncertain destiny. It is not clear what Shell's strategy will be, now with the face of RAISEN as described below. It has always had access to national crude, it is the second largest seller of oil derivatives in the country and its refinery is quite complex which allows it to process heavier and cheaper crudes than light ones.

(xii) Something about the prices of derivatives

On the other hand, there are tensions over the internal price of oil. The authorities and the oil companies had agreed on "market liberalization." That is to say that the prices of derivatives would evolve according to the variations of the international price of Brent crude (before Mr. Aranguren the international reference price for the Argentine market was the WTI oil, used in almost all Latin American countries, while Brent is used in Europe. Brent crude is between 5% and 15% more expensive than WTI and this favors local oil companies) and, for conversion to pesos, with the evolution of the exchange rate, but both factors are gradually increasing and this has repercussions on inflation rates. Ironically, oil prices in Argentina, in a free market, are inelastic to the fall and elastic to the rise of the two factors that determine them, that is, the value of international reference crude and the value of the US dollar. The situation is also different for companies that are only in Upstream, such as Pluspetrol, than those that are only in Downstream, such as Shell, or those that are in both sections of the production chain, such as YPF S.A. and AXION. The former are sellers of crude oil, and they do so at between US $55 and US $62 a barrel and complain because they should sell it at almost US $77, as could be the price of Brent. The latter complain because they have to buy the crude oil they refine at a high price while the consumers are unhappy about the increase in the price of the derivatives they sell and the third ones produce the crude oil they refine and how, the former ask to sell the derivatives at the international price. All this leads the Downstream companies to claim that the prices of their products are almost 20–30% behind. The conclusion is that the prices of both crude oil and derivatives cannot be freed up because the markets are strongly olygopolized and the State does not have its state-owned company such as YPF as the market regulator and must do so in the interests of present users and future generations. Tensions have even been generated between oil derivative distribution companies and retailers, which have led to the setting of quotas for the delivery of derivatives to stationers. The distributors complain because the service stations ask for more products than they need to store them and sell the more expensive surpluses due to the gradual price increases caused by this free market policy. In the past, stationers had to make a sworn statement of the volumes they had available to avoid these "extraordinary" profits. Now, the fuel tax is a fixed sum (which is revised) and not a percentage of the price, and then, the price increases, if the absolute value of the taxes is not corrected, go to the oil companies and the one who stops collecting is the State.

(xiii) The pipelines

As for the oil pipelines, the current rules and regulations allow what are known as "producers' pipelines," in which the operators themselves allocate resources for their construction and then have volume quotas in those systems.

Under this modality, the new oil pipeline that connects Loma Campana to the Pellegrini Lake area is currently being completed, promoted jointly by Tecpetrol, the oil company of the Techint Group, and the national flag operator YPF. The authorities plan to free this segment of the activity and that any company, not only the producers, can build oil pipelines and then charge a toll for the transport.

There is a company of these characteristics "Oleoductos del Valle S.A. (OLDELVAL)" that takes the crude oil from the Neuquén Basin to Puerto Rosales in Bahía Blanca. Pampa Energía has 21% of the shares, and the concession would end in 2028, with an option for 10 more years. So this modality is nothing new. In times of high production, there may be conflicts between the amount of crude oil transported by each company and this type of pipeline, when the producing areas belong to different companies. OLDELVAL was born in 1993 as a product of the privatization of YPF S.E. Managed by the state company, this type of problem is avoided because the areas are either exploited exclusively by that company or by partners or concessionaires who are obliged to deliver everything produced to the state entity.

OLDEVAL plans to increase the capacity of its pipeline to Puerto Rosales from 25,000 m^3/day to 50,000 m^3/day in 2024.

In early November 2018, Exxon bought Pampa Energía's shares in OLDEVAL with the idea of taking a more strategic position in view of the growth of its crude oil production in Vaca Muerta oil field where it operates in the areas of Bajo del Choique-La Invernada; Loma del Molle; Los Toldos Sur I; Los Toldos II Oeste; and Pampa de Las Yeguas I and participates as a partner in the Sierra Chata and Parva Negra Este areas.

(xiv) Crude oil exports

Due to the decrease in crude oil produced in the refineries (7% from 2015 to 2017) and the slight increase in Medanito-type crude oil production from Vaca Muerta oil field, exports of this type of oil would be enabled from November 2018. In other words, once again the interests of the producers would be considered, while the import of derivatives is permitted and idle capacity is generated in the refineries.

Thus, between January and September 2018, fuel exports were US $2226 million, with a year-on-year increase of 129%, while imports of the same type totaled US $5355 million, an increase of 26.9% compared to the same period in 2017.

The relative improvement is due to the fact that during 2018 there was an increase in exports, because the rise in the price of oil on the international market made it profitable for companies to extract more crude oil for export. Thus, external sales of crude oil alone were US $1103 million with an increase of 316.2%. This crude is produced mainly in the Gulf of San Jorge Basin, which due to its heavier characteristics is used less in the local refinery park.

On the other hand, energy product imports in the first nine months of the year showed a 25.6% increase in dollars, due in the higher prices (increased in 27.8%), even though there was a 1.7% reduction in volumes.

However, it is noteworthy that, with respect to September, INDEC observes that among the largest increases in total imports (US $100 million more) are purchases mainly of gas oil, naphtha, gasoline excluding aviation fuel and natural gas, the latter in last place.

This indicates that, despite the fact that local refineries are operating with idle capacity, as the public prices of fuels are approaching international prices, the tendency to import diesel and gasoline is increasing, due to the deregulation of the market.

(xv) Gasoline and diesel oil balance

It is interesting to see how this market has evolved.

Only the case of gasoline and diesel oil is analyzed because they are the largest volume among the oil derivatives (Tables 2.5 and 2.6).

Table 2.5 Gasoline balance—(10^3 TEP)

	Production	Import	Export	Sales (**)
2014	5469	337		5793
2015	6250	11		6299
2016	6334	185		6457
2017	6675	312		7022
2018 (*)	3414	782	24	3604

(*) First six months
Sources
2014 to 2017 National Energy Balance Sheet
2018: From Dynamic Tables of the Secretariat of Energy and Mining of Argentina

Table 2.6 Diesel oil balance (10^3 TEP)

	Production	Import	Export	Sales
2014	9610	3235	261	10958
2015	10160	3337	100	9736
2016	9849	3603	75	11264
2017	9937	3071	120	13887
2018(*)	5075	886	1.3	5805

(*) First six months
(**) Does not include sales to electricity generators
Sources
2014 to 2017 National Energy Balance
2018: From Dynamic Tables of the Secretariat of Energy and Mining of Argentina

There may be some inconsistencies between the data in the balances and the data for 2018 in the dynamic tables of the Secretariat of Energy.

Thus, it is striking that, in relation to September, the INDEC notes that among the largest increases in total imports (US $100 million more) are purchases mainly of diesel, gasoline, gasoline excluding aviation and natural gas, the latter in last place.

Another observation is that due to the strong increases in the tariffs of these derivatives in the first nine months of 2018, there would have been a significant drop in consumption.

According to the Chamber that groups the retailers (CECHA), fuel consumption in September 2018 recorded the worst indicators in several years. The drop in gasoline and gas oil sales was 6.65%, when comparing 2018 values with those of the same period in 2017. And if you compare what happened month by month, from August to September sales fell 10.20%. There was a collapse in the Premium gasoline dispatches with a drop of 27.50%. Diesel oil dropped 7.84% and super, 4.95%, always compared to September 2017.

"Until now, people have been restricting the quality of the product, migrating from the Premium to the super, which is why the general demand has remained relatively stable," explained Carlos Gold, president of the Confederation of Hydrocarbon Trade Entities of Argentina (CECHA).

But in August, consumption began to fall to the rhythm of an average 40% increase. In September, there were three major increases of over 10% and the increase was close to 75% in the first nine months of 2018.

(xvi) Tax and customs benefits

They are granted by Decree 1049 of November 2018 for what it calls new ventures in the activities related to hydrocarbons. These activities are the extraction of crude oil and natural gas; services related to the extraction of oil and gas, except prospecting activities; and geological and prospecting services.

This provision does not appear to be necessary. The oil companies are obtaining good profits (it can be seen so in the balance sheets), and it can be confirmed that a good part of the activities in Vaca Muerta oil field already have profits and one of them is the subsidized price of natural gas.

(xvii) The companies

From the business point of view, in the last five years, relevant events have taken place:

The Bridas Corporation group, owned by the Bulgheroni family and the Chinese company Cnooc, through its subsidiary AXION, completed the purchase of the assets of ExxonMobil (ESSO) in Argentina, Paraguay and Uruguay.

The company completed its landing at ESSO service stations in Argentina, one year after announcing the global operation for about US $700 million.

The owners decided that the assets would henceforth replace the ESSO name with AXION.

Thus, the Campana refinery, the fuel distribution terminals, airport terminals and the fuel transportation trucks will bear this new name.

Bridas Corporation is a company that integrates in equal parts Bridas Energy Holdings Limited and the Chinese Cnooc International Limited. Cnooc International Limited is controlled by Cnooc Limited, a company listed on the Hong Kong and New York stock exchanges, and is the largest offshore oil and gas producer in China and one of the main global competitors in the oil and gas sector.

In September 2017, Bridas and British Petroleum announced the merger of PAE and AXION ENERGY to create a new integrated energy company: Pan American Energy Group (PAEG).

The company will have as shareholders Bridas and BP, each with a 50% share package.

PAEG will become the main producer, employer and private investor in the oil sector in Argentina.

Pampa Energía sold its fuel refining and distribution business to TRAFIGURA for an estimated US $90 million. The oil company will begin to distribute its Puma products—the brand under which it operates in Argentina—in the network of 250 service stations of the former Petrobras.

TRAFIGURA is a group based in Singapore, where it is a market leader in the world industry of commodities and will be in charge of retail outlets, a refinery in Bahía Blanca, a lubricant factory and a fuel storage and dispatch terminal in Caleta Paula in the province of Santa Cruz.

It plans to invest US $450 million in a storage and distribution terminal in the city of Campana, Buenos Aires Province.

The company OIL Fuels of Cristóbal Lopez declared in bankruptcy by justice for a debt with the National State for 21 billion Argentine pesos between capital and interest for fuel tax, which the company charged to customers, but did not turn to the Treasury, disappears. At the end of October 2018, the situation was defined and the partnership between YPF S.A. and DAP S.A. kept their assets.

The offer of Destilería Argentina de Petróleo (DAP S.A.) and YPF consists of taking over the fuel supply of the entire network of OIL's flag service stations at market prices. In addition, the offer includes the exploitation of all the logistic assets in OIL's river terminal. The San Lorenzo refinery and direct access to the navigable Paraná River is one of YPF's main interests. On the other hand, Comercial del Plata, owner of 30% of Compañía General de Combustibles, has purchased 50% of DAP S.A. so that it also appears involved in the OIL Fuels business. Of the facilities, the fate of the San Lorenzo distillery, which seems doomed to be dismantled, is uncertain.

In August 2016, Shell Argentina sold the Dock Sur distillery and its service stations, that is to say its activities in Downstream, to RAISEN, a company owned by Shell and the Brazilian company Cosan. This company is a giant in Brazil and is 50% owned by Shell and 50% by the Cosan group.

In other words, the RAISEN group is Shell with another face.

In short, ESSO disappears and the face of Shell changes as more significant facts, because of the history of both companies in Argentina.

The presence of TRAFIGURA will imply that the business of Downstream will now have four big ones: YPF S.A., AXION, RAISEN and TRAFIGURA, which will oligopolize the market.

It has also reappeared the Pérez COMPANC oil business, which in the last three years has begun a process of restructuring that includes a change of image and brand.

Now, the group is preparing for the next step in this commercial strategy that is oriented toward diversifying its service offer in the three business segments in which PECOM operates.

In that objective is the public interest to keep the actions that the National State controls in TRANSENER, in addition to renewable energies.

The bet is based on the potential that PECOMP executives see in the country's energy business based on the government's plans to promote alternative energies with the aim of diversifying the country's energy matrix.

They are also preparing to invest in interconnection and transmission of electrical energy by participating in future tenders for the construction of high and medium voltage networks.

Always focusing on services, these plans are complemented by the consolidation of agreements with the country's major hydrocarbon producers, especially in the Vaca Muerta area. In the area, PECOM already provides services to YPF in the pipeline between Loma Campana and Lago Pellegrini, which is 80 km long and connects the production of Vaca Muerta oil field.

Although it has no direct relation with the oil activity, in March 2018 the merger of ENARSA with the binational company EBISA took place, giving rise to IEASA (Argentine Energy Integration), a decision accompanied by the sale of diverse assets of Enarsa for 1000 million dollars. A few years ago, Enarsa had delegated the purchase of imported liquid natural gas to YPF S.A.

The issue of pricing policy and taxes on oil derivatives will be dealt with in a specific section together with the natural gas and electricity tariffs.

2.7.2 Natural Gas

2.7.2.1 Some Data

Proven reserves do not include non-conventional natural gas reserves. According to the International Energy Agency's report for 2013, the resources, not the reserves, of this type of gas would be equivalent to 67 times the proven reserves of Argentina in 2012, that is, 21,654 billion m^3 (approximately 6.7%); it would give recoverable reserves of 5694 billion m^3, that is, 18 times the country's proven reserves, the sixth in the world (see Nicolás Di Sbroiavacca, 2013, "Shale oil y shale Gas en Argentina," Working Document of Fundación Bariloche, August 2013). In Spanish. In English in this volume.

It should also be borne in mind that the production mentioned in Table 2.7 must be reduced by the consumption of natural gas at wellheads, losses and vented gas (Tables 2.8, 2.9 and 2.10).

2.7.2.2 Some Thoughts

(i) Argentina is a "gas-bearing" country

Natural gas represents almost 54% of the gross energy supply and this share has been growing over time. In 2007, it absorbed 51% of the aforementioned supply.

Table 2.7 Natural gas reserves and production

Year	RC	PBP	PMP	PAP	PT	PNOC	PCON
Unit	10^6 m^3						
2015	350,484	27,236	11,636	4022	42,894	6684	36,210
2016	336,669	26,784	12,606	5600	44,986	9381	35,605
2017	325,000(+)	26,046	12,841	5707	44,594	11,461	33,133
2018(*)					11,252		

Source Ministry of Energy and Mines of Argentina; IAPG
Notes
CR: Proven Reserves
PBP: Low Pressure Production
PMP: Medium Pressure Production
PAP: High Pressure Production
PT: Total Production
PNOC: Non-Conventional Production
PCON: Conventional Production
(*) January to March

Table 2.8 Sales—(10^6 Mm^3)

Year	VR	VC P	VI	VEE	VT
2015	10,229	1764	12,632	14,916	43,570
2016	10,835	1847	12,084	16,002	44,625
2017	9637	1705	12,488	17,257	44,695

Source Ministry of Energy and Mining of Argentina
Notes
VR: Residential Sector Sales
VCP: Commercial and Public Sector Sales
VI: Sales Industrial Sector
VEE: Electric Power Sector Sales
VT: Total Sales (includes Other Sectors, essentially CNG)

Table 2.9 Natural gas imports and exports—(10^6 m^3)

Year	MGNG	MGNL	MGNT	XGNG
2015	5956	5314	11,270	1.4
2016	6221	4650	10,871	28
2017	6870	4799	11,669	
2018 (*)	1715	2619 (*)(*)	1715	41

Source Ministry of Energy and Mines of Argentina; IAPG
Note
CNG: Imports of Natural Gas from Bolivia
MGNL: Liquefied Natural Gas Imports
MGNT: Total Natural Gas Imports
XGNG: Natural Gas Exports
(*)(*) Between April-July 2018. There will be 50 ships, brought mostly by TRAFIGURA at prices higher than in 2017, due to the increase in Brent oil and the devaluation of the peso. The prices will be 7.56 US $/MMBTU for those that re-gasify in Escobar and 7.12 for those that do it in Bahía Blanca
(*) January to March

Table 2.10 Trade balance natural gas—(10^6 USD)

YEAR	MGNG	MGNL	MGNT	XGNG	Scale commercial
2015	1366.7	2141	3507.7	1.01	−3506.69
2016	863	936	1799	17.4	−1781.6
2017	1258	1013	2271		−2271
2018 (*)	346		346	5.2	−340.8

Source Ministry of Energy and Mines of Argentina; IAPG
Note
CNG: Imports of Natural Gas from Bolivia
MGNL: Liquefied Natural Gas Imports
MGNT: Total Natural Gas Imports
XGNG: Natural Gas Exports
(*) January to March

(ii) The fall in production

But, just as this participation has increased, natural gas production has fallen and Argentina, from being a net exporter until 2006, has become a net importer and in increasing quantities, especially of liquid natural gas.

In the last two years, the price effect has decreased consumption in all sectors except electricity generation, which is responsible for the slight increase in natural gas sales.

These facts, together with the lack of investment in exploration in conventional fields, have led to a fall in production between 2006 and 2014.

The need not to neglect the explorations in non-conventional areas would be confirmed by the announcement, in May 2018, of the English company Echo Energy PLC of the discovery of gas in the beginning of the drilling of the first of the four consecutive exploration wells in Fraction C of its concession in the maritime coast of the province of Santa Cruz.

(iii) The implementation of the "Gas PLUS Plan"

The PLAN GAS PLUS is a continuation of the "Plan Gas" of the Kirchnerist era. The production of deposits of the Total Austral Company in the Southern Marine Basin, the contributions of the non-conventional gas of Vaca Muerta oil field (essentially in Fortín de Piedra of Tecpetrol) and essentially almost US $7.5 per million BTU with which the Vaca Muerta natural gas subsidized explain the increases in the first quarter of the year 2018 compared to the same period of the year 2017 (see Table 2.11). The production values for May 2018, with 127.2 million m^3/day, confirm the growing trend.

In 2017, the total production of natural gas would have been 122 million m^3/day of which 25% would have been non-conventional natural gas essentially from tight formations.

On typical days in winter 2018, 158.7 million m^3/day were injected, of which 67% was domestic and 33% imported.

The composition of imported gas was: 57% liquid natural gas; 34% from Bolivia and 9% from Chile.

(iv) **Recent production**

Total production in July 2018 reached 132.1 million m^3/day, almost 5% more than in July 2017. In other words, it had been recovering, essentially due to shale and tight gas, which in July 2018 represented 35.5% of the country's total. Two fields explain almost 92% of the increase: Fortín de Piedra of Tecpetrol in Vaca Muerta

Table 2.11 Evolution of gas production by operator in millions of m^3—January–March period

Empresas	2017	2018	Dif.%
YPF S.A.	3804.8	3733.2	−1.87
Total Austral S.A.	2777.9	2891.0	4007
Pan American ENERGY	1368.8	1282.3	−6.32
Tecpetrol S.A.	329.7	549.7	66.74
Petrobras Argentina S.A. (Pampa Energía)	393.8	441.7	12.15
YSUR Energía Argentina YPF)	322.1	385.4	19.63
General Fuel Company	234.6	333.9	42.33
PLUSPETROL S.A.	272.9	269.1	−1.36
ENAP SIPETROL Argentina S.A.	205.4	260.3	26.71
SINOPEC	188.7	168.8	10.53
Others	979.5	936.1	−4.43
Total	10,875.5	11,252.0	3.43

Source Ministry of Energy of Argentina

oil field and East Indian Field of Compañía General de Combustibles in Santa Cruz, essentially due to the subsidies to the natural gas production. It is that in front of a Henry HUB (almost international marker that is applied in the US market) of US $2.97/MMBTU, the price that the subsidized ones receive during 2018 is US $7.5. The Argentine internal reference value that affects the residential tariffs (entrance price to the transport system PIST) is of 4.68. The high value of the tariffs that the families pay would have caused a fall of the consumption of this type of users in the period January–July 2018 of 7.2%, compared with equal period of year 2017.

From 2019 onward, producers will sell natural gas to households at US $3.72 per million BTUs and to electricity companies at US $3.40.

(v) **Looking ahead**

It is very likely that self-sufficiency in natural gas will be achieved.

The Vaca Muerta oil field would be the main supplier, without neglecting the Southern Basin.

With the mentioned subsidy of US $7.5 per million BTUs that is being granted, especially to the so-called Gas Nuevo, essentially from Vaca Muerta oil field, the incentives to invest and extract are the highest in the world, but the Argentine domestic market is very small.

It should be mentioned that the US $7.5/MMBTU (value that is not known where it came from) was higher in 2017 than the marginal cost (which would be the highest price of imported natural gas). Thus, in 2017, the gas imported from Bolivia came out about 5.5725 (composed of 5.4614 for the gas, 0.0502 for fuel and 0.061 for transportation to the border) and imported liquid natural gas at 5.72. It is even estimated that both values will increase after 2018. The Bolivian value is US $6.20; LNG, including regasification, 7.4 and Chile's, in the gaseous state, 10.50 (since it is re-gasified in Chile). As the subsidy is paid in US $, it would then be "convenient" (following the neoliberal positions) to import rather than produce, but this would affect the enormous profits of the companies that operate with gas in Vaca Muerta.

After 2022, "new" natural gas would be paid for, the price that "sets" the market. In June 2018, the price received by the producers was 4.68. So the subsidy they received at the wellhead was about US $2.82/million BTU.

The idea of the government, or a part of it, is that in the future the distributors and CAMMESA will bid for the natural gas prices they acquire with the idea of lowering prices. This is because the values are dollarized and are affected by devaluation processes, in a market with a floating US $ and this affects the rates paid by users, who are condemned to permanently increasing rates, because even if the price in dollars drops, the peso will continue to devaluate.

The budget adjustment proposals in September 2018 cast doubt on whether the production achieved by the "Gas Plan" would continue to be subsidized.

The fiscal impact, foreseen by the previous Energy Minister, was to fall from US $600 million in 2018 to US $400 million in 2019. But the increase in production to be subsidized would leave that figure far behind, bringing it up to almost US $1.1 billion in 2019.

One thing was a subsidy of US \$3 per million BTUs over 13 million m^3/day and another over 21.

Then, it would be thoughtful to limit the volume of subsidized natural gas, and this would affect the considerable profits of Tecpetrol (Fortín de Piedra) and CGC (Santa Cruz).

This situation shows the lack of coordination between the actions of the different ministries and secretariats and the lack of planning. Although the previous government had chosen to establish the subsidy, up to paying US \$7.5/million BTU, to encourage the production of natural gas essentially in Vaca Muerta oil field, the current government did not put a limit to the volumes reached by the measure and consequently favored extraordinary profits of the companies.

At the end of September 2018, the government was thinking about not extending the benefits of the "Gas Plan" to new exploitations.

How would the situation be different if the state-owned YPF had been in charge of developing the new gas?

(vi) **Exports**

The Argentine market being "small," it has been decided to export, first to Chile through the existing and idle gas pipelines since the 1990s. But the Chilean market is also quite small.

A report by HUB Energía Consultores points out that production in the Neuquén Basin could grow, according to an optimistic scenario, from 76 to 130 MM m^3/d between 2018 and 2022 and then stabilize until the end of the series in 2027. For the Southern Basin, on the other hand, the expectation in the best scenario is to go from 29 to 44 MM m^3/d. In the first case, by 2027, 90% of production would be of non-conventional natural gas. In the Southern Basin, almost all the gas would be from offshore formations.

This assumption is based on the fulfillment of an intense investment plan, essentially in Vaca Muerta oil field.

There is also the idea of transporting it in a gaseous state to the sea (Bahía Blanca), where it would be liquefied into liquid natural gas and the destination would be the "world."

According to ECOJURNAL of October 10, 2018, YPF S.A. would contract a barge from the Belgian firm Exmar equipped with a floating gas liquefaction unit. This would be a project that would allow small volumes of gas to be exported by ship from the dock of the petrochemical company MEGA in Bahía Blanca. The plant will be operational in 2019.

The vessel called Caribbean Flag has the capacity to process some 2.2 million cubic meters per day (MM^3/day) of gas, of which around 10% is used for own consumption.

YPF will now sign a lease agreement that would cost up to US \$30 million per year. The project also foresees the rental of a "storage ship" to collect the liquid natural gas.

It is estimated that the cost of liquefaction on the Exmar barge would be around US \$2.50 per million BTU of gas processed. With a wellhead cost of US \$3.50, it could be in a position to export liquid natural gas at US $6 per MMBTU plus freight and transportation costs.

In turn, to facilitate the economic viability of the project, the Ministry of Finance would agree to reduce the withholding taxes on gas exports. It would be a reduction of up to 50% of the export duties in force today (instead of charging 4 pesos on each dollar exported, the Treasury is willing to charge 2 or 3 pesos).

In October 2018, a company from Qatar (a leader in liquid natural gas) appeared interested in this project and would be willing to invest US $1.5 billion to put the plant into production in 2024. But this is nothing tangible yet.

These alternatives have their problems because there is a world over-supply of liquid natural gas, especially due to the appearance of the USA, which will become the world's leading producer of liquid natural gas.

This situation makes some potential investors doubt whether or not to turn to Vaca Muerta oil field and similar formations in the country.

The relative advantage that Argentina has is its location in the Southern Hemisphere, so that when in the Northern Hemisphere it is winter and there is a shortage of gas, in the South is summer and "spare," relatively. Its main competitors would then be: Australia, Nigeria, and Indonesia and the Asia-Pacific market. To this should be added Argentina's relative distance from that market with the consequent higher freight rates.

There are currently seven areas with "Gas PLUS" subsidies that ensure producers a sustainable price at the wellhead in non-conventional fields. There are five projects already approved that are charging this price, another six that were approved between May and June 2018, and eight that will receive the subsidy before June 2019. Of the total, 16 areas are located in Neuquén, two in Río Negro and one in Santa Cruz. During 2018, the subsidy will amount to more than 1 billion dollars. The government assures that the fiscal adjustment will not affect the mass of public resources.

Five projects have already been approved for the Gas Plan: Fortín de Piedra (Tecpetrol), Aguada Pichana Este/Rincón La Ceniza (Total), Campo Indio Este El Cerrito (CGC) and Estación Fernández Oro y La Rivera I/II (YPF). Four of these projects are in Neuquén and one in Río Negro (Estación Fernández Oro).

Among the areas pending approval are Loma Negra (CAPEX), Las Tacanas (YPF), Cerro Las Minas (YPF) and Rincón del Mangrullo (YPF).

Two other projects would soon be approved: La Calera (Pluspetrol) and Punta SEMILLOSA (Tecpetrol).

Another set of areas will be incorporated into "Gas Plus" before June 2019. Four of these projects are in charge of YPF; these are Plaza Huincul, El Orejano, Aguada de la Arena and Río Neuquén.

The remaining projects that will enter the plan are Aguada Pichana Oeste/Aguada de Castro (PAE), Agua de Cajón (CAPEX) and Centenario Centro (Pluspetrol).

In summary, by October 2018, in addition to the contracts already subsidized, there will be another 8 or 9 to reach a total of 14 beneficiaries under the provisions of Resolution No. 46 of the then Ministry of Energy and Mines. The budget for 2019 includes a limit of US $700 million for these purposes.

By mid-November 2018, Wintershall was authorizing exports from Vaca Muerta oil field to Brazil.

The volumes would be 1.2 million m^3/day at a price of US $5.65 per million BTUs (of which US $1.5 would be the cost of transport along the old Mercosur pipeline, as it would run for more than 2300 km).

The average price agreed for shipments to the southern zone, the Bio-Bío region in Chile and the Santiago de Chile metropolitan area, is US $4.20 per million BTUs.

Until November 2018, there were 18 export contracts, 17 to Chile and 1 to Brazil for a total of 21.2 million m^3/day, to which it would be necessary to add, in 2019, 2.5 million to YPF S.A. exportable as liquid natural gas by the port of Bahía Blanca thanks to an agreement with the Belgian company Exmar Energy, by ten years, to which a ship would be rented that would liquefy the gaseous fuel. In total, there would be then about 23.6 million m^3/day that still would be inferior to those imported from Bolivia to the gaseous state and like liquid natural gas by Bahía Blanca and several ports on the Paraná River.

(vii) **The Argentine domestic market**

Sales reached 122.45 million m^3/day and imports 31.96. In other words, almost 26% of what is sold is imported.

Then, it is recapped, the size of the Argentine market is very small.

The Chilean market is even smaller (20 million m^3/day).

Thus, as it has already been mentioned, the attractiveness of the wellhead price of US $7.5/million BTUs for large companies would only be justified by a massive export of liquid natural gas through the port of, for example, Bahía Blanca. But there is an oversupply of liquid natural gas in the world caused essentially by the appearance of the USA as a producer of Shale Gas and an exporter. In addition, CIF prices for liquid natural gas from Japan are at US $8.10/million BTUs.

Consequently, we must be very careful not to think that Vaca Muerta oil field is the unique solution to Argentina's problems.

If we look at the Argentine domestic market, we see that 40% of the population does not consume natural gas from the grid. It is consumed by 8.2 million homes and not 5.4.

In other words, there would be an additional potential market, assuming equal average consumption of 26.4 million m^3/day.

The Northeastern Argentina (NEA) pipeline, the construction of which has been halted, could have absorbed, at best, some 4 million m^3/day.

But not all homes, currently not supplied, have homes suitable for installing network gas.

YPF S.A. has expressed interest in building a large petrochemical project in Bahía Blanca from the gas coming from Vaca Muerta and is looking for a partner.

The Down with which YPF S.A. is developing the El Orejano area in Vaca Muerta oil field has been mentioned as a possible partner. Down has a very bad record as a real petrochemical producer in Argentina, and it seems that its projects were intended to block the Argentine petrochemical development, which led to the installation of the state-owned polo in Bahía Blanca.

Also, as said, YPF S.A. is thinking of making seasonal exports of liquid natural gas from Bahía Blanca for which it should build a liquefaction plant.

In addition, other trunk pipelines would have to be built and distribution networks expanded in many urban areas of the country.

To this end, the gas collection pipelines in the Neuquén Basin must be expanded, and the NEUBA II pipeline that reaches Bahía Blanca and currently has idle capacity must be extended.

For the moment, it would seem that the most immediate action would be to return exports to Chile.

Spain's Gas Natural Fenosa (GNF) is investing some US $50 million to increase the capacity of the Gasoducto del Pacífico, the strategic gas interconnection between Chile and Argentina that was built twenty years ago and runs through Vaca Muerta oil field.

The Gasoducto del Pacífico was built after an investment of US $320 million. It is 543 km long and connects the Argentinean region of Neuquén with the Chilean region of Bio-Bío. The infrastructure was designed to export gas from Argentina to Chile, and so it operated until 2005 when the country used it only to cover its internal needs and stopped selling to its trans-Andean neighbor. Now, thirteen years later, GNF has decided to reactivate the infrastructure and is building a new compressor on the Argentine side, just where the branch to Chilean territory begins, which will increase its capacity from the current 7.5 million cubic meters per day to 12.5 million.

These additional five million cubic meters are already contracted in equal parts by YPF and ExxonMobil.

There are also two other gas pipelines planned in the region, for example, the Gas del Sur Transport Company, which would build a gas pipeline with a treatment plant with an investment that could reach US $800 million.

But for now the real thing is that during the winter of 2018 some 3 million m^3/day would have been imported from Chile injected through the Pacific and Gas Andes pipelines, for which the Chilean ENAP signed a contract with the Argentine IEASA.

(viii) **Imports from Bolivia**

According to the contract, during the winter season, Bolivia had to deliver a minimum of 20.9 106 M^3/day and a maximum of 24 to Argentina but an average of 18 were sent.

There would have been several factors that impacted the reduction to ten that was taking place in late October 2018: increased production of Vaca Muerta oil field and other natural gas fields, a greater provision of energy from hydroelectric sources and nuclear power plants, and it would seem that this would continue during the summer months and Bolivians expect that in the winter of 2020 shipments would grow again, since for now the consumption of Northern Argentina would not be supplied with the gas of Vaca Muerta. In any case, future contracts for the export of Bolivian gas would be under discussion.

(ix) **The debt and commitments with the companies for the natural gas**

There are two types of debt.

One for the so-called Accumulated Daily Difference (ADD) which allowed extra charges at the end of each semester to be made to users in the event of significant differences in the exchange rate agreed for the acquisition contracts and the actual exchange rate. This debt would be paid in installments during the years 2019 to 2021 and would amount to approximately US $350 million. There would be negotiations so that 50% would be absorbed by the distributors and the other 50% by the State. The main beneficiary companies would be Metrogas, Natury (ex Gas BAN), Camuzzi and Ecogas l.

This nonsense of the DDA comes from the laws and decrees of deregulation of the business of the gas of 1992 and so that the claims of the distributors are not repeated; it would be necessary to repeal that legislation.

The other debt is for the "Gas Plan" that ended on December 31, 2017, and the creditors are gas-producing companies and the State would pay it from the beginning of 2019 until June 2021. From the beginning of next year and until the end of June 2021, the State would settle US $1583.2 million with YPF, Pan American ENERGY and Total Austral, among others.

As for the "Gas Nuevo" in 2019, subsidies would be given to companies in Vaca Muerta and the Southern Basin for US $700 million.

(x) **Something about the tariff increases**

The increases in "normal" rates, which were expected to start in October 2018, will be greater for users with lower consumption. At the national level, households in category R1 (up to 500 cubic meters of gas per year) increases range from 60% to 63% and for households with higher consumption, R3 4 (from 1801 cubic meters of gas per year) increases range from 19 to 21%. For example, in the Buenos Aires metropolitan area (the most densely populated area of the country), where 70% of households belong to the lowest consumption categories, the monthly increase would be around $400. Since the beginning of the current government's term, the increase in natural gas rates in the AMBA area (city of Buenos Aires and 40 counties in Greater Buenos Aires) would have been an average of 1.848%.

This structure makes no logical sense and shows the lack of social sensitivity from the government.

To a great extent, the dollarization of the tariffs has considerably aggravated the situation.

(xi) **The Problem of the Gas Plan**

During the administration of the former Minister of Energy and Mining, the first one of Mr. M. Macri administration, the Gas Plus Plan originated in the government of Cristina Kirchner was continued in order to increase gas production, which, as mentioned, was in sharp decline.

To this end, Mr, Aranguren established a subsidy essentially for the production of non-conventional gas, based on recognizing prices of US $7.5/MMBTU in 2018;

7.0 in 2019; 6.5 in 2020; and 6.0 in 2021. Then, the subsidy ranging from US $4.0 to US $3.0 would disappear and producers would receive the price set by the "market," which would be between US $3.5 and 4.0/MMBTU.

These measures proved once again the lack of coordination between the policies of the different ministries.

With the mentioned incentives, essentially Tecpetrol in the Fortín de Piedra Field strongly increased the production of natural gas, taking it to 11 MMm^3/day in the middle of 2018 and hoping to reach 19 by the end of that year or in the first semester of 2019.

These values implied a high disbursement of the government in subsidies.

But the economic and social crisis in Argentina is "declared," it is decided to go to the International Monetary Fund as a lender and this requires an adjustment plan.

Consequently, the government decided to reduce the subsidy to the production of non-conventional gas.

The budget proposal for 2019 limits the amount of this subsidy to US $700 million, which would be equivalent to a production of about 17 million m^3/day, much less than what Vaca Muerta would deliver (only Tecpetrol would keep all the subsidy). But it seems that a Trust fund would be created with the idea of contributing resources above the mentioned US $700 million if the production to be subsidized was increased.

Currently, there are six concessions recognized as deserving of the subsidy and eleven with the specifications approved by the province of Neuquén awaiting endorsement by the National State (Orejano; Río Neuquén; Aguada La Arena; Octagon Fiscal: North Dorsal: Dadin and Rincón Mangrullo, all from YPF; Centenario Centro and La Calera from Pluspetrol; Punta SENILLOSA from Tecpetrol; Los Todos I Sur from Exxon; el Mangrullo and Sierra Chica from Pampa Energía).

In other words, a real problem. With what criteria do we distribute the now limited and scarce amounts of subsidies? How do we avoid those considered "acquired rights" by the companies? How do we "attract new investors in the face of permanent changes in the rules of the game"? How do we make this compatible with the government's hope that Vaca Muerta's exports and harvests will bring the country out of the crisis?

It has been estimated that the total cost (development cost plus operating costs) at the wellhead of non-conventional gas is no more than US $1.7/MMBTU (according to the Ministry of Energy, this corresponds to the production of Orejano and Fortín de Piedra), but it is linked to very high investments that some estimate at no less than US $10–15 billion per year. So, was it necessary to have a large subsidy? It is estimated that US $3.5/MMBTU would be sufficient.

What would have happened if the exploitation of Vaca Muerta had remained in the hands of the totally state-owned YPF?

In any case, it would seem wise, so as not to repeat the history of the 1990s, to reserve the existing natural gas, even that from non-conventional fields, for future generations and to industrialize it in the country by adding value to it, for example, through the aforementioned petrochemical plants and the production of fertilizers.

The situation can even cause the reduction of natural gas production from Vaca Muerta, and YPF S.A. has made it known (they say that with US $4/MMBTU is not profitable to extract the Tight Gas and would allocate more resources to oil production) and the Total Austral would be willing to curb its investment in Vaca Muerta. But this of YPF S.A. seems contradictory with the announcements of contracting a Belgian company to install in Bahía Blanca a ship that liquefies the natural gas of its property to export it later.

(xii) **Gas pipelines and treatment plants**

In the Neuquén Basin, especially in the Vaca Muerta area, the proliferation of producers and the relatively rapid increase in production has generated a bottleneck due to the lack of sufficient capacity to evacuate production. This has motivated some concessionaires like Tecpetrol from Fortín de Piedra to project the construction of a gas pipeline to the Gran Rosario area.

TGN and TGS would join to use a new gas pipeline that would connect Vaca Muerta with the town of San Nicolás.

The project initiated by the Techint Group, TGN's controller, would constitute a joint venture with the oil company Pampa Energía, majority partner of TGS, together with Compañía General de Combustibles (CGC).

The 1040-kilometer gas pipeline would depart from Vaca Muerta and would first reach TGS's compression plant in Saturn. This is a strategic point for distribution to both the greater Buenos Aires and liquid natural gas projects.

From there, the pipeline would continue to San Nicolás, very close to Rosario, in the heart of the agro-industrial developments.

The system would have a transport capacity of no less than 15 million cubic meters of natural gas per day, but could be extended to reach 39 million. The work would cost between US $1.2 and 1.4 billion and would be completed in 30 months (Río Negro Newspaper, October 8, 2018).

That is to say, what has already been mentioned is repeated for the subject of oil pipelines and the proposal of the authorities is the same, that is to say that any company, not only the transporters and distributors of natural gas, can construct gas pipelines (also treatment plants) and rent the service to the producers. This situation is also a legacy of the privatization of Gas del Estado in the 1990s since this company operated the pipelines very efficiently and coordinated with YPF S.E. the construction of the pipelines from the production fields. The fact is that this type of policy underlies the modality of the business units and the "atomization" of the activities linked to the hydrocarbon industry.

At the Mega Company's Treatment Plant in Loma de la Lata, ethane natural gas, propane, butane and gasoline are extracted, but the increase in the production of Vaca Muerta filled the plant's capacity and the companies that own it (YPF, Down and Petrobras are studying its expansion).

Gas from dozens of producing blocks enters the plant to the point that it represents 50 to 55% of the production of the Neuquén Basin. The plant separates the rich gases that are liquefied and sent through a polyduct to Bahía Blanca, and the methane is injected into the country's main gas pipelines such as NEUBA II, Centro Oeste 1 and Centro Oeste 2, thus reaching a large number of homes throughout the country.

One-third of the gas consumed in the entire country passes through the Loma La Lata plant, but with a design with a processing capacity of 36 million cubic meters per day, the plant is currently operating with 39 million cubic meters through a careful control of pressures.

The impossibility of processing a greater amount of gas means that the networks that go to homes, businesses and industry are including without separating the rich gases, losing opportunities for export such as liquid natural gas or processing ethanol to produce plastics (opinion of the Manager of Mega, Mr Alejandro Fernández appeared in the article "Por Vaca Muerta queda chica Mega," Río Negro Newspaper, 26 October 2018).

(xiii) **The Institutional**

At the institutional level, the majority shares (70%) of the gas distributor Metrogas, the largest distributor in the market, are currently for sale to YPF S.A. The process is being performed by the Citibank. It is argued that since YPF S.A. is a producer of natural gas, according to National Law 24,076 it could not also be a controller of a distributor. It would be expected that the same criteria would be applied to other companies that act in various links of the natural gas chain (e.g., Transportadora de Gas del Sur, which participates in trunk gas pipelines and is building a gas pipeline in Vaca Muerta, of 92 km, which will cross the areas of Bajada de Añelo, Bajo del Choique, La Invernada, Pampa de las Yeguas I and II, Parva Negra Este y Oeste, La Escalonada, Rincón La Ceniza, Los Toldos Norte, Sur, Este and Oeste, La Calera, El Orejano and Sierra Chata).

Four companies are interested: CGC, the oil company of Corporación América, the holding company headed by Mr. Eduardo Eurnekian; Integra, a firm headed by Mr. José Luis Manzano; Camuzzi, controlled by DISVOL of Alejandro MacFarlane (Camuzzi is also a distributor); and Gas BAN, the distributor of the Spanish group Gas Natural Fenosa.

It is interesting to mention that Metrogas notably improved its operational and economic performance, to the point that in 2017 it registered a net profit of 775 million pesos due, among other things, to tariff increases.

Metrogas is a company that emerged in 1992 within the framework of the privatization of Gas del Estado. It provides distribution services to 2.4 million customers within its service area, 60% of which are located in the autonomous city of Buenos Aires.

It is announced that Distribuidora Gas del Centro (Córdoba, Catamarca, La Rioja) will absorb Gas de Cuyo (Mendoza, San juan, San Luis), both constitute Ecogas in the hands of Techint and attend almost 18% of the natural gas users in the country.

(xiv) **Natural gas exports to Chile**

The objective was not to harm the companies producing Vaca Muerta and to generate an immediate Chilean market for them.

In mid-August 2018, the new Secretary of Energy and Mining stated that Vaca Muerta Gas exported to Chile would not receive a subsidy.

Four types of contracts were established with the characteristics of supply interruption and non-interruption of supply.

In the winter months and if the Argentine domestic market required it, the supply would be cut off.

When the capacity of the domestic market is exhausted, the contracts would not be interrupted.

Basically, the companies that have the subsidy of the Gas Plan (which starts as already mentioned with a remuneration of US $7.5 per million BTUs) will not be able to export that gas, but they will be able to export the non-subsidized one.

For now, the amount of natural gas exported would have been limited to 6 million m^3/day.

The idea is to export up to 22 million m^3/day in the future (through the Del Pacífico 12.5 pipeline, which has been extended by Gas Fenosa, and through the Andes 9 pipeline).

At the end of August 2018, the first three contracts for exports to Chile that were not for Vaca Muerta natural gas were signed.

The gas is destined for the METHANEX plant in Chile and comes from the Santa Cruz Austral Basin (Compañía General de Combustibles) from the Marina Austral Basin in Tierra del Fuego and from the Argentine State (Pan American Energy and Total Austral).

The price for this transaction was agreed at US $4.08/MMBTU, and the successive ones will have a floor of 3.45 and a ceiling of 4.50, in all cases far from the 7.5 that the companies recently benefited from the Gas Plan, while the price of natural gas imported in winter from Chile would be around US $10.

The guidelines set a maximum shipment of 750,000 cubic meters each, and the three shipments can reach a maximum export of 2,250,000 cubic meters per day and extend to 2020.

In October 2018, approval of new orders to export from the Neuquén Basin to Chile with YPF up to a maximum of 1.5 million m^3/day and with Exxon up to 400,000 m^3/day was imminent. In both cases, the buying company would be INNERGY and the selling price of the Gas would be 4.20 USD/MMBTU.

PAE requested and was approved in October 2018, 1.5 million per day to supply thermal power plants in the Metropolitan Region and the V Region of Chile, at a price of 4.05 USD/MM BTU.

Wintershall requested and was approved in October 2018 for Nehuencó and Candela thermal power plants to export 1.5 million per day at US $4.20/MM BTU.

CGC requested and was approved in October 2018, for the same destination as Wintershall, another 1.5 million per day at 4.20 USD/MM BTU. This export was approved at the end of October 2018.

The natural gas was injected into the Gas Andes pipeline.

Other companies are waiting for the approval of their orders.

Due to the problems of lack of evacuation capacity and the fall in demand due to high tariffs, around 15 million m^3/day of natural gas would have been left unplaced.

For this reason, YPF S.A. and Tecpetrol would resort to the storage of gas in "depleted" fields such as Los Bastos in the Neuquén Basin.

They would inject 20% of the surplus production or 2.5 million barrels/day during five to eight months in summer and would take it out in four months during winter.

As of November 2018, there were 11 export contracts: seven from the Neuquén Basin to Bio-Bío and Central Chile; one from Salta to Antofagasta via the North Andean Gas pipeline; and three from Santa Cruz and Tierra del Fuego. The contracts were agreed at prices between US $4.2 and 4.3/million BTU with a retention of 4 pesos per dollar.

(xv) **The Argentine market and gas from Vaca Muerta**

The contribution to the Argentine domestic market is limited by the current idle capacity of the NEUBA I and NEUBA II pipelines, which together reach 9 million m^3/day.

Tecpetrol, as mentioned, has been the company that has put into production more quickly its Fortín de Piedra Field, in Vaca Muerta oil field.

In August 2018, it has reached a production of 11 million m^3/day, and it is in a position to increase, but it is limited by the evacuation capacity of the existing pipelines in the area and is not authorized to export to Chile.

Consequently, it has decided to build a new gas pipeline from Allen (Río Negro Province) to San Nicolás (Buenos Aires Province) to supply the high industrial consumption in the area.

This pipeline would have a capacity of 22 million m^3/day and would cost US $1.3 billion that Techint would be willing to finance. This is because with subsidized prices until 2021 and with the programed expansion of production at Fortín de Piedra (it would reach 17 million in 2019 and then 22 million from 2020 onward), its income would be, between September 2018 and December 2022, US $7159 million and investments (gas pipelines, treatment plant and wells) US $4370 million. For subsidies, in the same period, it would receive US $2019 million from the State.

The issue of subsidies under the Gas Plus Plan is complicated by the policy of lowering all types of subsidies being implemented by the government with a view to reducing the budget deficit.

As of August 2018, nineteen projects had been submitted for the benefits of the Gas Plan, seven had been approved and twelve were pending.

The complication lies in the fact that a limit would be set for the total benefited production of between 50 and 60 million m^3/day, when a single project (Fortín de Piedra of Tecpetrol) would absorb 22 million.

How would the quota to be subsidized be distributed, in order of appearance?

The other decisions linked to Vaca Muerta oil field would be:

To paralyze the works of the Northeastern Argentina (NEA) gas pipeline (GNEA), the flow of natural gas would be inverted, which would not come from Bolivia but from Vaca Muerta oil field. To do this, works must be completed south of Santo Tomé (province of Corrientes), to the Bermejo River, and only the north of Santa Fe and Chaco Provinces would be supplied, leaving Corrientes, Formosa and Misiones behind.

Renegotiate contracts for importing natural gas from Bolivia whose cost, it is argued, would be very high and lower the volumes committed, especially in the summer months. Request that is not well seen by the Bolivian Government. In this sense, there would be negotiations with the brother country in order to establish two periods: winter between May and September with volumes of 18–19 million m^3/day and summer between October and April with about 12 million. It should be remembered that the supply contract runs until 2026.

The increase in the production of natural gas from Vaca Muerta led to the termination of the contract that allowed the import of liquid natural gas through Bahía Blanca (it was gasified in the rented ship), but due to the problems of lack of infrastructure for the treatment and transport and the disappearance or reduction of the subsidies to the producers (the mentioned US $7.5/MMBTU) it would seem that it would be necessary to contract another ship to replace the shortages of liquid natural gas, in special for the power stations, in the winter months—in other words, a new sign of the lack of planning and policy coordination still within the Secretariat of Energy and Mining itself.

(xvi) **The profits of the companies in the natural gas chain**

"The gas companies" have recorded large annual profits and even some of them posted higher profits in the first half of this year than last year's total. All the companies that make up the gas complex are benefiting from the tariff shock policy applied by the current government. Not only are they charging for the public service of gas supply a disproportionate value in relation to the income of the majority of the population and the state of small and medium businesses and industries. They have also managed to socialize the negative impact of the devaluation in their results. This is done by analyzing each of the balance sheets of companies in the gas complex presented on the stock exchange. The balance of 2017 is impressive.

The analysis of the financial statements of the distributors (there are nine that make up the national network) that are listed on the exchange allows us to know the numbers of the respective financial years that recorded total profits of 4379.4 million pesos.

The two carriers (TGS and TGN) together earned 3635 million pesos.

Of the six dominant producers in the market, four (YPF, PAE, Tecpetrol and Pampa) added a net profit of 21,462 million pesos (the other two, Total Austral and Wintershall, are not listed on the local stock exchange).

In total, last year, the gas complex that can be analyzed with public access balance sheets noted an overall gain of 29,476.4 million pesos.

The impressive magnitude of these profits is exposed when compared to those of the previous year. The distributors had lost a total of 805.5 million pesos; the transporters had gained just 671 million; and these four oil companies had lost 21.558 million pesos. The 2016 balance sheets of Metrogas—controlled by YPF—Camuzzi Pampeana, Camuzzi Sur, TGN and YPF largely explain these negative balances; the rest had profits. In 2016, the tariff increases began to impact, which explains the relative improvement in most years, and then record fabulous profits in 2017.

In the first half of 2018, despite the devaluation, all companies (except Metrogas) in each of the three links of the gas complex have recorded profits and several above the previous year's total. They also demanded compensation for the devaluation of the peso, established by resolution of the former Minister of Energy of the current government that dollarized the tariffs, a request that was granted.

The impact of the tariffs is therefore very important, as a 50% increase was established as of October 2018.

With this increase already approved, the gas increase accumulates 1600% from the end of 2015.

"The cut in the social tariff was also announced and the 10% bonus for residential users who could reduce their consumption by 20% with respect to the same period last year was eliminated" (taken from the article "El Imperio del Gas" by Alfredo Zaiat published in Página 12 newspaper, Buenos Aires, October 14, 2018).

These enormous profits of the companies, product of the great tariff increases, would not be reconciled with the lack of investment in works such as Ecogas, which distributed profits in the 2017 fiscal year for a total of 617.1 million pesos and only invested 150.7 in works, when it should have contributed to the expansion of networks 278.2.

2.7.3 Vaca Muerta Oil Field

2.7.3.1 General

Something about Vaca Muerta has already been mentioned in the paragraphs on oil and natural gas.

We will refer to Vaca Muerta because it is the Argentine formation where, since 2010, activities of the so-called non-conventional hydrocarbons are being carried out, especially those located in the province of Neuquén. The resources of the Neuquén Basin also extend to the provinces of Río Negro, La Pampa and Mendoza. In this last province, there is a strong popular resistance against the possible environmental impacts of the application of the fracking techniques.

It is not intended here to describe what fracking is, nor what Vaca Muerta actually means, nor its future possibilities, nor its environmental impacts.

At least three papers can be consulted for this purpose.

Di Sbroiavacca, Nicolás, 2013. "SHALE OIL and SHALE Gas in Argentina; States of Situation and Perspectives." Working Documents of Fundación Bariloche. In Spanish, In English in this volume.

Bravo, Víctor. 2013. Working Documents of Fundación Bariloche, September 2013. Particularly analyzed in this paper are the environmental impacts of fracking.

Bertero, Raúl, 2013. "Expected Evolution of Vaca Muerta Field Production." University of Buenos Aires and National Academy of Engineering.

Here, we will refer to how the current political power is handling this formation for which it will be necessary to give some specific information.

The start of the research and production activities began during the Kirchnerist Government, especially through the controversial contract between YPF S.A. and the Chevron Company.

Those interested in knowing something about this contract can consult the work:

Bravo, Víctor. 2015. Analysis of National Law 27,007, called the Hydrocarbon Law, and the Hydrocarbon Policy for the period 2003 to 2014." Working documents of Fundación Bariloche, in Spanish. In English in this volume.

It is interesting to note that the existence in Loma de la Lata of an enormous potential for natural gas, at greater depths, was known by the geologists of the State YPF at least in the 1970s.

In the formations where petroleum energy and natural gas are hosted in rocks of very low porosity and permeability, or in very compact clays, even in the same bedrock, they are called "SHALE OIL or SHALE Gas" (oil and natural gas present in clays) or tight oil or tight gas. Oil and natural gas from compact sands).

Tables 2.12 and 2.13 show the unproven technically recoverable natural gas and non-conventional oil resources.

In other sources, higher values appear because they are estimated with recovery factors for the largest volumes in situ.

In the work of Nicolás Di Sbroiavacca, values of the recovery factors are taken based on the experiences of the exploited deposits in the USA.

The values are estimated with a recovery factor of 6.5% on the in situ resources.

Vaca Muerta would represent six times the proven natural gas reserves of Argentina by 2015 and 3% of the world's reserves.

Table 2.12 Non-conventional gas resources—(10^9 m^3)

Basin	Unsubstantiated technically recoverable resources
Neuquén	3833
VACA MUERTA	(2031)
St. George	769
Southern Magellanic	1064
Chaco Paranaense	28
Total	5694

Source "SHALE OIL and SHALE Gas in Argentina, Situation and Perspectives" Ing. Nicolás Di SBROIAVACCA. Working Documents of Fundación Bariloche August 2013

Table 2.13 Non-conventional oil resources—(10^6 m^3)

Basin	Unsubstantiated technically recoverable resources
Neuquén	3160
Vaca Muerta	(2520)
San.Jorge Gulf	79
Southern Magellanic	1043
Chaco Paranaense	1.6
Total	4285

Source "SHALE OIL and SHALE Gas in Argentina, Situation and Perspectives" Ing. Nicolás Di SBROIAVACCA. Working Documents of Fundación Bariloche August 2013

Total resources would represent eighteen times Argentina's proven natural gas reserves in 2012.

The values are estimated with a recovery factor of 5.6% over the in situ resources.

Vaca Muerta would represent six times the proven oil reserves of Argentina by 2015 and 0.3% of the world's reserves.

The total resources would represent eleven times the proven oil reserves of Argentina in the year 2012.

In this way, the exaggerated values indicated for Vaca Muerta are somewhat limited.

We will refer to this formation in this section, since it is the one that is being researched with pilot and development projects.

In any case, if these resources were to become reserves, the volumes would be really important. If only the Vaca Muerta reserves are considered, in oil, they would be greater than the proved reserves of Brazil, and in natural gas, they would be greater than those of Canada.

2.7.3.2 Vaca Muerta Oil and Natural Gas Production

Tables 2.14 and 2.15 include oil and natural gas production from non-conventional Vaca Muerta formations.

In May 2018, non-conventional gas production represented 37% of the country's total.

More than half of the oil and natural gas extracted in the province of Neuquén are non-conventional.

Table 2.14 Non-conventional oil production

Year	PT	PNC	PNC/PT
	10^3 m^3/day	10^3 m^3/day	(%)
2015	87.42	4.17	4.8
2016	84.28	5.43	6.4
2017	76.24	7.05	9.3
2018(+)	76.86	7.43(*)	9.7

Source Secretariat of Energy and Mines of Argentina
(+) First Quarter
(*) Estimated
PT: Total Country Oil Production
In May 2018, unconventional oil represented 12% of the country's total production

Table 2.15 Non-conventional natural gas production

Year	GNT	GNNC	GNNO/GNT
	10^6 m^3/day	10^6 m^3/day	(%)
2015	117.3	18.31	15.6
2016	123.25	25.7	20.9
2017	122.18	31.4	25.7
2018(+)	125.02	33.4(*)	26.7

Source Ministry of Energy and Mines of Argentina
(+) First Quarter
(*) Estimated
GNT Natural gas production Total Country
CNG Non-Conventional GN Production

2.7.3.3 Vaca Muerta Concessions

(i) Concessions

The process of granting the concessions consists of first granting the areas for five years, during which time the companies carry out what they call their "pilots," and then, concessions are granted for thirty-five years.

As of September 2018, there were thirty-two non-conventional concessions in Neuquén. These are deposits that include projects that exploit both Shale and Tight. Generally, this whole package is called Vaca Muerta oil field, although it is only one of the non-conventional geological formations in the Neuquén Basin.

In total, these 32 projects "promise" 156 billion dollars of investment over the next 30 to 35 years and the drilling of no less than 11,200 wells. By the end of September 2018, some 546 wells had been drilled and some $8.093 billion invested.

The Shale Vaca Muerta formation covers 30,000 square kilometers and is present centrally in Neuquén, but also occurs in Río Negro and Mendoza.

(ii) The companies

The main companies that were working in Vaca Muerta as of July 2018 were the following:

YPF, which has 35.3% of the area, or about 9605 km^2. It is the largest producer and is associated, among others, with Chevron (USA), Dow (USA), Statoil (Norway) and PETRONAS (Malaysia).

Gas and Oil (state company of Neuquén Province) with 5.8% of the surface (1575 km^2), associated with Exxon (USA), Shell (Holland-England) and Statoil. Its production is currently marginal.

PLUSPETROL (Argentina) with 4.3% of the area (130 km^2).

Wintershall (Germany) has blocks with a portion for 2.3% of the surface (194.5 km^2).

Total: It occupies 4.8% of the surface of Vaca Muerta (1316 Km^2). It is one of the giants of the industry and it is composed of French capitals.

Exxon MOBIL: It has participation in 3% of the surface (565.2 km^2).

Pan American Energy (PAE) (Argentina, China and Great Britain): It has 4.8% of the surface (1298 km^2).

Shell: It has a portion of 2.9% of the surface (502 km^2).

Tecpetrol (Italy Argentina): It occupies 2.9% of the surface (778 km^2) and exploits Fortín de Piedra which already produces 10% of the fluid that is extracted in the province.

Pampa ENERGY with 4.2% of the surface (1140 km^2).

CAPEX with 1.2% of the surface (350 km^2).

ENERGY View with 2.9% of the surface 778 km^2.

Chevron, with 2.1% of the surface (558 km^2).

Together these thirteen companies occupy 77% of the Vaca Muerta oil field.

(iii) Vaca Muerta and the Gas Plan

It is important to mention the Gas Plan, which aims to encourage non-conventional gas production through the granting of subsidies.

As already mentioned, the subsidy consists of the granting of a decreasing support price that starts at US $7.5 per million BTUs (3 dollars more than the average price, since US $4.68 is recognized for electricity generation) in 2018 and ends at US $6.0 in 2021, and then, it would be governed by "market prices." But what would be these prices paid for electricity generation? It is not known.

There are differences with respect to what is understood by incremental production as some in the government estimate that it would be in addition to the previous year's production and the private sector from the moment that non-conventional production began. But what about conventional incremental production such as TOTAL Austral offshore and others in the Southern Basin? Twenty oil companies are waiting for the definition of the National Energy portfolio on the future of subsidies. The entry into the Gas Plan in Vaca Muerta requires the approval of the province of Neuquén in the first instance and the final approval of the National Energy board.

So far, Neuquén has approved fourteen projects, while the nation has only authorized the payment of the subsidy to seven. The others are still pending. Among the projects approved were: La Ribera I and II (YPF), Agua del Cajón (CAPEX), Aguada Pichana Oeste and Aguada de Castro (PAE), Rincón La Ceniza (Total), Aguada Pichana Este (Total) and Fortín de Piedra (Tecpetrol). The other problem is that the subsidies are in dollars, and with the devaluation of June–July 2018, the amount in pesos has increased almost 40% from 22,000 million to 30,800.

So far, TECHINT, with Fortín de Piedra, has been the most favored, while YPF S.A., which has most of the projects stopped (e.g., El Orejano) for the granting of subsidies, is the most damaged.

(iv) Investments made in Vaca Muerta

They are substantial and will reach US $3.438 billion by 2017 (about half of the amount raised nationwide in Upstream). By 2018, they are estimated at 4434 (approximately 55% of those projected for the entire country also in Upstream).

Of the $154 billion committed by September 2018, 10% would have been invested.

The three most important areas, in the "hot zone" of Vaca Muerta, are Fortín de Piedra (Tecpetrol); Loma Campana (YPF, Chevron); and Aguada Pichana Este (Total; YPF, WINTERSHALL, PAE). These areas are in massive development and only occupy 2% of the surface of Vaca Muerta oil field.

In September 2018, YPF and PLUSPETROL will add La Calera.

Year after year, investments in non-conventional areas are increasing, and some sources indicate that the development of Vaca Muerta would require no less than US $10 billion per year. This is quite risky, so it will be pointed out later.

(v) **Something about productivity**

As for oil, the reserve per well in Vaca Muerta can vary between 87,000 and 161,000 m^3, between half and twice as much as conventional wells.

Oil costs per development well vary between US $10 and 14 million, depending, among other things, on the number of fractures. These values are almost three times higher than those of conventional wells. It is estimated that in 2018 YPF S.A. would have a development cost, at wellhead, of about 12 US $/barrel (in 2019, it would drop to US $8 to which the exploration cost and operating expenses would have to be added. The cutoff point for making exploitation viable was estimated at US $43 per barrel. The highest initial productivity per well found in Vaca Muerta would not exceed 200 m^3/day.

Regarding natural gas, there is some information about the field with the highest production as of July 2018, Fortín de Piedra. The reserve per well is estimated at no less than 210 million m^3, with an initial production of 400,000 m^3/day and an investment cost per well of almost 12 million dollars. No wellhead costs per million BTUs are available but hardly exceed US $2.5 (for Fortín de Piedra, a development cost of US $1.54/MMBTU has been estimated) and the cutoff point that would make exploitation viable at US $2.5. Then, the profit is very large if they receive in 2018, US $7.5; in 2019, 7.0; in 2020, 6.5; and from 2021 about 6.0 dollars per million BTU. Then, the prices would be "free." In 2019, the owners of Fortín de Piedra (TECHINT) would receive about US $1.4 billion in revenue, so the committed investment of US $23 million would be recovered in less than three years.

Consequently, the hydrocarbon business in Vaca Muerta would be in natural gas rather than in oil.

This explains, along with the subsidies to the production, that most of the contracts have as objective the extraction of natural gas.

In spite of which there are concessions to produce petroleum, as for example the Coiron Amargo South East area, granted to PAE in association with the Neuquén provincial company G&P, with an surface of in 228 km^2.

The problem lies in the very high investments required and that has been mentioned would not fall below 10,000 million dollars per year.

(vi) **The Vaca Muerta train**

The routes that lead to Vaca Muerta, both from other towns in Neuquén and from the port of Bahía Blanca, are congested by the intense truck traffic. These essentially transport pipes, cement and sand for fracking operations.

The train (called "Nord Patagónico") would travel about 700 km from Bahía Blanca to the town of Añelo in Neuquén, with an estimated cost of about 780 million dollars through the system of Public–Private Participation and a construction period of four years.

One of the inputs used in the fracking, the siliceous sands, represents almost 10% of the cost of the well. Between 7000 and 15,000 tons of sand are used per well (depending on the number of fractures), and this gives an idea of the magnitude of the savings.

With this, the cost of drilling a well would be reduced by almost 10%.

Each formation, consisting of 30–40 wagons, would transport about 900 tons of cargo, circulating with a frequency of 42 min.

There would be some concern among the inhabitants where the train formations pass through because of the inconvenience that would be caused by the passage of the trains.

(vii) **The water problem**

A lot of water is consumed, almost 35,000 cubic meters per well with the addition of about 400 tons of chemicals diluted in water so that the solution is 2% chemicals and 98% water.

In short, a large amount of water, polluting and toxic chemicals and the expulsion of natural gas into the atmosphere, since not everything is recovered. This expulsion of natural gas increases the greenhouse effect, since the greenhouse effect power of natural gas is 23 times that of the main agent of this type, the carbon dioxide.

A large part of the injected fluid returns to the surface (between 30 and 50%), and this fluid is highly contaminating.

The one that remains in the subsoil can migrate toward water layers and contaminate them.

In Neuquén, the first 500 m of the vertical pipe will be required to be cemented in order to avoid the contamination of the underlying water layers.

In any case, due to the cracks generated in the rocks by the injections of water, sand and chemicals, the contaminated fluid can ascend to those layers.

The transport of water and materials and inputs causes a huge traffic of trucks, including tankers, which circulate on rural roads.

In Neuquén, it is estimated that the water would come from the artificial Lake Mari Menuco or from the Colorado, Neuquén or Río Negro Rivers. The consumption of water from underground water tables would be prohibited, and only surface water could be used.

The Neuquén Government has even announced that it will build a network of aqueducts in charge of the province to supply the needs of the oil companies.

The problem would not be water consumption but water contamination and the disposal of recycled sludge from the drilled wells.

Anyway, the volumes of water used, still reduced by almost 60%, must be treated and reinjected, and this is precisely the problem, in the treatment and disposal of these waters which is called "flowback."

This means that the main problem is not the consumption of water, which if reduced will be welcome, but the final disposal of the polluted water and this is no longer a problem of quantity but a problem of quality (source: Victor Bravo, 2013, "An opinion on fracking," Working Documents of Fundación Bariloche), in Spanish. In English in this volume.

"Vaca Muerta moves millions of liters of water per day, either in trucks on the roads or in aqueducts and hoses from the Neuquén River to the Australian tanks that inject it into the oil wells."

The cost is enormous. It is estimated at US $5.7 per m^3 of water.

According to industry estimates, it takes more than 1000 water trucks, with 35,000 L each, to drill an oil well. This implies a large investment, so operators are increasingly appealing to technical strategies to lower the costs of moving water. One that is in force is the use of the "FLEXI" pipe, a kind of hose that sucks liquid from the river and carries it for several kilometers to the well cisterns. Sources close to YPF determined that the use of this method allows for a 50% reduction in the costs of transferring water to the wells in the various non-conventional basins in the area. These are hoses that are pressurized and carry the water to begin the first stages of fractures in the locations.

For now, this system does not completely replace the one needed by Vaca Muerta for the fracking water, but little by little it intends to move fewer trucks on the routes, as it happens with the comparison of what will be the irruption of the NORPATAGONIC TRAIN, which will move sand from Bahía Blanca to Añelo to use it in the hydraulic stimulation. In the eagerness to lower costs, there are service companies that have already started to draw up investment projects, so that operators can take advantage of them. This is the case of the "Agua Distribuida" project with an aqueduct.

The projected aqueduct is 68 km long, from the La Calera deposit to Parva Negra, with a diameter of 20 in. (500 mm). The idea is to transport about 30,000 cubic meters of water per day from the Neuquén River to supply some of the Vaca Muerta operators.

According to the project, which is still being evaluated, a water intake will be installed from the Neuquén River at an altitude of 430 m above sea level. Then, a pumping station will be built for 8 km long, up to 554 m above sea level. That is, the water will rise to a height of 124 m, where it will be stored in a cistern. From there, the water will go down by unevenness for about 60 km, to have connections to about 12 areas, such as La Calera, Bajada del Añelo, Aguada de la Arena, Pampa las Yeguas Block II, Sierra Chata, Parva Negra, El Orejano, Pampa Las Yeguas Block II Northwest, Bandurria Norte, Bandurria Centro and Aguada Pichana Este. (source: La Mañana de Neuquén, newspaper, October 14, 2018).

(viii) **Other environmental impacts**

Although it was mentioned earlier that the development of this topic could be consulted in the publication "An opinion on fracking," Víctor Bravo, 2013, Working Documents of Fundación Bariloche (in Spanish; in English in this volume), we want to refer here to a report of the United Nations Committee on Economic and Social Rights that points out some concerns about the extraction of hydrocarbons in Vaca Muerta and its impact on global warming.

Among its proposals, the committee expressed "concern about the plans for large-scale exploitation of non-conventional fossil fuels" in Neuquén, considering that "the total exploitation," with the hydraulic extraction method it uses, "would consume a significant percentage of the world's carbon budget to achieve the stipulated 1.5 degree Celsius warming goal and recommends increasing the contribution of renewable energy sources."

They also refer to the role of native peoples, suggesting that the State should guarantee "systematic" consultation prior to "granting concessions to State-owned companies or third parties, for the economic exploitation of the lands and territories traditionally occupied by them." They proposed using "the protocols developed and agreed with the indigenous peoples, which take into account the specifications of each people and each case."

2.7.3.4 Questions and Concerns

These considerations raise questions and concerns about what Vaca Muerta and non-conventional hydrocarbons in general may mean for the development of Argentina.

The aim is to demonstrate that Vaca Muerta oil field will save the country by providing an annual growth rate of between 0.3 and 0.4% for Argentina as a whole and an accumulated growth rate of between 4.5% and 4.8% until 2030, according to a joint analysis by the Ministries of Energy and Mining and Production.

It has been said, and the numbers seem to confirm it, that Vaca Muerta oil field is essentially important in gas resources.

The production of both oil and natural gas, despite the recovery in the first months of 2018, especially in gas, is currently not sufficient to meet the requirements of the domestic market.

If the rhythm of investments of the companies that are working in Vaca Muerta continues and intensifies, undoubtedly not beyond the next 5 years the self-supply of gas would be reached. Oil supply can also be achieved, but not only by Vaca Muerta but also by the exploration and development of conventional fields, substitution and rational use.

Against this can play the government's idea of only subsidizing the production of gas until 2021, and then, prices would be those of the "market," i.e., lower than those in force in 2018. This could reduce the rate of investment of all or some of the companies that have not yet managed to receive the subsidies. Despite the fact that costs are much lower than the prices of 4.5 or 5 dollars per million BTUs that the market would have after 2021.

In any case, for reasons of the magnitude of resources and production subsidies, the companies, many of which are foreign, are more interested in gas than in oil. But they fear that because of the adjustment plans the government will end subsidies for companies that have not yet obtained this benefit.

As the Argentine market is small, despite the fact that almost 60% of families do not consume natural gas, and so is the Chilean market, there is still the expectation of exporting to the rest of the world. But here the disadvantage, beyond the strategy of some of the foreign companies that are in Vaca Muerta, is the current almost saturation of the world market by the emergence of US liquid natural gas, competing with that of Qatar and Australia and gas networks in Russia. Thus, the panorama of exporting LNG does not appear as the additional outlet for Vaca Muerta's surpluses, let alone consider this venture to be the country's "salvation."

The investments required for the development of these fields are enormous (the decline of the wells is very high and this requires, as the experience of the USA shows, the drilling of a large number of wells to maintain and expand production), and despite the strong presence of YPF S.A. (associated in turn with multinationals), it would seem that the amounts needed would be provided by this last type of company. Then, their strategies will prevail, which do not necessarily, but rather the opposite, coincide with those of the country.

So, beware of the "Dutch disease," which has always been suffered by two major exporters like Nigeria and Venezuela, and it seems that there is no strategy as of Norway.

The Dutch disease originated in the 1960s in Holland due to the appearance of the great gas resources in the North Sea, and the Dutch thought that this would bring the bonanza to the country. The great increase in its external income caused the appreciation of its currency, the destruction of its industry and agriculture, and the impoverishment of vast sectors of the country. Since then, the phenomenon of expansion of foreign exchange income due to booms in the export of natural resources (oil and minerals) that, without adequate measures, generates revaluation of the local currency and contraction and loss of jobs in other productive sectors and final negative effects on the economy as a whole, has been called "Dutch disease."

Norway, unlike, for example, Nigeria, in the face of the same fact, created a compensation fund with the income that was essentially regulated by the State and its state oil company. The case of Nigeria is tragic, since beyond being one of the most corrupt countries, largely due to the actions of a multinational, it exports oil and imports its derivatives for domestic consumption.

It has already been said of the presence of the multinational companies in Vaca Muerta that their strategy of development of the resources will be the one fixed by the head office (since they will have freedom to dispose of the production according to their convenience and this can be that of REPSOL in the 1990s, to export to the maximum without importing the depletion of the resources nor the supply of future generations with the consequence of the current dependence on imported natural gas and oil) and the strategy of the national companies, without a state company as the old YPF was, will also be that of maximizing their income. The fact is that a large number of Argentine businessmen are "rentiers," their aversion to risk is very high and many have grown up as contractors or state concessionaires in public services, a task that has a very low level of irrigation and a safe demand. This characteristic comes from the time of the colony, from the country of cattle raising and then agriculture, where the large latifunds enjoyed the quality income of the humid Pampa and the risk was that it would not rain. So, what will be left for the country? Only the royalties, the occupation of labor and the income tax, as it happens with every extractive activity.

Another of Vaca Muerta's problems is related to the lack of infrastructure, essentially oil and gas pipelines (beyond the idle capacity of some of the existing ones) and heavy drilling equipment. The construction of oil and natural gas evacuation infrastructure requires time, as well as investment, and this clashes with the current government's idea that Vaca Muerta will be the solution for the contribution of the enormous amount of foreign currency that the country requires almost immediately. On the other hand, the multinational companies want the state to build much of the missing infrastructure, especially roads, railways, hotels, homes, hospitals, running water and other services. These "externalities" would then have to come out of the state coffers, and the big oil companies would be dedicated to extracting and transporting the oil and natural gas. How far from the role played by the State YPF in creating and sustaining the villages around its operations!

Perhaps the solution to the Vaca Muerta case would be a controlled exploitation, with the consequent social license (due to the environmental effects of fracking), conserving the resources so that future generations should not have to import the hydrocarbons. Essentially industrialize primary production by converting it into petrochemical goods and fertilizers, using the surplus income to contribute to the change in the productive structure of the country through industrialization and technological development. But this would only be possible if Vaca Muerta's hydrocarbon resources were managed by the National State and the provinces through a federal company as YPF should have been. However, as of May 2018, almost 80% of Vaca Muerta has been granted in concession for 35 years, so that the companies holding the concessions already have acquired rights and it would be very costly, except for breach of contract, for these areas to revert to the Provincial States.

What will be left for the Vaca Muerta Country? Royalties, some taxes paid in the provinces, income tax, low impact on the local equipment industry (as duty-free imports are allowed), low-quality local inputs, sand mining and little else. Today, there are almost 70,000 poor people in Neuquén, and Vaca Muerta does not seem to be thinking about them.

So, will Vaca Muerta be another lost opportunity?

2.7.4 Liquid Petroleum Gas (LPG)

2.7.4.1 Some Data

At the date of preparation of this document, no information was available for 2017 covering the main numerical aspects of the liquefied gas industry.

The author generated in 2015 a study on this energy source called "A reflection on the Home Plan," Working Document of Bariloche Foundation, May 2015, about a plan that will promote the "Kirchnerist" Government with the aim of responding to the LPG problem, especially regarding the so-called social carafes.

Table 2.16 LPG supply and demand—(10^3 TEP)

Year	2014	2015	2016	2017
Total production	2882	2841	2865	2835
NG production	1551	1875	1691	1730
Refinery production	1331	966	1174	1105
Exports	981	940	967	980
Own consumption	28	29	29	34
Total sales domestic market	1889	1873	1898	1783
Residential	1360	1348	1366	1284
Commercial and public service	227	225	228	214
Agricultural	94	94	95	89
Industries	208	206	209	196

Source Energy Balance Sheet, Secretariat of Energy and Mining of Argentina

The observations made in the above-mentioned document are still valid, and even the shortcomings have worsened along with the social situation (Table 2.16).

2.7.4.2 Some Thoughts

(i) **Argentina is a net exporter of LPG**

With approximately 33% of the production, there would be no problem to satisfy the internal consumption of LPG and less for the 2.5 million users that would constitute the universe of "poor subsidized" without access to natural gas by networks. This situation led to the implementation of the so-called Plan Hogar from 2015 during the government of Cristina Kirchner.

(ii) **Production**

Most of the LPG comes from natural gas fields and the rest from distilleries. In both cases, it is a by-product of the process.

(iii) **Concentration of sales**

Almost 70% of the sales in the domestic market are concentrated in the residential sector.

(iv) **Consumption**

It is assumed that, depending on the area and number of household members, these users would consume between 500 and 750 thousand tons of LPG annually, equivalent to 50 or 74% of total residential LPG consumption. It is estimated that there are 4.5 million residential LPG users, of which 2.5 million would be subsidized.

(v) **Subsidies**

The amount of the planned subsidy would have been 3 billion pesos in 2015, equivalent to 1.6% of the country's total economic subsidies in 2014 and 11.6% of social subsidies.

(vi) **The Household Plan**

In fact, the implementation of the Plan Hogar, which was an excellent idea, clashed with the shortage of containers and with the virtual impossibility of accessing subsidized carafes. This situation worsened after the current government took office.

Periodically, the LPG prices foreseen for the Plan Hogar are updated and the price of the kilocalorie of this fuel is 2.6 times greater than the natural gas by networks for the non-subsidized users and 22% greater for the subsidized ones. So fuel for the "poor" is the most expensive.

(vii) **The users**

It is estimated that there are about 5.3 million residential LPG users and almost 50% should have access to the Plan Hogar, but in almost all the cities of the country the subsidized cylinders are very scarce and the users end up paying not 240 or 260 pesos per 10 kg container but over 400 pesos.

(viii) **Packaging**

It would then be necessary to have much more 10–15 kg containers available and to ensure the supply at home of the users, increasing the amount of subsidies since fuel is a necessary and indispensable good. Otherwise, the substitute is firewood which, if it cannot be freely appropriated from the forest, is the most expensive useful calorie for the poor.

(ix) **Caloric consumption**

Thus, LPG is a fuel that essentially supplies the caloric consumption of the families, who use it because they do not have natural gas by networks.

The "Home" Program would work with much less inconvenience if the State Gas Company were recreated and assigned the role of executor of the plan throughout the country. It is very difficult to want to convert, through regulations and inspectors, the current LPG market, whose sellers are governed by profit, into a Public Service as residential LPG consumption really is. The Gospel says that it is very complicated "to serve two masters at the same time."

2.7.5 Compressed Natural Gas (CNG)

Argentina is one of the most advanced countries in terms of CNG use for automotive transport. In addition, it has an industry developed for the manufacture, installation and maintenance of CNG equipment in vehicles and is even an exporter.

2.7.5.1 Some Data

Some data of the Live Vehicle Fleet (i.e., circulates) will be presented to see the incidence of vehicles converted to CNG. The data refer to the years 2015, 2016 and 2017.

2.7.5.2 Some Thoughts

(i) **CNG vehicle fleet**

As it can be seen in Tables 2.17 and 2.18, the fleet converted to CNG has increased from 1,742,762 vehicles in 2015 to 2,035,301 at the end of 2017 (Table 2.19).

The 47% of the fleet is registered in the city of Buenos Aires and the province of Buenos Aires.

54.2% of the live fleet is less than nine years old.

Table 2.17 Characteristics of the argentinean automotive fleet 2015–2017

Year	Automobiles (%)	Light commercial vehicles (%)	Heavy commercial vehicles (%)	Total (number)
2015	86.3	10.7	3.6	12,012.150
2016	85.8	10.7	3.5	12,505.920
2017	85.3	11.1	3.6	13,302.620

Source Association of Argentine Component Manufacturers (AFAC)

Table 2.18 Vehicle fleet by fuel type 2015–2017

Year	A Naphtha (%)	A diesel (%)	A CNG (%)	Hybrids (*) (number)
2015	50.0	35.5	14.5	300
2016	49.0	35.8	15.5	300
2917	49.0	35.7	15.3	400

Source Argentine Association of Component Manufacturers (AFAC)
(*) To Gasoline and Electricity

Table 2.19 Evolution of CNG Consumption: 2015–2017

Year	CNG consumption (10^6 m^3)	Total NG sales (10^6 m^3)	Share of CNG consumption in total NG sales (%)
2014	2843.2	42755.3	6.6
2015	2980.8	43570.6	6.8
2016	2826.6	44685.5	6.3
2017	2553.6	44696.0	5.7

Source IAPG

(ii) **Evolution of consumption**

The relative increase in the price of natural gas with respect to naphtha explains the fall in consumption of CNG from 2105 to 2017.

The above situation has been reversed so far in 2018, and there has been an increase in conversions of gasoline vehicles to the use of CNG. It is that in the first eight months of the year 2018 the price of gasoline has increased by almost 50% while CNG has done so by 22%.

It is that the cost of converting vehicles to CNG ranged in August 2018 between 17,000 and 34,000 pesos, with a lifetime of fifteen years. With a price of premium petrol of $39.88/liter, traveling 15,000 km would be 58,465 pesos, and with a cost of CNG m^3 of US $14, the same journey would be 21,000 pesos, with a saving of 37,465 pesos, amortizing the cost of conversion in less than a year. Of course, the relative prices naphtha/CNG will continue to be varying and with them the tendency to convert or not.

(iii) **Conversion equipment**

An element to consider is that the imported component of the conversion equipment reaches almost 55% of the total cost (even in mid-2018 were lacking inputs due to the increase in conversions), and this disadvantage should be compared with the lower environmental impact of CNG with respect to the naphtha and the development of a major industry of these teams in the country that also exports.

For these reasons, neither would be advisable, as it arises from some information, the idea to enable the import of vehicles converted to CNG use, and on the contrary, it would be necessary to facilitate (which does not imply to subsidize) this activity that creates local work and to foresee the substitution of the mentioned imported components, which are essentially tubes and electronic parts.

2.8 Electric Power

2.8.1 Some Facts

First, some data will be included for the years 2014 to 2017.

Information on the following concepts is included:

Hydroelectric potential;
Installed power;
Power available;
Maximum gross power demand;
Demand;
Generation;
Demand by user type;
Generation by renewable energies;

Billing consumption by user type;
Number of users;
Invoicing by province;
Fuels used for the generation of public service;
Energy used for generation;
MONOMIC spot market price per hour;
Length of transport lines;
Transformer power;
Transport network failures (Tables 2.20 and 2.21).

The only hydroelectric exploitation of certain installed power, in the period under analysis, has been the AH Punta Arena in San Juan with 62 MW (Table 2.22).

The current government mentions that it will build seven hydroelectric power plants of more than 50 MW, which are included in Tables 2.23, 2.24, 2.25, 2.26, 2.27, 2.28, 2.29, 2.30, 2.31, 2.32, 2.33, 2.34, 2.35, 2.36, 2.37, 2.38, 2.39, 2.40, 2.41, 2.42 and 2.43.

2.8.2 Some Comments

(i) **The electrical system**

Table 2.20 Hydroelectric potential, December 2017

Situation	MW	GWH
Installed (*)	11,880	36,870
Inventory not installed (**)	22,853	10,5358
Total	34,733	14,2228

Sources
(*) Wholesale Electrical Market Administration Company, CAMMESA Yearbook 2017
(**) Guillermo V. MALINOW: "Potencial y Desarrollo Hidroeléctrico Argentino", June 2013

Table 2.21 Status of future hydroelectric projects

Status	N° of projects	MW	GWH
Under construction	1	62	296
Basic project	12	7763	36,866
Feasibility	5	1619	4871
Pre-feasibility	22	7824	38,724
Inventory	36	5585	24,601
Total	76	22,853	105,358

Source Guillermo V. MALINOW: "Potencial y Desarrollo Hidroeléctrico Argentino", June 2013

Table 2.22 Payroll of hydroelectric plants in operation as of september 2017

Name	Company	Department	Province	System	Installed power (MW)	Owner
YACYRETA	E.B. YACYRETA	ITUZAINGO	CORRIENTES	MEM	2746	Binational
Valle Grande	EJSEDSA	VALLE GRANDE	JUJUY		0.128	Private
VALLE COLORADO	EDET		TUCUMAN	Aislada	0.334	Private
Urugua-í	EMSA	IGUAZU	MISIONES	Interconectada no MEM	120	Provincial
ULLUM	HIDROTERMICA SAN JUAN	ULLUM	SAN JUAN	MEM	42	Private
TILCARA	EPSE	ZONDA	SAN JUAN	MEM	63.28	Private
TAFI DEL VALLE	EDESA		SALTA	AISLADA	0.04	Private
SANTO TOMAS	COOPERATIVA DE ELECTRICIDAD DE BARILOCHE		RIO NEGRO	Interconectada no MEM	0.36	Provincial
SAN ROQUE	EPEC GENERACION	SANTA MARIA	CORDOBA	MEM	24	Provincial
SAN FRANCISCO	EJSEDSA	VALLE GRANDE	JUJUY		0.118	Private
ALTO D LA LOMA	C.H. SALTO DE LA LOMA SIEYE	JACHAL	SAN JUAN	MEM	0.6	Private
SALTO ANDERSEN	CENTRAL SALTO ANDERSEN	PICHI MAHUIDA	RIO NEGRO	MEM	8	Provincial
SALTITO II		CAINGUAS	MISIONES	Interconectada no MEM	0.64	Cooperative
Saltito I	COOP.2 DE MAYO	CAINGUAS	MISIONES	Interconectada no MEM	0.64	Cooperative

(continued)

Table 2.22 (continued)

Name	Company	Department	Province	System	Installed power (MW)	Owner
S. GRANDE ARG	CTM SALTO GRANDE	CONCORDIA	ENTRE RIOS	MEM	945	Binational
Río Corralito	EDESA SA	ROSARIO DE LERMA	SALTA	Interconectada no MEM	13.2	Private
RIO REYES	HIDROELECTRICA REYES EJSEDSA	DOCTOR MANUEL BELGRANO	JUJUY	MEM	7	Private
RIO PICO	EDERSA		RIO NEGRO	AISLADA	0.106	Provincial
RIO HONDO	HIDROELECTRICA RIO HONDO SA	RIO HONDO	SANTIAGO DEL ESTERO	MEM	15	National-private concession
RIO GRANDE	EPEC GENERACION	CALAMUCHITA	CORDOBA	MEM	750	Provincial
UEBRADA ULLUM	QUEBRADA ULLUM—EPSE	ULLUM	SAN JUAN	MEM	45	Provincial
PUNTA NEGRA	EPEN		NEUQUEN	Interconectada no MEM	0.065	Provincial
Punta Negra	EPSE	PUNTA NEGRA	SAN JUAN	MEM	63.28	Provincial
PUERTO MORENO	EPEN		NEUQUEN	Interconectada no MEM	0.12	Cooperative
PUEBLO VIEJO	HIDROELECTRICA TUCUMAN SA	MONTEROS	TUCUMAN	MEM	15	National-private concession
PIEDRAS MORAS	EPEC GENERACION	TERCERO ARRIBA	CORDOBA	MEM	6.3	Provincial
PIEDRA DEL AGUILA	HIDR. PIEDRA DEL AGUILA S.A.	COLLON CURÁ	NEUQUÉN	MEM	1400	National-private concession
PICHI P. LEUFU	PICHI PICUN LEUFU	COLLON CURA	NEUQUEN	MEM	285	National-private concession

(continued)

Table 2.22 (continued)

Name	Company	Department	Province	System	Installed power (MW)	Owner
Paso de las Carretas	SER BEEF SA	CORONEL PRINGLES	SAN LUIS		1.2	Provincial
Pampichuela	EJSEDSA	VALLE GRANDE	JUJUY		0.077	Private
. BANDERITA	HIDR. CERROS COLORADOS S.A.	CONFLUENCIA	NEUQUEN	MEM	472	National-private concession
NIHUIL 4	HIDR NIHUIL IV (EMSE SE)	SAN RAFAEL	MENDOZA	MEM	18	Private
NIHUIL 3	HIDROELECTRICA LOS NIHUILES SA	SAN RAFAEL	MENDOZA	MEM	42	National-private concession
NIHUIL 2	HIDROELECTRICA LOS NIHUILES SA	SAN RAFAEL	MENDOZA	MEM	110	National-private concession
NIHUIL 1	HIDROELECTRICA LOS NIHUILES SA	SAN RAFAEL	MENDOZA	MEM	72	National-private concession
MUTQUIN	EPEN		NEUQUEN	AISLADA	0.06	Provincial
MCH S.GUILLERMO	C.H. SAN GUILLERMO SIEYE	CALINGASTA	SAN JUAN	MEM	0.1	Private
Manzano Amargo (La Fragua)	EPEN		NEUQUEN	Interconectada no MEM	0.4	Provincial
LUJANITA	PAH LA LUJANITA—ENARSA	LUJAN DE CUYO	MENDOZA	MEM	1.7	Private
LUJAN DE CUYO	PAH CT MENDOZA—ENARSA	LUJAN DE CUYO	MENDOZA	MEM	1	Private
Los Toldos	EDESA SA	SANTA VICTORIA	SALTA		0.426	Private

(continued)

Table 2.22 (continued)

Name	Company	Department	Province	System	Installed power (MW)	Owner
LOS REYUNOS	H. DIAMANTE SA	SAN RAFAEL	MENDOZA	MEM	224	National-private concession
LOS QUIROGA	HIDROELECTRICA RIO HONDO SA	BANDA	SANTIAGO DEL ESTERO	MEM	2	National-private concession
LOS MOLINOS 2	EPEC GENERACION	SANTA MARIA	CORDOBA	MEM	4.5	Provincial
LOS MOLINOS	EPEC GENERACION	SANTA MARIA	CORDOBA	MEM	52	Provincial
LOS MADERAS	HIDROCUYO S.A.	EL CARMEN	JUJUY	MEM	30.6	Private
LOS ALERCES	EPEN		NEUQUEN	AISLADA	0.024	Provincial
Loma Atravesada	EDERSA	BARILOCHE	RIO NEGRO	Aislada	2.3	Private
Las Loicas	MUNICIPALIDAD DE MALARGUE	MALARGÜE	MENDOZA	Aislada	0.03	Municipal
LA VINIA	EPEC GENERACION	SAN ALBERTO	CORDOBA		16	Provincial
LA POMA	EPEN	ALUMINE	NEUQUEN	Interconectada no MEM	0.4	Private
LA MOSCA	EJSEDSA		JUJUY	AISLADA		Private
La Cíenaga	HIDROCUYO S.A.	EL CARMEN	JUJUY	Aislada	0.45	Private
LA CARRERA	EJESA		JUJUY	Interconectada no MEM	0.18	Provincial
LA CALERA	EPEC GENERACION	COLON	CORDOBA		4.4	Provincial
JULIAN ROMERO	CENTRAL JULIAN ROMERO 5 SALTOS	GENERAL ROCA	RIO NEGRO	MEM	6.2	Provincial

(continued)

Table 2.22 (continued)

Name	Company	Department	Province	System	Installed power (MW)	Owner
Ing. Emilio Frey	COOPERATIVE			Interconectada no MEM	1.6	Cooperative
GRAL. ROCA	EMP DE ENERGIA DE RIO NEGRO SA	GENERAL ROCA	RIO NEGRO	Interconectada no MEM	2	Provincial
FUTALEUFU	HIDROELECTRICA FUTALEUFU SA	FUTALEUFU	CHUBUT	MEM	472	National-private concession
FITZ SIMON	EPEC GENERACION	CALAMUCHITA	CORDOBA	MEM	10.5	Provincial
. AMEGHINO	HIDROELECTRICA AMEGHINO SA	GAIMAN	CHUBUT	MEM	46.8	National-private concession
ESCABA	HIDROELECTRICA TUCUMAN SA	JUAN BAUTISTA ALBERDI	TUCUMAN	MEM	24	National-private concession
EL TUNAL	AES JURAMENTO	METAN	SALTA	MEM	10.6	National-private concession
EL TIGRE	H. DIAMANTE SA	SAN RAFAEL	MENDOZA	MEM	14	National-private concession
El Cajón	DGSP	FUTALEUFU	CHUBUT	Aislada	0.36	Provincial
DIVISADEROS	APELP	PUELEN	LA PAMPA	Interconectada no MEM	10	Provincial
CUSI CUSI	EJSEDSA	SANTA CATALINA	JUJUY		0.072	Private
CTA. DEL VIENTO	HIDR.CUESTA DEL VIENTO – EPSE	IGLESIA	SAN JUAN	MEM	10.6	Provincial
CRUZ DEL EJE	EPEC GENERACION	CRUZ DEL EJE	CORDOBA	MEM	1.1	Provincial

(continued)

Table 2.22 (continued)

Name	Company	Department	Province	System	Installed power (MW)	Owner
CIPOLLETTI	CENTRAL HIDRAULICA CIPOLLETTI	GENERAL ROCA	RIO NEGRO	MEM	5.4	Provincial
CIENEGA (CIENEGA DE PAICONE)	EJSEDSA	SANTA CATALINA	JUJUY	Aislada	0.02	Private
Chuscha	EDESA SA	SAN CARLOS	SALTA		1.195	Private
CHOCON	HIDROELECTRICA EL CHOCON SA	CONFLUENCIA	NEUQUEN	MEM	1260	National-private concession
CHIQUILIHUIN	DGSP		CHUBUT	AISLADA	0.179	Provincial
CH SAN MARTIN	GENERADORA ELECTRICA MENDOZ.SA	LUJAN DE CUYO	MENDOZA	MEM	6.6	Private
CH LS CORONELES	GENERADORA ELECTRICA MENDOZ.SA	SAN RAFAEL	MENDOZA	MEM	6.6	Private
CH CARRIZAL	CONSORCIO POTRERILLOS	RIVADAVIA	MENDOZA	MEM	17	Private
CESPEDES	CENTRAL HIDRAULICA CESPEDES	AVELLANEDA	RIO NEGRO		5.2	Provincial
CENTRAL NAZARIO BENAVIDEZ	PLASTICOS GASA SA	CAPITAL	SAN JUAN	Interconectada no MEM	0.35	Private

(continued)

Table 2.22 (continued)

Name	Company	Department	Province	System	Installed power (MW)	Owner
CENTRAL BARREALITO	BARREALITO SRL	CALINGASTA	SAN JUAN	Interconectada no MEM	0.42	Private
CASSAFOUSTH	EPEC GENERACION	CALAMUCHITA	CORDOBA	MEM	16.2	Provincial
CASA DE PIEDRA	CENTRAL CASA DE PIEDRA	PUELÉN	NEUQUEN	MEM	60	Provincial
CARACOLES	C.H.LOS CARACOLES—EPSE	ZONDA	SAN JUAN		121.4	Provincial
CADILLAL	HIDROELECTRICA TUCUMAN SA	TAFI VIEJO	TUCUMAN	MEM	12.6	National-private concession
CABRA CORRAL	AES JURAMENTO	LA VIÑA	SALTA	MEM	100.5	National-private concession
C.H.CACHEUTA	CONSORCIO POTRERILLOS	LUJAN DE CUYO	MENDOZA	MEM	120	Private
C.H. A.CONDARCO	CONSORCIO POTRERILLOS	LAS HERAS	MENDOZA	MEM	61	Private
BUTACO	DGSP		CHUBUT	AISLADA	0.176	Provincial
BENJAMIN REOLIN	EPEC GENERACION	CALAMUCHITA	CORDOBA	MEM	33	Provincial
AUQUINCO	ENERGIA DE CATAMARCA SAPEM		CATAMARCA	AISLADA		Provincial
ARROYITO	HIDROELECTRICA EL CHOCON SA	CONFLUENCIA	NEUQUEN	MEM	127.8	National-private concession

(continued)

Table 2.22 (continued)

Name	Company	Department	Province	System	Installed power (MW)	Owner
ALUMINE	ENERGIA DE CATAMARCA SAPEM		CATAMARCA	Interconectada no MEM	1	Provincial
ALICURA	AES ALICURA	LACAR	NEUQUEN	MEM	1050	National-private concession
ALBERTO GRAFFIGNA	ALBERTO GRAFFIGNA		SAN JUAN	Interconectada no MEM	0.5	Private
AGUA DEL TORO	H. DIAMANTE SA	SAN RAFAEL	MENDOZA	MEM	150	National-private concession
Total					11,880	

Source Ministry of Energy and Mining of Argentina

Table 2.23 Hydroelectric developments to be built, according to the current government

Central	Río	Power (MW)
NESTORKIRCHNER-Jorge COPERNIC	Santa Cruz	1350
El TAMBOLAR	San Juan	70
CHIHUIDO I	Neuquén	637
YACYRETÁ –Apipe	Con Paraguay	240
Aña CUÁ	Con Paraguay	135
Portezuelo del Viento	Grande Mendoza	90

Source Secretariat of Energy of Argentina

Until the early 1990s, practically the entire Argentine electricity system was in the hands of the State, with the companies Agua y Energía and HIDRONOR standing out. Subsequently, almost all the plants were privatized, leaving only the national government in charge: the binational Yacyretá (shared with Paraguay) and Salto Grande (shared with Uruguay), as well as the three nuclear plants (Embalse, Atucha I and Atucha II).

TRANSENER is a mixed company that deals with the transmission of electricity at high voltage. The distribution is divided between private companies such as EDENOR, EDESUR and EDELAP that operate in the city of Buenos Aires and in the city of La Plata and provincial companies that distribute in territories of the provinces.

CAMMESA is a mixed company that operates and dispatches the Interconnected Electric System. The Patagonian System has also been integrated into the National System.

The ENRE is a public agency in charge of regulating the national electricity system.

Large Users are grouped into three categories:

Large Wholesale Users (GUMA);
Major Retail Users (GUME);
Large Private Users (GUPA).

There are interconnections with Chile, Paraguay, Uruguay and Brazil, and from 2019, there will possibly be one with Bolivia.

"It is worth remembering that the main hydroelectric works built in our country were developed by the National State through the specialized agencies Water and Electric Energy and HIDRONOR following perfectly defined study stages as regards their scope and content, such as: Inventory, Technical-Economic Prefeasibility, Technical-Economic Feasibility, Preliminary or Basic Project and finally the Executive Project" (Malinow 2013, "Potencial y Desarrollo Hidroeléctrico Argentino)

(ii) **The hydroelectric potential**

Development of Argentine Hydroelectric Power.

Table 2.24 Installed power in (MW)

Year	Turbine gas (TG)	Turbine steam (TV)	Cycles Combi-nados (CC)	Diesel groups (DI)	Sub total thermal	HIDRO	Nuclear	Wind	Solar	Total
2012	4071	4451	9205	1052	18,779	11,148	1005	112	6	31,050
2013	4074	4441	9205	1074	18,794	11,096	1010	164	8	31,072
2014	4036	4451	9191	1388	19,065	11,106	1010	187	8	31,376
2015	4451	4039	9227	1226	18,943	11,108	1010	46	8	31,115
2016	4451	5251	9227	1293	20,225	11,170	1755	187	8	33,345
2017	4451	6006	10,436	1825	22,718	11,243	1755	227	8	35,951

Source CAMMESA Statistical Yearbooks Years 2012 to 2017

Notes

In the year 2014 in Diesel Equipment (DI) 329 MW of Mobile Plants are included

In the year 2016, 558 MW of Mobile Diesel Plants must be added

In the year 2017, 200 MW of Mobile Diesel Plants must be added

Table 2.25 Installed and available power—Year 2014—(MW)

	Gas turbines (TG)	Steam turbines (TV)	Combined cycles (CC)	Subtotal thermal	HIDRO	Nuclear	Rest (DI; wind; photovoltaic) total	Total
Power available	2374	2738	7131	12,243	10,551	963	1169	24,926
Installed power	4035	4451	9191	17,677	11,106	1010	1583	31,376
Power not available	1661	1713	2060	5434	555	47	414	6450
Maximum demanded power								24,034
"Reserve"								892

The power "reserve" as of 2014 was 3.5% of the available power
Source Based on CAMMESA 2014 Statistical Yearbook

Table 2.26 Installed and available power—Year 2017—(MW)

	Gas turbines (TG)	Steam turbines (TV)	Combined cycles (CC)	Subtotal thermal	HIDRO	Nuclear	Rest (DI; wind; photovoltaic)	Total
Power available	3277	4834	8995	17,106	10,681	703	1806	30,296
Installed power	4451	6006	10,434	20,891	11,243	1755	2060	35,949
Power not available	1174	1172	1434	3785	562	1052	255	5653
Maximum demanded power								25,628
"Reserve"								4668

Source Based on CAMMESA 2017 Statistical Yearbook

Table 2.27 Available power over installed power and reserve—Year 2017—(%)

	Gas turbines (TG)	Steam turbines (TV)	Combined cycles (CC)	Subtotal thermal	HIDRO	Nuclear	Rest (DI; wind; photovoltaic) total	Total
Available power/installed power	73.6	80.5	86.2	81.9	96.6	40	87.7	84.2
Reserve over maximum power								18

Source Own elaboration based on Table 5.1.7

Table 2.28 System power reserve

Year	Power available	Power maximum	Reserve (*)	Maximum power reserve
Unit	(MW)	(MW)	(MW)	(%)
2014	24,926	24,034	892	3.7
2015	25,794	23,949	1845	7.7
2016	27,348	25,380	1968	3.8
2017	30,296	25,628	4668	18.2

Source CAMMESA Yearbooks
(*) Own elaboration based on CAMMESA's Yearbooks

Table 2.29 Demand by type of destination—GWH

	2014	2015	2016	2017
MEM agents	126,421	132,100	132,970	132,436
Exports	6	55	329	69
Pumping	485	578	465	401
Own consumption and losses	4293	4136	4306	4294
TOTAL	131,205	136,870	138,070	137,200

Source CAMMESA 2017 Yearbook

Table 2.30 Generation—GWH

	2014	2015	2016	2017
Thermal	83,265	86,625	90,349	88,838
Hydraulic	40,683	41,464	38,012	41,280
Nuclear	5258	6519	7677	5716
Wind and solar	629	608	561	632
Imports	1390	1655	1470	734
Total	131,205	136,870	138,070	137,200

Source CAMMESA 2017 Yearbook

Table 2.31 Other renewable generation—(GWH)

	2014	2015	2016	2017
Biodiesel	2	0	1	0
Biomass	114	155	193	243
Wind	613	593	547	616
Hydroelectric < or = 30 MW	1457	1624	1820	1696
Solar photovoltaic	16	15	14	16
Biogas	103	84	58	64
Total	2305	2470	2633	2636
Percentage of MEM demand covered by Renewables (%)	1.8	1.9	2.0	2.0

Source CAMMESA 2017 Yearbook

Table 2.32 Wholesale electricity market demand by type of user—GWH

	2014	2015	2016	2017
Residential	51,444	55,424	57,067	56,889
Intermediate consumption	35,995	37,351	38,541	38,291
High demand	39,028	39,337	37,503	38,255
Total	126,467	132,110	133,111	132,436

Source CAMMESA 2017 Yearbook

Table 2.33 Electricity invoicing by consumer sector (GWH)

Sector	2014	2015	2016
Residential	40,386.6	42,078.5	43,391.4
Commercial	19,494.2	20,442.2	20,976.6
Industrial	38,024.7	39,470.5	36,848.8
Health services	1230	1227	1286
Public lighting	4036.3	4163.4	4211.4
Transport	563.9	565.8	341.9
Irrigation	1049.4	964.6	919.5
Official	4003.7	4209.2	4403.3
Rural	1224.7	1272.6	1256.7
Others	1831.3	1887.3	1322
TOTAL	111,844.8	116,281.1	114,957.6

Source Secretariat of Energy and Mining of Argentina Statistical Yearbooks

Table 2.34 User number (No.)

	2014	2015	2016
Residential	13,382,765	13,583,050	13,861,632
Total	15,572,140	15,794,362	16,090,317

Source Secretariat of Energy and Mining of Argentina Statistical Yearbooks

Table 2.35 Turnover by province (%)

	2014	2016
Capital Federal y Gran Bs As	35.4	35.9
Rest of Province Bs As	15.0	14.0
Santa Fe	9.4	9.4
Córdoba	7.1	7.4
Rest of provinces	33.1	33.3
Total country (GWH)	111,844.817	133,469.976

Source Author's elaboration on data from the Secretariat of Energy and Mining of Argentina

Table 2.36 Consumption of fuels for the generation of electricity by the public service. Price of Fuels (%)

	2014	2015	2016	2017	Price (USD/MMBTU)2017(*)
GN	47.5	44.2	48.2	64.5	11.5
FO	29.1	30.6	26.5	15.7	10.4
GO	16.9	19.5	20.9	15.0	8.4
CM	6.5	5.7	4.4	4.8	5.2
Biodiesel	–	–	–	–	
Total (10^3 TEP) (*)	9154.4	9884.7	9800.6	8036.4	
Average specific consumption (Kcal/KWH) (*)	2019	2029	2006	1949	

Source (*) CAMMESA 2017 Yearbook
The percentages are based on data from the CAMMESA 2017 Yearbook

Table 2.37 Power generation for public service electricity (10^3 TEP)

Energy	2014	2015	2016	2017
Hydraulic	3558	3526	3248	3454
Nuclear	1280	2204	2224	1745
Coal	542	526	393	353
Wind	53	51	47	53
Solar	1	1	1	1
Natural gas	12,071	12,380	13,282	14,330
Diesel oil	1554	1930	2051	1204
Fuel oil	2664	2860	2598	1261
Biodiesel	0	38	0	0
Total	21,723	23,516	23,844	22,401

Source Energy Balance Sheet Secretariat of Energy and Mining of Argentina

As of December 2017, it was 34,733 MW, with a capacity to generate 142,228 GWH. The hydroelectric potential was equivalent to 97% of the country's total installed capacity in 2017. Of the hydroelectric potential, 66% was in varying degrees of progress for its use; 0.3% of them was under construction; 34% with a basic project; 7% with a feasibility study; 34% with a pre-feasibility study; and 24% with an inventory. The installed hydroelectric power at the end of 2017 was only 34% of the detected hydroelectric potential (see Tables 5.1.1 and 5.1.2). Table 5.1.3 shows the list of hydroelectric plants in operation as of September 2017.

The hydroelectric power plants in file by the current government for their construction (Table 5.1.4) would be about 2522 MW until 2025, but almost all of them are largely delayed in some cases due to conflicts between provinces (Portezuelo del

Table 2.38 Energetics for the generation of electricity for the public service (%)

Energy	2014	2015	2016	2017
Hydraulics	16.4	15.0	13.6	15.4
Nuclear	5.9	9.4	9.3	7.8
Coal	2.5	2.2	1.6	1.6
Wind	0.2	0.2	0.2	0.2
Solar	0.0	0.0	0.0	0.0
Natural gas	55.6	52.6	55.7	64.0
Diesel oil	7.2	8.2	8.6	5.4
Fuel oil	1203	12.2	10.9	5.6
Biodiesel	0.0	0.2	0.0	0.0
Total	100	100	100	100

Source Own elaboration based on Table 5.1.17

Table 2.39 Public service generation and self-production of electricity—(10^3 TEP)

	2014	2015	2016	2017
Public service	10,693	11,322	11,480	11,255
Auto production	1222	1184	1178	1268
Total	11,955	12,506	12,658	12,523

Source Energy Balance Sheet Secretariat of Energy of Argentina

Table 2.40 MONOMIC price of the hourly spot market without transport charges ($ current/MWH)

	2014	2015	2016	2017
Energy component	538.8	642.2	1041.5	1052.4
Component power and reserve	10.6	10.3	13.3	121
Monomic price	549.4	653.5	1054.8	1173.4
Distributors' revenue	95.2	95.3	312.0	572.2

Source CAMMESA 2014-2017 Yearbooks
As of August 2018, the MONOMARY price was $2284.3 per MWH and the amount collected from distributors was $1343

Table 2.41 Length of transport lines—(km)

	Tensión (KV)	2009	2014	2015	2016	2017
High voltage	133 a 500	11,853	14,392	14,756	14,756	14,756
Trunking distributors	33 a 330	17,080	19,061	19,532	19,550	19,713

Source CAMMESA Yearbooks

Table 2.42 Evolution of transformer power—(MVA)

	2009	2014	2015	2016	2017
High voltage + high voltage in reserve	14,450	20,900	23,916	24,700	26,640
Trunk distributors	11,872	13,756	14,575	14,897	15,268

Source CAMMESA 2017 Yearbooks

Table 2.43 Annual failures in transport networks. N° of faults/100 km—year

	2009	2014	2015	2016	2017
High voltage	0.5	0.5	0.5	0.5	0.5
Trunking distributors	2.4	2.1	2.3	2.7	2.7

Source CAMMESA 2017 Yearbooks

Viento), but essentially due to financing problems. The only one that seems to be advancing is the binational Aña Cuá. The country's budget for the year 2019 only registers the contribution of some funds for Chihuido I and the two in Santa Cruz. The others are subject to the Public–Private Partnership Program (PPP), which will be discussed later.

In recent years, the execution of new hydroelectric projects has been neglected, with the satisfaction of demand falling on thermal power stations, with the consequent negative effect on the energy trade balance, since, as will be seen, the weight fell on natural gas, which had to be imported in increasing values for this purpose.

(iii) **Power**

See Tables 5.1.5 to 5.1.9.

In 2017, the installed power was mainly thermal (58%), and within it, although combined cycles predominated, the inefficient equipment of gas turbines, which did not always operate at the top of the load curve, was still important. The three nuclear power plants only contributed 2.3% of the available power and 4.9% of the installed power (due to the maintenance shutdown of the Embalse plant), and the wind turbines and photovoltaic panels were insignificant.

The power available in 2017 was approximately 18.2% higher than the maximum with a reserve margin of 4668 MW (in 2015, the reserve reached 1845 MW).

Between 2009 and 2014, the maximum power grew at a rate of 4.2%, while between 2014 and 2017 it grew at a cumulative annual rate of 2.2%, showing the effects of the recessive economic process.

Between 2015 and 2017, the maximum power grew by 1679 MW, while the available power grew by 4502 MW. In any case, supply problems are essentially due to deficiencies in transmission and transformation.

It should also be noted that the works entering a period have been the result of decisions and executions made in the previous period.

It should also be noted that the works entering a period have been the consequence of decisions and executions made in the previous period.

As regards the rate of installation of new power, to satisfy the demand of a developing country and not a stagnant one, the values should not be less than 1000 MW per year, plus a power that ensures an adequate level of available reserve, with an appropriate mix of plants that supply the peak and base of the system.

According to the Secretariat of Energy and Mining, between 2017 and July 2018, 4100 MW (almost 95% in conventional thermal power plants) will be installed in 41 power plants.

(iv) **Energy demand**

See Tables 5.1.10 to 5.1.15.

Total electricity demand by destination is concentrated in the wholesale electricity market whose consumption grew between 2009 and 2014 at a cumulative annual rate of 3.9%, and between 2014 and 2017, it decreased to 1.6% cumulative annual.

Generation by type of energy source showed the growing predominance of fossil fuels, which every year displaced hydraulic and nuclear energy, which showed the neglect of the abundant hydroelectric exploitation and the practical paralysis of nuclear energy, despite the entry of the third nuclear power plant in 2014, after thirty years of the second reservoir.

The import of electricity was insignificant.

Renewable energies (except for hydroelectric power of more than 30 MW) contributed barely 2% of the needs of the wholesale electricity market. The greatest contribution was made by small hydroelectric plants, and there was a growing supply of wind power, among the renewable energies detected.

The invoicing by type of user, in the year 2016, was concentrated in two sectors: the residential (37.7%) and the industrial (32.1%), increasing the participation of the residential and diminishing that of the industrial as sample, perhaps, of the process of deindustrialization. The spatial concentration of the population and of the country's productive activity was evidenced by the fact that the city of Buenos Aires, Greater Buenos Aires and the provinces of Santa Fe and Córdoba absorbed almost 67% of the electricity turnover.

Almost 99% of the population had electricity service.

In the first nine months of the year 2018, compared to 2017, the demands decreased sharply, in the residential sector 5.5%, in the commercial sector 6.9% and in the industrial sector 5.3%. These data are in line with the recessionary process, rate increases and fall in real wages, so far in 2018. Even the distributors mentioned the increase in the number of delinquencies and in electricity thefts.

(v) **Energy used for electricity generation**

See Tables 5.1.16 to 5.1.18.

The tables show the contribution of fuels to generate fossil thermal electric energy, as well as the contribution of the different energies that allow obtaining electricity from the public service. In both cases, the high participation of natural gas, the scarce relevance of wind and solar energy and the decline of hydroelectric power are evident.

(vi) **Generation of electricity from the public service and self-production**

Table 5.1.19 includes the generation of public service and self-production of electricity, the latter representing between 10% of the total. Surely with the implementation of the "distributed energy" policy, this percentage will increase in the future. This issue will be discussed later.

(vii) **The MONOMIC price of the spot market**

The gap between CAMMESA's collections and MONOMIC prices fell from 86% in 2015 to 51.3 in 2017 and 41.3 in August 2018, essentially as a result of tariff increases.

The hourly spot market MONOMIC price without transport charges showed a very significant difference with what was collected by the distributors (see Table 5.1.20), covered by increasing subsidies. In 2014, they barely covered 17% of the MONOMIC price and the rate increases brought it up to 49% in 2017. When the problem of energy prices and tariffs (oil derivatives, natural gas and electricity) is analyzed, we will return to the subject.

Wholesale market prices were out of step with the marginal costs of electricity generation. The application of the Resolution of the Secretariat of Energy 240/03 allowed that the prices were sanctioned as if there was full supply of national natural gas, without any type of restrictions, and in addition, a maximum price was fixed for energy in the wholesale electricity market of 120 $/MWK, and both things caused an important distortion with respect to the real costs. This, in turn, led to a substantial increase in electricity subsidies provided by the State to keep the electricity system running. Thus, by February 2015, CAMMESA accumulated a debt with the generators of more than 55 billion pesos.

The different components of the MONOMIC price vary according to the volume of thermal generation required, depending mainly on the hydroelectric supply, on the price of natural gas and in a mitigated way on the value of liquid fuels since their value is included in the price as a surcharge (SCTD).

The MONOMIC price presents seasonality throughout the year, being higher in the winter months, related to liquid fuels.

The Ministry of Energy and Mines Resolution No. 20/2017 modified the seasonal prices by establishing new prices and new user categories.

The plaintiffs' payments are not sufficient to level out the actual costs that are covered by treasury contributions (CAMMESA Yearbook 2017).

The wholesale cost of electricity would decrease in 2018 to US $66/MWH due, essentially, to the fact that CAMMESA instead of paying natural gas at 5.20 million BTU will pay it at 3.40 and to a lesser extent by the generation income with renewable sources.

Despite this, the price paid by users will continue to increase, due to the strong devaluation of the peso as tariffs are dollarized.

From August to October 31, 2018, users will pay about $1400/MWH at a cost of $3000/MWH.

So if the cost was US $66/MWH (2450 $/MWH), the subsidy, by the State, would be 40% of the cost and the user would pay the remaining 60%.

It is estimated that the amount to be subsidized in 2018 would be more than 65 billion pesos.

On the other hand, the 2019 budget establishes a subsidy to the electricity sector (not only CAMMESA), of about 99,484 million current pesos.

(viii) **Transportation; transmission and system failures**

By 2017, all transport systems were interconnected. Tables 5.1.22 to 5.1.24 present figures referring to the transmission system, the transformation system and the faults detected.

The electrical system showed faults in distribution and transformation that were recorded with the supply cuts, especially in the areas served by EDENOR and EDESUR in spite of the very large increases in tariffs received since 2016.

When the temperature exceeds 30 °C in summer, the cuts begin.

This situation led the ENRE to issue Resolutions 198 and 199 that modified the service quality standards.

For example, at EDENOR, between March and August 2017, there were 63 faults in medium voltage feeders and between September 2017 and February 2018, 143.

EDESUR also verified 63 and 143 in the same periods.

The ENRE warned that the number of service interruptions, and their duration, greatly exceeded the established guidelines.

As a result, fines of between 330 and 600 KWH per affected user were applied.

The ENRE also established that the above-mentioned companies should return about 500 million pesos to the affected users.

As an example, the cut at the end of August 208 affected 300,000 users of EDESUR out of a total of 2.2 million and EDENOR affected 20,658 on the same date.

It would seem that it is cheaper for the companies to pay the fines than to make the necessary investments.

EDENOR had planned to invest some 5211 million dollars during 2018 and EDESUR some 4400 million dollars.

(ix) **The case of TRANSENER**

The sale of the state's participation in TRANSENER to private shareholders, which is a consequence of Decree 882 of 2017, is not convenient. The fact is that TRANSENER is essentially profitable, so why privatize it?

According to Mr. Jorge La Peña's calculation, the state's stake in the electricity transmission company (which is 26%) was worth between US $350 million and US $400 million until a few months ago, but in October 2018 it fell to around US $120 million.

The decree was published in the Official Gazette on November 1, 2017, when the electricity transmission company's share on the Buenos Aires Stock Exchange was at $42.55, or US $2.39, at an exchange rate of $17.80 to the dollar at that time. The share rose to a maximum of $65.05 on February 1 this year (US $3.31 at an exchange rate of $19.65 per dollar) and fell back to $43.90 by October 16, 2018 (US $1.19, at $36.90 per dollar). This means that the dollar stock lost 50% of its value in almost a year and 64% from its peak in February 2018. Nor did it make sense to leave the sale process of these shares in the hands of the consulting firm Price Waterhouse & Co Asesores de Empresas S.R.L (PWC Argentina), which won the bid for $13.3 million.

It would seem that one of the main interested parties in the purchase is PECOM from Pérez COMPANC.

The State had bought the shares at $7 in mid-2016 came to quote $65 and in October 2018 was $50 (Santiago Spaltro, El Cronista newspaper, Buenos Aires).

2.9 Mineral Coal

2.9.1 Some Facts

Unlike what happens in Europe, in Latin America, with the exception of Colombia and in certain regions of other countries, coal has not been used by families and barely by industries to generate steam; its use is limited essentially to the generation of electricity and the steel industry.

Table 2.44 summarizes the data on coal.

Table 2.44 Data on Coal—10^3 TEP

Year	Production	Imports	Exports and bunker	Electricity generation public service	Electricity generation self-production	Steel	Industry
2014	33	1464	17	542	12	807	6
2015	20	1432	8	526	10	819	5
2016	14	1154	7	393	12	617	26
2017	12	1087	0	353	12	687	26

Source Ministry of Energy and Mines of Argentina Energy Balances

2.9.2 Some Comments

(i) **Reserves**

The Measured Coal Reserves reached 477.9 million tons in 2016, which added to the Indicated Reserves of 206.5 and inferred of 67.8 gave a total of 752.3 million tons.

(ii) **Production**

The current production of Río Turbio, the only field in operation, in 2017, reached 20,340 tons. The maximum production peak was reached in 1979 with 1,326,254 tons, and since then, the drop has been permanent, as practically, after the privatization of Yacimientos Carboníferos Fiscales (YCF) and despite the investments made by the previous government in the Río Turbio mine.

In 2016, the ratio of Reserves to Production Measures reached 23,500 years, showing the waste of this resource.

(iii) **The markets**

Since Río Turbio coal is not suitable for conversion into coke economically, the almost only consumer market is thermoelectric power plants. As practically the only generating plant was San Nicolás, the drastic decrease in production can be explained. Most of the coal consumed is imported, suitable for producing coke and essentially destined for the steel industry to feed the coking plants. The rest of the coal is used by the San Nicolás thermal power station. The permanent decrease in consumption is essentially due to the crisis in the country's steel industry.

(iv) **The problem of the Boca de Mina plant**

In 2018, a 240 MW capacity power plant was being completed at the Río Turbio Mine, which will consume 1,200,000 TN per year of raw coal. But until the mine's production, which did not exceed 20,300 TN/year, is reactivated, the plant should operate with natural gas or imported coal. Important investments have been made in the mine that could, with an additional effort, reach the production required by the plant, while the plant is practically finished. The current government does not seem willing to complete the work for budgetary reasons. In the Río Turbio area, the population considers that closing the mine is condemning the population to disappear, while environmental groups oppose the use of coal to generate electricity for environmental reasons. The issue is complex and the worst way out is to do nothing.

(v) Most of the coal consumed is imported, of a type suitable for producing coke, and essentially destined for the steel industry to feed the coking plants. The rest of the coal is used by the San Nicolás thermal power station. The permanent decrease in consumption is due to the crisis in the country's steel industry.

2.10 Nuclear Energy—Uranium

2.10.1 Some General Information

Argentina has three nuclear power plants.

All of them use natural uranium and heavy water.

Atucha I, which came into operation in 1974, has 357 MW of power.

Embalse came into operation in 1984 and has 648 MW of power. It is being repowered to bring it to 683 and is also a source of production of Cobalt 60 used for radiological purposes.

Atucha II entered into operation in 2014 and has 745 MW of power, being the largest machine in the Argentinean electrical system.

The fuel assemblies of the Atucha I and Embalse nuclear power plants were manufactured until 1995 with uranium of national origin. For this purpose, the CNEA exploited various deposits in the provinces of Salta (Don Otto), Chubut (Los Adobes), Córdoba (Los Gigantes) and Mendoza (Huemul in Malargue and Sierra Pintada in San Rafael).

In 1995, due to a sharp fall in uranium prices on the international market, the activities of the Sierra Pintada deposit in the province of Mendoza (the only one in operation) were suspended and uranium concentrate was imported, and the rest of the transformation processes were carried out in the country.

The country imports all the foodstuffs it consumes.

In terms of reserves, Argentina has certified 7000 tons of uranium among the Sierra Pintada and Cerro Solo deposits, while the inferred and predicted reserves are an additional 3000 tons.

On the other hand, the current consumption of uranium is 215 tons per year counting the operation of Atucha II.

Consequently, if the fuel elements were to be produced again in the country, the duration of the 7000 TN certified uranium reserves would be thirty-three years.

When analyzing the electricity system, the participation of nuclear energy in installed power (4.8%, Table 2.26) and in generation (5.5%, Table 2.30) was mentioned.

Table 2.45 shows the origin and destination of nuclear fuels in Argentina.

Table 2.45 Nuclear fuels (10^3 TEP)

Year	Import	Consumption power plants of the public service
2014	1280	1280
2015	2204	2204
2016	2224	2224
2017	1745	1745

Source Ministry of Energy and Mines of Argentina Energy Balances

2.10.2 Some Problems

(i) **The paralysis of the nuclear plan**

An article by Andrés J. Kleine is transcribed, the concepts of which are shared by the author.

"Throughout almost 70 years, Argentine society has managed to generate a scientific-technological-industrial conglomerate that has placed our country in an important and respected position at the international level in the field of peaceful applications of nuclear and perinuclear technology. The main driving force has been the State through its investments and its great purchasing power, managing to organize first-rate Science and Technology institutions (CYT) such as the National Atomic Energy Commission (CNEA) but also a conglomerate of public companies, such as Investigación Aplicada Sociedad del Estado (INVAP), owned by the State of Río Negro; Empresa Neuquina de Servicios de Ingeniería Sociedad del Estado (ENSI), partially owned by the State of Neuquén and the CNEA, with its Heavy Water Industrial Plant (PIAP); DIOXITEK S.A. and Nucleoeléctrica Argentina (NASA). In addition, it has been able to convene national private companies to form joint ventures such as Combustibles Nucleares Argentinos (CONUAR) and many other large, medium and small companies that have bet on the development of this technology as qualified suppliers.

It is worth mentioning that Argentina opted, in the first half of the 1970s, for the line of natural uranium and heavy water reactors in order to become independent from the expensive and proliferating technology of uranium enrichment, which we still do not possess at the industrial level. Our three power plants work safely with this technology, which today we master after many years of work and investment and which also has a great potential for innovation. The government has just dismissed 250 workers from NASA's permanent plant as a first consequence of the cancellation of the project of the fourth heavy water plant, with very convenient financing assured, of enormous interest for our country, and talks about a turnkey purchase of a Chinese reactor in 2022. This is the worst possible combination. The fourth plant was to be financed 85% by China and our country's share was to be dominant (70%) with a perspective of continuity for many small, medium and large companies that had bet and invested in technology and installations. We want to avoid the cancellation of the project of the fourth plant because of the devastating consequences it will have and is already having. The destruction of NASA's management unit, where there are highly qualified personnel who participated in the very worthy construction and start-up of ATUCHA II and who are participating in the extension of the life of the Embalse plant, is not only an unjustifiable squandering of very valuable human capital, but is also a risk for the operation of existing nuclear power plants that require local knowledge and design capacity. This human capital will be taken advantage of in other latitudes where many nuclear power plants are being built, forming a new chapter in the brain drain of our country. We should also remember that the cancellation contravenes National Law 26,566, which explicitly declares the construction of the fourth plant to be in the national interest.

In addition to this desolate situation, DIOXITEK, the company in charge of manufacturing the base compound for nuclear fuels, has been dismissed.

These decisions will destroy thousands of highly qualified jobs and seriously jeopardize the continuity of the sector, squandering the capital accumulated during almost 70 years of continuous and fruitful work that has positioned our country among the most advanced in nuclear technology.

Andrés J. Kleiner, Senior Researcher at CNEA-Conicet, Full Professor UNSAM. (Página 12 newspaper, 20 August 2018)."

(ii) **Paralysis of the heavy water manufacturing plant (HWMP)**

"One of the biggest and most successful investments to sustain this line was the construction of the PIAP, a real technological jewel that we need to provide very high-quality heavy water to our present and future plants and to the research reactors that INVAP manufactures and exports. But it would seem that the funds to pay the salaries of the PIAP workers have run out (taken from A.J. Kleiner, cited above)."

"PIAP is controlled by the Neuquén Engineering Services Company (ENSI), a partnership between the province of Neuquén (51%) and the CNEA (49%). This is where the heavy water that moderates and cools the nuclear reaction in atomic power plants using natural uranium is produced. The company halted production in May last year due to a technical shutdown, and had initially promised to restart in August, but that never happened.

PIAP produced last year a stock of 23 tons of heavy water that could be sold to the state-owned NASA, the firm in charge of operating the three nuclear power plants, and thus get out of this delicate situation at least for a few months. First it was argued that NASA did not have enough resources to pay for the production because they also cut its funds and had their tariffs frozen. At the end of February, the government raised, through Resolution 73/18, the remuneration that the company receives for the sale of energy, but NASA did not buy the heavy water anyway.

One of the alternatives to keep the PIAP working is to change the target and instead of heavy water it would produce fertilizers.

2.11 Renewable Sources

2.11.1 Some General Aspects

Here, we will analyze the biomass energies (essentially biodiesel and bioethanol with some mention to others), wind and solar energy.

Something has already been mentioned when analyzing electricity, and it is important because a large part of the consumption of these sources is destined for the generation of electricity.

Large hydroelectric plants (more than 30 or 50 MW) that were also included with electric power will not be discussed here.

Table 2.46 Renewable energy potentials

Energy	Unit	Value
Solar (*)	KWH/m^2 day	5
Wind (**)	MW	5000 a 20000
Biomass (***)	10^6 TN	148
Small hydroelectric inventory (****)	MW	430

Sources
(*) Grossi Gallegos, Hugo "Red SOLARIMÉTRICA del Servicio Meteorológico Nacional
(**) MR Consultores "2nd National Communication of the Government of the Republic: Evaluation of the renewable energy market in the Argentine Republic—October 2005. The Gross Potential would reach 20000 MW
(***) According to FAO WISDOM model
(****) PROINSA Proyectos de Ingeniería S.A. "Study to improve the knowledge and promotion of the Hydroelectric offer in Small Developments", 2006

Special analysis will be made of the renewable and distributed energy plans.

The contribution of renewable energies in the installed electrical power and in electricity generation is shown in Table 5.1.12.

According to the law, this contribution should reach 8% of total electricity generation in 2017, but in 2014 it was barely 1.5% and in 2017 it was 2%.

(i) **Potential**

The potentials are shown in Table 2.46.

(ii) **The energy balances of these sources**
(iii) **Contributions to power and electricity generation**

The contribution of renewable energies in the installed electrical power and in electricity generation is shown in Table 2.21.

2.11.2 Some Data by Renewable Energy Source

(i) **Solar Energy**

The Solar Radiation Maps of the Secretariat of Energy allow us to deduce that 11 of the 24 Argentine provinces present annual average values above 5 KWH/m^2 -day which makes them suitable for the installation of photovoltaic panels.

Almost all provinces are suitable for solar thermal use.

The total installed power in 2017 was 8Mw Peak, and the energy generated was 16 GWH.

(ii) **Wind Energy**

The total installed power in 2017 was 227 MW, and the energy generated was 616 GWH.

(iii) **Biomass Energy**

In 2017, the electricity generated was as follows:

Biomass: 243 GWH
Biogas: 64 GWH

Table 2.47 includes the balance of firewood, cane bagasse and biomass waste.

Table 2.48 shows the balance for bioethanol and biodiesel.

In the year 2017, 50% of the anhydrous ethanol was obtained from corn and the remaining 50% from sugarcane.

Table 2.47 Energy balance of firewood, bagasse and biomass waste (10^3 TEP)

	Production	Electricity generation auto production	Coal companies	Residential consumption	Commercial consumption	Industrial consumption
Año 2014						
Firewood	869	179	475	86	43	86
Bagasse	866	106				760
Waste	405	405				
Total	2140	690	475	86	43	846
Año 2015						
Firewood	913	208	495	84	42	94
Bagasse	804	146				658
Waste	365	365				
Total	2082	719	495	84	42	752
Año 2016						
Firewood	834	181	445	84	42	84
Bagasse	830	104				726
Waste	292	292				
Total	1956	577	445	84	42	810
Año 2017						
Firewood	833	199	421	86	43	86
Bagasse	962	114				848
Waste	386	386				
Total	2181	699	421	86	43	934

Source Ministry of Energy and Mines of Argentina Energy Balances

Table 2.48 Bioethanol energy balance. Biodiesel (10^3 TEP)

	Production	Export	Power plants public service	Transport
2014				
Bioethanol	340			340
Biodiesel	2290	1426		864
2015				
Bioethanol	413			413
Biodiesel	1604	702	37(*)	865
2016				
Bioethanol	451			451
Biodiesel	2367	1447		920
2017				
Bioethanol	560			560
Biodiesel	2556	1469		1087

Source Ministry of Energy and Mines of Argentina Energy Balances
(*) CAMMESA does not put Biodiesel consumption in Electric Generation in the year 2015

By 2017, mostly biodiesel was obtained from soybean oil and a small amount from waste edible oils.

Ethanol and biodiesel were mixed up to 10% with naphtha and up to 12% with diesel, respectively. There were intentions to increase these percentages.

(iv) **Small Hydroelectric Developments**

The electricity generated in 2017 was 1696 GWH.

2.11.3 The Plans

The plans formulated to develop the renewable sources are the RENOVAR, PROBIOMASA and MATER. From many years ago comes the PERMER.

As it has to do, essentially, with the renewable energies also will be included, in this paragraph, the PPA contracts; the PPP contracts and the distributed energy.

2.11.3.1 Renewal

(i) **The electricity framework and first actions**

"In accordance with the general regulation of the Argentine electricity sector (National Law 24,065 of 1992), the dispatch and operation of the national grid, together with the administration of economic transactions in the wholesale market, are the responsibility of Compañía Administradora del Mercado Eléctrico Mayorista

S.A. (CAMMESA). The Argentine government, generators, distributors, transmission agents and large users are represented on the Board of Directors of CAMMESA. Each of them holds 20% of the company's assets. At the end of 2015, the Argentine Government approved the Renewable Energy National Law 27,191, which established the basic principles for a new legal framework for promotion.

This National Law was regulated by Decrees N° 531/16 and N° 882/16.

Final consumers will be able to comply with the prescribed objectives through the purchase of electricity from distributors and/or directly from CAMMESA. Wholesale market users with an average annual demand greater than 300 kW may meet the objectives through the signing of supply contracts in the private market (either directly with independent power producers or with energy marketers) or through self-generation projects.

As a first step to comply with the Renewable Energy National Law 27,191, in May 2016, the Argentine Government launched the RENOVAR program, an open call for proposals that contemplates a series of tax benefits and financing mechanisms, as well as regulatory and contractual improvements.

Between 2016 and mid-2018, Rounds 1, 1, 2 and 2 were carried out under the RENOVAR Plan.

All the projects committed to achieve commercial qualification within the next 18 to 36 months. If this is not done, the contracts will expire.

(ii) **The legal framework**

Regulatory Decree No. 531/16: Regulates National Law 27,191 (March 2016).

Regulatory Decree No. 882/16: Introduces certain amendments to National Law 27,191 and sets quotas for tax benefits (July 2016).

MINEM Resolution No. 71/16: Preliminary version of the Terms and Conditions for the National and International Open Call for Proposals of the RENOVAR Program—Round 1—Public Consultation (May 2016).

MINEM Resolution N° 72/16: Procedures to grant fiscal benefits under National Law 27,191 (May 2016).

Joint Res. 123/16: MERCOSUR Import Tariff Program (MINEM and Ministry of Production, July 2016).

MINEM Resolution N° 136/16: Definitive version of the terms and conditions for the National and International Open Call for Proposals of the Renovar Program - Round 1 (July 2016).

MINEM Res. N° 147/16: Approval of the FODER Trust Contract (August 2016).

MINEM Resolution No. 202/16: Regulatory framework for projects included in previous regimes.

(September 2016).

(iii) **National Law 27191; Renewable Energy Law**

The established objectives were to reach 8% of the national electricity system supply from these sources by 2017 (2% had been reached by September 2018).

The Contracts

The contractual framework of the RENOVAR Program is based on two contracts that are articulated to provide all the elements commonly found in a renewable energy supply contract. Both contracts are governed by Argentine law and provide for the possibility of international arbitration. The project development companies awarded enter into a 20-year renewable energy supply contract with CAMMESA, which acts as a buyer on behalf of distributors and users in the wholesale market. Under the supply contract, the project development companies assume the obligation to build and to reach the commercial qualification date within the term established by each bidder in the offer. 100% of the electricity generated by the plant is paid for at the price awarded, which is set in US $ and adjusted annually. Project developers are required to deliver a minimum amount of electricity annually, and any shortfall is subject to a period of compensation and/or penalty, as appropriate. The traditional provisions necessary for financing projects without resources have been incorporated into the model contract from the beginning.

Along with the supply contract, the project development companies will enter into an Agreement of Adherence to the FODER Trust, by virtue of which they will have the character of beneficiaries of the FODER Trust. The FODER is a structured public trust with two main trust accounts (financing and guarantee) and a series of sub-accounts with specific purposes. In the framework of the rounds carried out, the main objective of the FODER is to provide guarantees of payment for energy (liquidity) and payment for termination (solvency).

The Prices

The prices of the RENOVAR Program rounds were offered in nominal terms. The projects were awarded on the basis of the price offered, adjusted for network losses and the deadline for the commercial qualification date. Additionally, the national component declared in the selection process was considered. As for the supply contract, the nominal price of the awarded project is subject to adjustment for two concurrent factors, namely (i) the Annual Price Adjustment Factor and (ii) the Price Incentive Factor. In both cases, these are coefficients that increase the annual price offered by the bidder, since it is not permitted by current legislation to adjust prices with indexes in foreign currency 1.7% increase.

In all cases, the Price Incentive Factor constitutes, for the government, a positive net benefit that increases the economic return of the project and allows a faster recovery of the investments.

The technical arguments reaffirming the advantage of applying the Price Incentive Factor are as follows:

It favors the immediate installation of awarded electricity generation plants.

Improvement of the current value of the projects.

Increases the net liquidity of the project in the first ten years of activity which, in a competitive scenario, could result in a lower average price offered.

Strengthens the financial profile of the projects by developing their profitability and financial conditions.

As can be seen, these are all measures in favor of private bidders, most of which are foreign companies.

Some points of the National Law 27,191 enacted in October 2015.

In other words, the basis of the provisions related to the promotion of renewable energy sources comes from the previous government.

The law consists of 22 articles divided into nine chapters.

Essentially, it modifies the provisions of National Law 26,190 of 2006, which had not generated substantial results since the decisions of the previous government were dominated by actions aimed at traditional energies, both hydrocarbons and electricity.

The objective of the law was to reach 8% of electricity generation with renewable sources by 2017 and 20% by 2025.

It defines what is understood by "Renewable Sources: These are the renewable sources of non-fossil energy that are suitable to be used in a sustainable manner in the short, medium and long term: wind, solar thermal, solar photovoltaic, geothermal, tidal, wave, marine current, hydraulic, biomass, landfill gas, sewage treatment plant gas, biogas and biofuels."

The power limit established by this law for hydroelectric power plant projects shall be up to fifty megawatts (50 MW).

To facilitate, the development of these sources establishes a series of tax benefits (Early return of VAT, accelerated amortization for income tax; tax certificate if the integration of the electromechanical equipment of national origin is 60% and if there is no national production of the equipment a bonus equivalent to 20% of the value of national equipment, etc.) and may not be taxed by any type of national, provincial or municipal tax, except royalty for use of fiscal land.

A trust fund called FODER is created to finance the investments, essentially from the National Treasury and equivalent to no less than 50% of the savings that these sources produce from fossil sources.

By December 31, 2017 and December 31, 2025, consumers of all types of electricity should be supplied with no less than 8% and 20%, respectively, of their electricity from renewable sources.

Large consumers (with more than 300 KW) must contract similar percentages to those mentioned for the remaining users of electricity or self generate with renewable sources and the prices of the contracts cannot be higher than US $113/MWH in equivalent local currency.

Electromechanical capital goods may be imported free of taxes.

For dispatch, intermittent renewable sources will be considered as if they were run-of-river hydropower plants (taken directly from the Secretary of Energy and Mining)."

(iv) **The Results**

Up to October 2018, three rounds had been held: I, I, 5 and 2.

A total of 147 projects have been presented with the powers and types of energy shown in Table 2.49.

Table 2.49 RENEWAL Projects: 1, 1.5 and 2

Source	Power (MW)
Wind	2466
Solar	1732
Biomasses	158
Biogas	65
Small HIDRO	32
Waste biomass	13
Total	4466

Source Ministry of Energy and Mining of Argentina

Table 2.50 Contract prices by source

Source	Price (USD/MWH)
Wind power	18.9–66
Solar	41–60
Biomass	110–143
Biogas	118–135
Small HIDRO	89–105
Waste biomass	128–130
Total (average)	54.72

Source Ministry of Energy and Mining of Argentina

The total power of the projects is 4466 MW and the average price is US $54.72/MWH.

Table 2.50 includes the minimum and maximum prices of MWH by type of source.

CAMMESA's MONOMETIC price was US $61.73/MWH in August 2018.

It is interesting to note that the cost of MWH generated by small diesel groups operating four or five hours a day is between US $300 and 400 much higher than, for example, small solar energy facilities.

(v) The Mini Round 3

On November 14, 2018, Resolution No. 100 of the Secretariat of Energy and Mines called for an open national and international tender for contracting in the wholesale electricity market (MEM) for electricity from renewable sources.

The resolution granted benefits to the bidders, different according to the type of technology, which were equivalent to 45% of investment per KW set as a reference.

It also requested the bidders to disaggregate the components of the equipment to detect those of national origin.

No special privileges were indicated for domestic bidders.

The idea of this Third Round would seem to be to look for smaller-scale projects than in the previous rounds.

The reason would be that the available capacity in the medium voltage networks and the transport restrictions in the high and extra high voltage networks would be two of the main reasons for targeting smaller initiatives. This, they say, would contribute to greater network stability.

The reference values for investments in some technologies were also modified downward. Only the wind energy projects—US $1400 per KW—and solar photovoltaic projects—US $850 per KW—were kept at the same amount as in the last round.

Unlike the renewable energy supply contracts, the guarantee period would now be halved from 180 to 90 days as the programs, according to the Secretariat, would be consolidated.

2.11.3.2 The PROBIOMASS

The project for the promotion of energy derived from biomass, PROBIOMASA, is an initiative of the Ministries of Agriculture and Planning through the Ministries of Agriculture, Livestock and Fisheries, and the Secretariat of Energy. And it has the technical assistance of the United Nations Food and Agriculture Organization (FAO).

Its main objective is to increase the production of thermal and electric energy derived from biomass at the local, provincial and national levels to ensure a growing supply of clean, reliable and competitive energy, and at the same time, open new agro-forestry opportunities, stimulate regional development and contribute to mitigating climate change.

The Resolution 25 of March 15, 2017, creates the program in the orbit of the SUBSECRETARIAT OF BIOINDUSTRY, of the SECRETARIAT OF VALUE ADDED of the MINISTRY OF AGRICULTURE. With the aim of developing and promoting a series of activities related to the project, such as to promote and consolidate an institutional network of BIOENERGIES, it generates a geographic information system on availability, accessibility and consumption of biomass that expresses the BIOENERGETIC potential at a national level and that allows a provincial zoning, identification and selection of the areas that, due to their characteristics, can satisfy demands for the BIOENERGETIC development of the country.

To advise and assist in the implementation of bioenergy projects.

At the end of 2017, there were about 63 projects detected, most of them at the idea level, and to a lesser extent in operation, dedicated to the energy use of Agricultural, Livestock and Urban Waste, for the production of Biogas, but essentially for the cogeneration of electricity.

2.11.3.3 The MATER

This is the name given to the Term Market for Renewable Energies.

Resolution 281 of 2017 establishes a regulatory framework so that large users of the wholesale electricity market (MEM) and large demands that are customers of public distribution service providers or distribution agents, with power demands equal to or greater than 300 kW, can adapt to National Law 27,191.

"Law 27,191 guaranteed the right of Large Users of Electricity to choose their renewable energy supplier and freely negotiate supply conditions. The former Ministry of Energy published Resolution 281/2017 creating MATER.

The new resolution regulates the operation of the Term Market for Renewable Energy (MATER).

In this term, Market, Large Users (shops and industries with relevant electricity consumption) may negotiate the purchase of renewable energy from generators and traders. They can, if they wish, secure up to 100% of their electricity consumption from renewable sources. Or they can purchase only the minimum steps required by law, which are 8% in 2018, to reach 20% in 2025.

If the consumer decides not to take any initiative, the State itself will supply it. Although it will not meet the legal minimums in the first years (it is estimated that the State will only reach 8% by 2020), the consumer has the possibility of delegating this management. The State, with the Renewal Program, is signing 20-year contracts, with average prices close to US $70 MWH (above the US $61.73 that it pays to CAMMESA as of August 2018). To this value, charges and costs will have to be added, some of which are pending definition.

If, on the other hand, the consumer decides to exercise this right on his own, he can sign Supply Contracts with a Generator or Marketer; or opt for Autogeneration. The best known case consists of installing solar panels in their own factory or premises, and producing part of the energy they consume on site. This has interesting benefits because of the savings in transmission costs, distribution, taxes, charges, etc.

In Argentina there are already a large number of Large Users (the total list exceeds 8000) who are making good progress in their renewable energy purchase processes. With the new regulations, this process will undoubtedly accelerate. There are advanced negotiation processes where a Generator or Marketer installs the solar plant on the Client's premises, and the Client only has to pay for the energy that he actually receives (taken from Juan Boch, President of SAESA, August 3, 2017)."

The following are some additional comments to Resolution 281 of 2017 made by Mr. José Carlos Cueva, energy lawyer for the Revista Energía Estratégica in dialogue with Guido Gubinelli.

The provision establishes that as of 2019 these large users must begin to consume 8% with renewable energies, whether they generate their own, purchase them from the company that administers the wholesale electricity market (CAMMESA) or obtain them under contract with a private party, as enabled by the resolution.

Also, if the law is not complied with, Large Users will be fined according to the percentage not complied with.

It should be noted that the law establishes that by 2020, the percentage of consumption with renewable energies must be 12%, gradually incorporating 2% every two years until, as of 2026, Large Users end up incorporating 20% of clean energy, the final mandate of the law.

Table 2.51 Marketing fees

Marketing fee	2017–2018	2019–2020	2021–2022	2023–2024	2025–2030
Obligation under Law 27,191 (%)	8	12	16	18	20
Maximum: Maximum charge associated with the obligation in USD/MWH	0	6	10	14	18

Source Resolution No. 281 of 2017

Table 2.52 Administration fees

Administration fee	2017–2018	2019–2020	2021–2022	2023–2024	2025–2030
Obligation under Law 27,191 (%)	8	12	16	18	20
Charge associated with the obligation (in USD/MWH)	0	0.05	0.05	0.05	0

Source Resolution No. 281 of 2017

In Cueva's opinion, the national government established some parameters to make the purchase and sale of renewable energy among private parties more auspicious than with CAMMESA, which would generate an expansion of the market.

Large Users that define not to leave the Joint Purchases to CAMMESA will have to pay Marketing and Administration charges, which will have to be paid as from the economic transactions of January 2019, for the percentage of demand that corresponds.

In the case of marketing, the charge will be US $6 per MWH as from that year, which will increase as operators comply with the expansion of the supply of renewable energy, until it reaches US $18 in 2025-30, as it is shown in Table 2.51.

As for the administration charge, it will be five cents per MWH from 2019 to 2024, as shown in Table 2.52.

When asked about the estimated value that will be set for Large Users who decide to buy energy from CAMMESA, Mr. Cueva explained that "it is still not clear how the weighted cost formula will be and how it will be transferred to the entire universe of users. These values will also have to be added to the Incentive and Adjustment Factor, which increase the final price per MWH awarded in the RENEWAL Program Rounds by about 20%."

On the other hand, the specialist considers that equality in the dispatch priority is important.

The priority will be governed by the earliest commercial qualification period of the projects and, in the event of a tie, the undertaking with the highest capacity factor.

"It was a positive measure to equalize the projects, because otherwise it seemed like a competition between the undertakings of the RENOVAR Program and those that would fall under this Resolution," Mr. Cueva pointed out.

One aspect that the energy lawyer emphasizes is that Large Users who initially decide to buy energy from CAMMESA but later decide to do so with a private company may opt out of the contract at two times during the year: in February or August, stating their intention 3 months in advance.

But Mr. Cueva warns that they will only be able to buy from CAMMESA again after 5 years. That is, "when they leave (the Large Users) will have to have at least one contract tied to 5 years, without risk of termination."

Finally, in the case of the user who opts for self-generation, the resolution indicates that the surplus produced by the clean energy source, exceeding in this first stretch 8% of renewable energy, may be sold in the SPOT market (taken from Guido Gubinelli, in Energía Estratégica).

2.11.3.4 PPA (Power Purchase Agreement)

Power Purchase Agreements (PPAs) are private energy marketing agreements that allow a utility to sell the electricity produced by its facility to a customer at a fixed price, usually over the long term. In this way, they make it possible to launch generation projects with stable financing prospects and outside of subsidy policies.

Private PPPs could be a way of meeting the limit of renewable energy consumption, as the terms and conditions of PPPs can be freely agreed between the contracting parties. The only exception is that the price cannot exceed US $113/MWH.

The first private PPA contract was made at the end of 2017, between GENNEIA and Loma Negra. The Rawson III wind farm, with a capacity of 24 MW, will provide energy to the cement company. Other users who also signed PPA contracts were Toyota Argentina and ALUAR.

2.11.3.5 The PERMER

The Renewable Energy in Rural Markets Project (PERMER) was set up in 2000 with the aim of facilitating access to energy in dispersed rural populations, far from the distribution networks.

The program subsidizes the provision and installation of:

(A) The provision of electrical energy and communication

Individual photovoltaic and/or wind systems

Mini-grids (hydraulic—solar/wind—hybrid).

(B) Solar systems for thermal purposes (parabolic cookers, solar ovens, solar water heaters)
(C) Photovoltaic systems for pumping drinking water.
(D) Photovoltaic systems with higher power for productive projects.

The implementation of each project is carried out in a coordinated manner between the Project Coordination Unit (UCP) of the Ministry of Energy and the provinces, guaranteeing its federal implementation.

The PERMER allows access to energy in a clean and renewable way, thus constituting a social inclusion policy that generates greater equality of opportunities and an improvement in the quality of life of the population (Secretariat of Energy and Mining).

2.11.3.6 Comments

As it was said, the idea of promoting the development of renewable sources came from the previous government, but in fact it did very little about it. Perhaps the achievements of the PERMER should be highlighted in some provinces.

The current government, on the contrary, has given a great impulse to the subject and the results in terms of the power under construction and the contracts signed show it.

As the program did not contemplate the forward and backward effects of a massive income, it has generated the opening, almost unrestricted, of the market to foreign companies essentially from Europe and China that have developed a capacity to produce very large equipment and to deliver it in short periods of time. This has dislocated national metalworking companies that cannot compete in terms of quantity and deadlines. This goes beyond some weak clause, since the law allows practically the import of all the metal mechanical part of the equipment and exempts them from any import duty.

"According to information from CAMMESA in the case of the wind energy projects awarded, the weighted national integration component was 10.7% in round 1, 9.5% in round 1.5 and 37.2% in round 2.

In the García del Río project (Buenos Aires), national integration was proposed at just 5.9%, in Vientos del Secano (Buenos Aires) at 7.9%, in Los Meandros (Neuquén) at 9.2% and in Cerro Alto (Río Negro) at 9.4%. In the Villalonga wind farm in GENNEIA, the national integration component specified is 16.7%. Pampa Energía's Corti wind farm Park (Buenos Aires) declared a national integration of 6.18%, while in the six other projects awarded in Round 1 the national component does not exceed 15%: PAE's Garayalde project (21%), Central Puerto's La Castellana (13.4%), Greenenergy's KOSTEN (8%), SENVION's Vientos Los Hércules (14.3%), GENNEIA's Chubut Norte (12.8%) and Parque Arauco II de Arauco SAPEM (7.1%).

National integration is low because most of these initiatives obtained the declaration of "critical project" under Resolution 4 of December 2017, which allowed them to continue importing wind turbines with a capacity of more than 700 KW at zero tariff. In the case of Arauco SAPEM, the 7.1% of national integration that it put in the specifications ended up being higher because it bought national towers.

In Renovar 1.5 the low percentage of local integration was repeated. SINOHYDRO's Parque Pampa (Buenos Aires) declared 20.5% national integration; ISOLUX's Miramar (Buenos Aires) 10.5%; Central de la Costa's Vientos de

NECOCHEA (Buenos Aires) 7.5%; Grupo FRALI's La Banderita (La Pampa) 22.2%; Pomona I (Río Negro) by GENNEIA, 9.4%; Parque Eólico del Bicentenario (Santa Cruz) by Petroquímica Comodoro Rivadavia, 7%, while the remaining four wind farms set national integration at between 3.5 and 11.9%.

In round 2 of the Renovar In this case the percentage of weighted national integration rose to 37.2% because Resolution 1 of September 29, 2017, published jointly by the Ministry of Production and the Ministry of Energy, proposed for wind energy a path of increasing national integration that sets an initial goal of 35% until June 30, 2020, raises it to 45% until December 31, 2021 and finally brings it to 50%.

The problem with this resolution is that in order to reach these minimums, a weighting of parts and processes of the wind turbine was set and if a 35% national integration of the wind turbine is reached, the norm allows to consider it one hundred percent as a national good. For example, the wind turbine tower counts as 23% national integration, the nacelle assembly as 10% and the hub assembly as 3%. Therefore, if the tower is manufactured domestically and some basic assembly tasks are performed, it reaches 36% and the entire wind turbine can be computed as domestic. In this way, some projects have climbed to 90% local integration because the wind turbine alone represents about 70% of the electromechanical component of the wind farms. In addition, this allows them access to a bonus that guarantees tax benefits as if the product were national and not a mere assembly of mostly imported parts. Meanwhile, the national industry was practically marginalized from the business and continues to be underdeveloped" (taken from Fernando Krakowiak "Negocio a la medida de los importadores," Página 12 newspaper, Buenos Aires, 13 November 2018").

Consequently, Argentina is not repeating the successful experience of nuclear development that positioned it at a global level. In this case, the effort would have been much less, since we are not dealing with cutting-edge technologies.

The alternative would have been to equip at a more leisurely pace that allowed for the massive participation of domestic industry, which could even have generated the possibility of exporting to neighboring countries.

It seems like another lost opportunity.

As far as prices are concerned, they started out very high, but by October 2018 the average equipment is below the MONEY price paid by CAMMESA (54.72 for renewable sources compared to 61.73 for CAMMESA contracts). This is due to the predominance of wind and solar, which contribute 55% and 38%, respectively, of the power in operation and whose prices are below the aforementioned average.

As for biodiesel and bioethanol, there is still a "conflict" between the oil companies and the sugar, corn and oil companies over the prices established by the government for these biofuels and the percentage of participation of both in the respective blends with gasoline and diesel. In this last aspect, the producers of biofuels aspire to increase the participation in these mixtures, even in some cases until taking it to 100% in the case of the biodiesel.

Another not insignificant aspect is the magnitude of biodiesel exports, which by 2017 were reaching almost 58% of production. The fact is that Argentina has one of the most important oil complexes in the world and these exports bring added value as do not the beans and residues of this industry which represent one of the largest items of the country's export income.

PROBIOMASA is a good initiative where the participation and contribution of institutions such as INTA, which has many years of experience in the field, would be essential.

The MATER introduces a new complexity in the electrical sector and makes us miss the simplicity of the system when it was managed by State companies and they actively participated in the unified dispatch of charges and in the construction of energy works with the financing of resources from the taxes on energy consumption (essentially on liquid fuels) with the contribution, always a minority, of the International Development Banks.

As of October 2018, there are 16 facilities under the RENOVAR and MATER plans in operation, with a capacity of 348 MWE.

The first development began operating in mid-2017, and the last one was connected in October 2018 by YPF when the Manantiales BEHR plant was built.

Among the 16 developments in operation are from small hydraulic exploitations, through plants based on biogas and biomass, solar and wind farms. In addition to these, another 86 are under construction out of the 157 already awarded, while the construction of hydroelectric and nuclear plants of great size is delayed, affecting in this last case the Argentine nuclear development.

2.11.3.7 Public–Private Partnership (PPP) Contracts

(i) **What they are and where they originated**

The PPPs will be briefly analyzed, since the current government does not favor state investment in infrastructure works such as energy and more specifically those in the electricity sector.

They are essentially "an agreement between the public sector and the private sector in which part of the services or tasks that are the responsibility of the public sector is provided by the private sector under a clear agreement of shared objectives" (World Bank). In addition, the agreement regulates the responsibilities of both parties and the remuneration that the private party will receive for taking on that responsibility.

PPPs are contracts between the public sector and the private sector to carry out works or supply goods. In these contracts, the (private) contractor assumes responsibility for obtaining a substantial part of the financing for the work.

They are projects aimed at infrastructure, housing, public services, productive investments, applied research and technological innovation.

But there is no room for those destined only for the provision of labor, supply and provision of goods and those financed essentially with National Treasury Funds.

They are obtained by public tender or competition.

Payments can be made by directly affecting resources or transfers or through duly regulated trusts.

The Trust Agent is the Bank for Investment and Foreign Trade (BICE) and the trustees are the ministers or secretaries under whose jurisdiction the work is carried out.

The duration is for thirty years.

This figure originated in the United Kingdom in the seventies and was developed mainly in the transport sector with mixed results, to be benevolent.

(ii) **The legal part**

The National Law that regulates the operation is the 27,328 of November 2016 and the respective Regulatory Decrees (118 of February 2017 and 936 of June 2017).

These regulatory decrees are very interesting for the benefit of private contractors, since they eliminate the obligation to pay VAT for the funds they participate in, exempting them from income tax and the tax on eventual goods.

If the contractor is foreign and there are disputes, courts of countries with headquarters in member states of the Convention on the Recognition and Enforcement of Foreign Arbitral Awards act.

(iii) **Control**

This control is carried out by a Bicameral Commission for the monitoring of contracts, presided over by a member of the opposition and vice-presided over by a member of the government. It is composed of seven representatives and seven senators.

(iv) **Advantages, disadvantages and criticism**

"A study by the CECE Foundation", composed mostly of professors from the University of Buenos Aires (UBA), adds that this system, developed in the early 1970s in the United Kingdom, generally involves the direct provision of a service to the public sector by a private company, through a contract that includes the design, construction and then operation and maintenance of an infrastructure.

According to CECE's report, by economist Alejandro Einstoss, "the greatest advantage for the public sector is that the works are financed by the private sector, thus avoiding budgetary restrictions, without generating public debt and taking advantage of the private sector's management capacity." On the side of the private sector, the advantages "come from being able to participate and structure projects that would be out of scale without the participation of the public sector," where the state assumes certain risks or reinsured a certain amount of income.

However, PPPs can present different disadvantages in relation to traditional public works. Among them, Einstoss highlighted the possible failures in the preparation of contracts that result in cost overruns for users, errors in the choice of projects and in the distribution of risks, high financial costs and recurrent renegotiations.

The Argentine Government's plan includes the implementation of 60 PPP projects between 2018 and 2022 for US $26 billion. The 2018 budget had already specified 52 projects that include the construction of highways, prisons, hospitals, housing and improvements to the energy system. The Undersecretariat for Public–Private Participation was created for their execution, under the Ministry of Finance. There are legislative precedents for PPPs in 2000 and 2005, although none "was widely used" (taken from the report of the CECE Foundation of the University of Buenos Aires).

"The origin of the PPP is Margaret THATCHER's United Kingdom, and so this type of public-private partnership scheme was extended to other countries" (Spain, Chile, Brazil, Panama, Colombia, Peru and France).

However, the British National Audit Office (NAO) denounced the poor results of the system because the over costs were up to 40% in the works.

Furthermore, in Spain, the bankruptcy of the construction companies that participated in this modality pushed the Spanish administration to nationalize four highways built through PPPs, throwing a loss of 2.5 billion dollars to the Spanish State (taken from Conclusion, interview with Mr Pino Solanas, April 29, 2018).

(v) **The experiences in other countries**

Different countries in the region and the world have been applying this system for years. A report by the European Union's Committee of Auditors argues that PPPs allowed public authorities to acquire large-scale infrastructure, but increased the risk of insufficient competition and therefore put contracting authorities in a weaker bargaining position. In this regard, when analyzing a dozen PPP cases in Spain, France, Greece and Ireland, he criticizes that this mechanism had "widespread shortcomings and limited benefits."

Something similar is stated in a report by the National Audit Office of the United Kingdom (NAO). PPPs were introduced there more than 25 years ago, and there are currently more than 700 such projects under construction or operation. The study notes that the government's use of such models "has declined significantly" because of "concerns about cost efficiency and value for money."

In Latin America, this type of project has increased in the last decade. The CECE Foundation report shows, based on World Bank data, that PPP investments in the region increased from US $8 billion in 2005 to US $39 billion in 2015. But this study also warns about the changes in the region: According to OECD data, more than 70% of PPP contracts signed from the early 1990s to 2015 were renegotiated (taken from Martin Slipczuk, in Cheque, April 22, 2018).

(vi) **The English experience**

The British PFI, later modified by the PRIVATE FINANCE TWO (PF2), served as a means for the British financial sector to participate in the financing of public works, as a way of doing works of a mixed nature, where the private sector that was the creditor of the financing used the public works and services on which it operated to favor its business and practices concerned. The PF2 was an update of February 2012 extending the terms of the contract, shortening the approval period to 18 months, simplifying procedures, etc. The greatest British corruption of this century is to CARILLON's credit that he used the IFP and PF2 in his own best interests.

CARILLON and other construction and financial companies benefited because they charged the state more than 2 to 3.75% of the current rate for loans to build roads, schools and hospitals. These figures were very important for the Royal Treasury because of the difference it had in taking credit to co-invest and the payments for private loans. It produced fiscal deficits for the conservative government. A state-owned bank like the Royal Bank of Scotland denounced the PF as "fraud against the people" despite the fact that it had made large profits. Politicians, until now, have used the system without paying money in advance and thus constraining the level of public debt, even without placing them in the annual budget.

A conservative government—the current one—was forced to abolish the system that this political current had instituted: Financially, it caused damage to the State and promoted a system of "legalized" corruption. The bankruptcy of CARILLON on January 15, 2018, was the cause of the elimination of the PPP system; while maintaining the contracts in force, it eliminated the possibility of new agreements. It was Philip Hammond, conservative minister of economy (chancellor of the exchequer) who promoted the elimination of the system.

In England, for 25 years, the Conservatives and Labour maintained this system of corruption, (taken from an article by Félix Herrero communicated to the author of this work by mail on November 5, 2018, entitled "The Great Corruption also marks history").

2.11.3.8 Distributed Energy

Some General Concepts

First it is defined, it is indicated how it works and some advantages of the distributed energy are mentioned according to information source of Spain where it has been applied for a long time.

(i) **What is distributed energy?**

Distributed energy is the generation and management of electrical energy in a decentralized manner, as close as possible to its place of consumption through, mainly, renewable energy sources.

(ii) **How does it work?**

One or more sources of generation (solar, wind, biomass, biogas or others) and sometimes storage are installed and connected to the consumer's internal network, which can be either an individual or a company. Depending on the complexity of the system, the installation can be of three types:

Distributed generation: The user maintains his connection to the centralized grid, which he uses at times when there is no generation available due to the natural intermittency of renewable sources. The energy generated can be used for own consumption (self-consumption) or to be fed into the grid. This type of distributed generation project is the most common and is an increasingly frequent practice.

Distributed generation + Integration of storage systems: Electricity storage systems, such as batteries, are added to the previous level. This reduces the dependence on the network, and surplus energy can be managed by storing it for later use. This type of system is becoming increasingly common thanks largely to the dramatic reduction in battery costs in recent years. With an ambitious project of self-consumption combined with storage, the electrical network ends up being a mere backup system.

OFF-GRID installation: The user is not connected to the centralized network, as all the energy he consumes comes from renewable sources generated locally in a small network created in his facilities. This network requires, in addition to batteries for storage, a system capable of guaranteeing the stability and operability of the network. They are usually known as microgrids, because they have the same capacities as a large grid. These microgrids can supply a single consumer or even larger communities in remote areas, so they are currently one of the keys to electrification of rural areas in developing countries.

(iii) **What are the advantages?**

Greater energy efficiency: Losses from the transmission of electricity in highly advanced transmission and distribution networks such as those in Spain represent approximately 8% of the total energy generated. This figure is even higher in less sophisticated networks in countries with greater geographical dispersion. Distributed generation, by bringing the point of generation closer to the point of consumption, largely reduces these losses.

Reduces electricity supply costs: It allows the use of energy generated for self-consumption, proportionally reducing the consumption of electricity from the network and saving costs. In fact, distributed generation is considered in some countries as a measure of energy efficiency, since for the purposes of the net balance the consumption of the network decreases.

Promoting sustainable generation: Most distributed generation facilities use renewable sources. The development of these will allow a greater penetration of these technologies in the energy mix, reducing harmful emissions and contributing to a more sustainable and environmentally friendly system.

Reduced landscape and environmental impact: It reduces the need for large installations that affect the landscape and ecosystems, also taking advantage of the existing network, without obliging it to be reinforced. As they are smaller and less powerful generation facilities, they are better integrated into the environment.

Facilitates the economic independence of isolated regions: Up to 17% of the world's population does not have access to electricity. In developing regions with large land areas such as Africa or Asia, setting up centralized networks is a major investment. In this scenario, it is preferable to develop microgrids that respond to the growing demand in a more efficient and effective way in economic terms.

It is more flexible: It allows a more flexible operation of the electricity system, especially if a future with a multitude of microgrids that can be linked together, working in a coordinated way, as an evolution of the SMART GRIDS (source: www.norvento.com).

Energy Distributed in Argentina

(i) **National Law 27,424/2017**

The operation is regulated by National Law 27,424, of November 30, 2017, on the Promotion of Distributed Generation of Renewable Energy integrated into the public electricity grid.

"In effect, since 1998, the country has a legal framework that seeks to promote renewable energies, which is delimited by National Law 25,019/1998, which proposes a promotion regime for wind and solar energy; National Law 26,190/2007", which sought to complement the previous regulations by including other types of renewable energies (such as geothermal, tidal, hydraulic, biomass and biogas) and, finally, National Law 27,191/2015 on the promotion of the use of renewable sources of energy for the production of electrical energy, which proposes to reach, by the end of 2025, a participation of 20% of these sources in the generation of all electrical energy consumed locally.

Although investments in these energy sources have recently encountered some obstacles, among which the need to update and expand the existing electricity transmission system stands out. Investments in renewable energy sources are on the rise since 2011, thanks to the boost provided by the RENEW program.

In this context, the domestic promotion of Distributed Generation of energy (DG) from renewable sources is an additional strategy that goes in the same direction. As already mentioned, the general characteristics of DG are as follows: (i) the reduction of transmission losses in the networks—at least up to a certain threshold or limit that depends, among other factors, on the location and degree of penetration of DG; (ii) the injection into the network of the total or surplus of the energy produced; (iii) and the fact that they usually have lower powers and a scale of production that can be both commercial (where the total energy generated is injected) and individual (where only the surplus is injected). Also, DG, if it is supported by renewable sources, in addition to having a typically smaller generation, allows other benefits that have more to do with the environment and the possibility of reducing generation costs in the future.

In Argentina, DG is limited by National Law 27,424/2017 "to the generation of electricity by users of the public distribution service who are connected to the provider's network and who, in order to inject the surpluses produced, must meet the technical requirements established in the regulation" (Article 3, paragraph h).

Similarly, it is important to clarify that the appropriate scenario for the development of DG is that where there is what is known as grid parity, i.e., where the price of the energy sold by the distributor is comparable to the price of generating it on its own account and there is the possibility of recovering the investment made by the small producer in the short to medium term. For this reason, in the regulations, the authorities usually set the billing mechanism or tariff structure that best suits local electricity market conditions in order to promote DG (in Argentina, for example, the net billing balance was chosen). In fact, the two main obstacles that governments face when trying to promote DG are the lack of legislation (allowing small generators to inject electricity into the grid) and the lack of a market that ensures an adequate return for small-scale generation (and the subsequent recovery of investment) (taken from KGM "Potential Impact of the New Law on Distributed Generation in Argentina from Renewable Sources," June 2018).

(ii) **The regulation of the Act**

On November 2, 2018, Decree 986 regulated National Law 27,424.

The National Director of Renewable Energy of the Ministry of Energy and Mines, Mr. Maximiliano Morrone, had advanced some contents of the Regulation.

"There are 16 million users in Argentina, and 70% of energy consumption is concentrated in users connected to the public distribution system. Of that percentage, 30% is in the hands of industrialists and the remaining 40% is in the hands of residential users," said Morrone. "We are talking about a 70% chance that this generation is in the hands of those who consume this energy in Argentina," the official said and said that they are working with the Ministry of Modernization to make the procedures for grid connection "simple and transparent. The National Director of Renewable Energies said that an important novelty is that the measure will reach residential users and SMEs that wish to connect up to 500 kW. Although users above 300 kW were covered by National Law 27,191, on renewable energies, to generate their own energy through clean energy sources, the management procedure is more cumbersome, and by including them in National Law 27,424, they facilitate the process of self-generation. The official said that once the law is in place, industrial and commercial users who generate their own energy will be able to repay the investment within 5 to 6 years. However, he acknowledged that in the different latitudes of the country not only the renewable resource is variable but also the value of the tariffs is different, so the calculation could vary. Another point emphasized by the Director of Renewable Energy is the issue of incentives. He explained that through FODIS (fund for the promotion of renewable distributed generation), measures were studied to boost the activity. He said that, for residential users, soft credit lines will be allocated, understanding that this is their greatest entry barrier for the acquisition of this equipment. For industrial and commercial users, tax certificate measures will be applied as incentives. In that line, the official anticipated that was included in the regulatory regime the "exemption of VAT and profits for those users of up to 300 kW" that begin to be self-generating through renewable energy sources. It should be noted that at first, the FODIS, created through Article 16 of the Act, will have an allocation of 500 million pesos for the first year of operation, from the National Treasury. Then, the fund will be fed by a series of mechanisms, such as half of the energy savings that renewable self-generation itself represents for the National State (based on the article published in Energía Estratégica of 9 May 2018: "Morrone: The Regulatory Decree of the Distributed Generation Law is ready").

The Net METERING

"For those user-generators whose service contracted with the distributor discriminates the price of energy within its tariff scheme in hourly segments, the injection of electric energy will be recognized and paid for at the price of each hourly band as appropriate". This system is known as net metering. And it allows the user-owner to compensate his energy consumption from the supplier with generated kilowatt-hours ("KWH") not consumed at that time and, consequently, injected into the grid. The "surplus" KWH can be consumed in the same distribution circuit, at the nearest load center that presents demand at that time. Therefore, the Net Metering scheme compensates this surplus energy at the same value as the final consumption tariff—a tariff that the electricity company would have charged for this consumption in the circuit, if it had supplied it itself from a central generating plant. The compensation will be valued in pesos and should impact the invoice for the period in which the injection was made (Centro de Desarrollo y Asistencia Tecnológica, CEDYAT).

Specifically, when the regulations come into force, both residential and commercial users will be able to generate their own energy and sell the surplus to the distribution company. Although sales prices for self-generated energy by users have not yet been established, this will, of course, lead to a reduction in the value of the bill to be paid to the companies, which will have to provide consumers with meters that allow them to see how much is being consumed from the electricity grid and how much is being injected into it.

Another point to highlight is the savings that will be produced by discounting the cost of transporting and distributing the electricity that arises when it goes from the power stations to the household socket, and means between 8 and 15% of the bill.

In addition, this regulation will bring a reduction in the initial investment of solar energy systems, since by implementing distributed generation the solutions do not need batteries, reducing the price and eliminating all types of maintenance for approximately 25 years, which is the useful life of systems of this type.

How much can a user save?

The cost of energy consumption in a house or business is relative; it depends on the province where it is located, the meters of the property and the use made of electricity, but according to official estimates of the electricity distribution companies, the average consumption of a typical family per month is equivalent to 350 KWH.

A solar energy system that generates 160 KWH per month, which allows an average private user to save 50% of their bill, requires an initial investment of about US $1909 + VAT (1.1 KW ON-GRID system) and is amortized in 5 or 6 years. With the new law, according to this example, the consumer will also be able to sell the surplus of self-generated energy in periods when he does not use it, such as when he is on vacation away from home (taken from M. Miodowsky in the article "How the law allowing users to generate their own energy will impact bills" published in the Ámbito Financiero journal, on November 7, 2018).

(iii) **Some concerns**

Because of economies of scale, electricity generated by large hydro and nuclear power plants has to be cheaper than electricity produced by small (less than 300 KW) residential or commercial generators or "large" generators (more than 300 KW) that are generally industrial (except in very special cases of cogeneration) and it does not seem that economies of series prevail over economies of scale.

If this were to become widespread and not a small portion of the transactions, it would atomize the electricity market and could raise generation costs and make the country less competitive.

On the other hand, if one of the advantages of this alternative was to eliminate the needs of the distribution networks, the transactions from the generator to the system could not be made when there was energy left over, nor could it be bought from the system otherwise.

In short, it seems interesting if it represents a small part of the transactions, for example, up to 10%.

In any case, foreign sellers of renewable sources equipment are happy.

2.12 Rational and Efficient Use of Energy (UREE)

2.12.1 Background of the Previous Government

Decree 140 of December 21, 2017, declared the UREE to be of national interest and priority, and approved the general guidelines of the National Rational and Efficient Energy Use Programme, setting out its objectives as follows:

To promote efficient energy use, taking into account that most of the energy used comes from non-renewable natural resources.

To reduce the impact on the environment by optimizing energy conservation and cost reduction as an essential component of energy policy and environmental preservation in the Argentine Republic.

To implement the international agreements on energy and climate change adopted by Argentina.

To recognize and promote the efficient use of energy as the most effective measure in the short and medium term, in order to achieve a significant reduction in carbon dioxide (CO^2) and other greenhouse gas emissions.

To contribute to the establishment of conditions that favor the sustainable development of the nation, the growth of employment and the increase of productivity, through the implementation of energy efficiency policies within a framework of environmental requirements, protection of natural resources and commitments to mitigate the emissions of greenhouse gases responsible for the process of global climate change.

Promote a change of culture in the efficient use of energy, through cultural-educational strategies, generation of structural changes based on the modification of individual behaviors through programs and plans that must be conducted by highly specialized agencies, in addition to the adoption of high-efficiency technologies.

Promote the desirability of the public sector assuming an exemplary role before the rest of society, by implementing measures aimed at optimizing energy performance in its facilities.

To prioritize the expansion of the scope achieved in the industrial sector in terms of energy efficiency through a specific medium -and long-term subprogram, bearing in mind that this sector represents 30% of the country's total energy consumption.

The decree had short-, medium- and long-term objectives, and the actions were divided into the following sectors and topics: Industries, Commerce and Services, Education, Cogeneration, Efficiency Standards and Labeling, Public Lighting and Transport, Housing, Climate Change and Clean Development.

In other words, there was a close link between the UREE and the development of renewable sources and environmental problems.

It can also be seen that UREE's policies came from the previous government.

Another thing is the real actions that were implemented in the previous government.

2.12.2 The Current Government

The Undersecretariat for Energy Saving and Efficiency was created in December 2015 by Decree 231/15 of the National Executive Branch as an area under the Secretariat of Strategic Energy Planning of the then Ministry of Energy and Mines. It is organized by thematic areas as follows: Industry, Residential, Commercial and Public, Transport, Planning, Measurement and Verification, Education, National and International Programmes and Communication and Dissemination.

2.12.2.1 The Actions

Information taken from the subsecretariat's 2017 Annual Report, citing only the main ones:

(i) Participation in the preparation of the energy plan 2030.

The Undersecretariat established the savings targets in the following sectors: Residential, Public Lighting, Industries, Transport and UREE policies, concluding that by 2030 the actions would imply a saving of 10.2% on total energy consumption.

(ii) Progress in the development of Energy Efficiency Indicators to measure the performance of consumer networks.
(iii) Participation in the elaboration of the Sectoral Plan for Climate Change.
(iv) In the Industrial Sector: Creation of the National Energy Efficiency Fund aimed essentially at SMEs. Energy audits.

(v) Tariff benefits for electro-intensive industries.

Development and distribution of a guide for the best operation of engines.

(vi) In the Transport Sector

Labeling for vehicle emission control.
Training tasks.
Pilot plan for efficient driving in the fleet of the Secretariat of Energy and Mining.
Efficient driving on driver's licenses.
Technical verification of efficiency in vehicles.
Preparation and distribution of a guide for the transport of loads.

(vii) In Education

Current and virtual training.
Work in Universities.
Diplomas at the University of Buenos Aires and National University of Jujuy.

(viii) In the Residential, Commercial and Public Sectors.

Efficient lighting. 94,000 luminaires were installed.
Energy Diagnosis of Public Buildings.
Labeling Program for Home Appliances.
Preparation of a National Household Energy Expenditure Survey.
Characteristics of Energy Efficient Social Housing Plans.
Development and Distribution of a Guide for Administrators and Building Managers.

In addition to this, a launch manual was to be published in early November 2018 with a series of suggestions related to how to drive the vehicles. It is called Efficient Driving Guide for Light Vehicles.

2.12.3 Comments

One of the variables usually taken into account in UREE's analyses and proposals is that of the absolute and relative prices of the various energy products.

Given that these are generally goods with low or very low price-demand elasticity, tariff incentives or disincentives will not necessarily serve to make UREE policies effective.

Another aspect, and this is more related to the industrial sector, when it does not involve electro-intensive activities, is that the impact of energy costs on production costs is usually very low, especially when compared to wage and financial costs. So on the side of the tariffs, this type of industrialist will not have too many incentives to apply UREE in their processes. Then, we must look for other types of advantages such as tax relief. Then, it will be the State that will have to bear the cost of the programs, under the assumption that a saving in the energy supply will be economically and environmentally favorable to the interests of the country.

Although the low price-demand elasticity of energy does not help to generate UREE policies in the residential sector, taking into account that these are necessary goods and in several cases not replaceable, very low relative prices, as occurred in much of the previous government, make UREE policies unviable.

But at the other extreme, excessive tariff increases such as those applied by the current government cause, when not accompanied by increases in wage income, an artificial incentive to UREE policies.

The situation is not sustainable over time and there is a risk that users will not be able to use the services, especially natural gas, LPG and electricity. In any case, the measures to promote the labeling of household appliances are welcome, but they are only valid for new equipment and it is unfeasible for the majority of the population to recommend the change of appliances due to the deterioration of the level of income.

Even more so when the so-called social tariffs practically disappear and the discounts for lower consumption compared to similar periods of time in the past are eliminated.

In these cases, UREE policies clash with the satisfaction of basic needs.

In the transport sector, the consumption of gasoline and diesel, traffic congestion in cities, GHG pollution, the absence of public transport of people and the virtual elimination of rail freight make this sector the most important for implementing UREE policies.

Measures to replace lamps in public lighting and projects to label homes as energy savers have also been very effective. However, this must go hand in hand with plans to build houses adapted to the country's climate.

In summary, the work of the current Undersecretariat for Energy Saving and Efficiency would appear to be very active and positive.

2.13 Energy Prices and Tariffs

At this point, the energy prices and tariffs will be analyzed.

On the one hand, the issue of fuel prices will be addressed, particularly in the case of fuel taxes, since the current government changed the application methodology from applying percentages to fixed amounts. Here, we will include something about the prices of bioethanol and biodiesel.

On the other hand, the natural gas and LPG tariffs will be presented and finally the electricity tariffs.

The energy price lists for some months from 2014 to 2018 will be presented.

But first some conceptual considerations will be made.

2.13.1 Tariff Policies, Structures and Subsidies

Firstly, energy is a necessary good, and without it, a country's social and economic system would not function. Although this is a statement, it could be proved right if the goods are divided into those that are necessary and those that are expendable.

You can live without drinking beer (unless you are member of a special community), but you cannot do so without electricity or fuels (from wood to natural gas depending on where you live, the level of income you have or the availability of one or another fuel).

An industrial or commercial activity or a service cannot function without these energies.

Another previous concept, which is not accepted by the neoliberal economy, is that energy is a public service.

It is like education and health are indispensable and the State has the obligation to supply them. It is being discussed whether this should be done through public companies as was the case in Argentina in the 1970s, or through private companies as has been the case in Argentina since the 1990s. YPF is a mixed company with a state majority.

Perhaps the best thing would be a state-owned company in the energy sector, but with well-managed companies as in the case of Electricité de France, where the directors enter through a tender process and must report annually to Congress on compliance with the plan's contracts.

Another characteristic of the energy sector is that it is impossible for it to function without planning and without taking into account the short, medium and long term, since the works that the sector requires to function imply large investments that mature in the medium and long term.

Tariffs are used by companies to cover their operating costs and part of their investments. In the case of large investments that usually take place over several years (power plants, refineries, pipelines, explorations and exploitations), it would not be possible to charge the users present for the entirety of the required investment that is displaced over time.

For this reason, another economic policy tool is financing (there is no energy company in the world that does not resort to financing to meet its investment needs).

In other words, the investment needs cannot come entirely out of the tariffs, and a part of them must cover the financial costs (e.g., the purchase in installments of an appliance. If you cannot pay its cost all at once, it is bought in installments. The cost is financed by a bank, or a supplier).

Subsidies such as taxes are economic policy tools, and their appropriateness or appropriateness will depend on how they are used.

Families who consume energy (e.g., electricity, natural gas and LPG) do not have the same level of income and consume in general in relation to that income level.

A family living in a 20–40 m^2 apartment will consume less than a family living in a 200 m^2 house.

Therefore, not only the average rate should be taken into account, but also a rate structure should be implemented. This structure can be progressive or regressive.

It is progressive when the price of the unit consumed (KWH or m^3 or KGR) is lower, the lower the consumption and reciprocally. In other words, those who consume less will have a lower unit rate than those who consume more. This generates what in economics is called a cross-subsidy, where those who consume more subsidize those who consume less.

The important thing is that the average rate covers all the operating costs and a part of the investment costs (not all of them, but the rate would be very high and the present generations would be financing future generations).

If the average tariffs were very low, future generations would either have no energy or they would have to finance companies with very high average tariffs if they want to have energy and they would be financing the past generations that had very low average tariffs.

The same should happen with the service and productive sectors. In this case, the final consumption sectors (commerce) should have higher unit rates than the intermediate consumption sectors (industry, especially SMEs) and public service providers (transport).

This should lead to the design of a tariff structure that has different unit rates according to whether the families consume little or a lot; sectors that are of final consumption or intermediate consumption and according to whether the temperature where they are located is more or less rigorous.

If the problem is analyzed as a whole, i.e., medium- and long-term plans are made, tariff structures will emerge that are equitable, socially, economically and from a generational point of view.

This should lead to the stratification of household consumption and even determine a reasonable minimum consumption (which does not imply having a 40 W light bulb per room or a consumption of gas only for cooking) that will have the minimum rate.

Of course everything would be solved if all the inhabitants of the country had a stable, dignified and fairly paid job. Then, the tariff structure could be different.

In summary, in recent years, in Argentina, there was an irrational tariff structure that led to generate a policy of indiscriminate subsidies (not for all provinces) that together with the delegation of investments in the private energy sector (the state companies had disappeared) led to supply problems in electricity and natural gas and the need to import fuels (this is essentially responsible for the policy of energy exports in the 1990s and early 2000s and the late reaction, only in 2012, with the particular form of nationalization of YPF).

But on the other hand, the current distorted magnitude and tariff structure, especially in AMBA, should not have been corrected in such a short term (and it is not yet finished) but gradually and elaborating a new tariff structure far from the neoliberal principle that says "that everyone should pay according to their costs," because this would inevitably lead to a regressive tariff structure that would make it impossible in many cases to pay a large part of the population, especially those with fixed incomes that are not updated at the rate of the tariffs.

The issue of energy saving or rational energy use can and should, if well managed, influence tariff structures, but far beyond a system of rewards and penalties. It seems that even the deductions from energy service payments for lower relative consumption in comparison with similar periods in previous years would have disappeared.

The application of the so-called social tariff would also have been greatly restricted.

But why did these extreme imbalances not occur before the 1990s?

Essentially because of the existence of the energy companies among them YPF, Gas del Estado and Water and Energy that were privatized in that decade.

The financing of energy investments came from fuel taxes (essentially gas and diesel) which together with contributions from national banks such as the Development Bank and the Banco Nación contributed almost 70 or 80% of the cost of the works, and the rest came from multilateral banks such as the IDB and from equipment suppliers.

The public energy companies set their tariffs and the political authorities sometimes used them to avoid increasing inflationary rates, which delayed the work.

Another aspect was that the public energy companies did not need the profit item, but the one that contemplated the expansion of the service, and this meant that the tariffs did not owe so high, because the "shareholder," to whom they had to be accountable, was the State, that is, the population.

Furthermore, energy was considered a public service and energy consumption a right.

The fact is that energy meets the needs of the productive sector and the population.

On the other hand, neoliberalism considers the energy sector as just another sector producing goods and services.

Therefore, for this conception, energy activities are self-contained and isolated business units and each of them must be self-sustaining.

Consequently, they do not accept subsidies, and ultimately, if a business unit gives "losses," it must import what it produces.

There is no connection between the energy sector and the rest of the sectors: in political terms, between the Ministry of Energy and the Ministries of Finance, Production or the Central Bank.

Then, autonomous decisions taken in isolation, for example, by the Energy Authority, with an exorbitant tariff increase, do not take into account the repercussions on the rest of the system (e.g., inflation, fall in consumption, standard of living of the most vulnerable population).

The then Minister of Energy was given the directive to increase tariffs in order to end subsidies and reduce the deficit, which was considered to be the sole cause of inflation. He fulfilled his task, the repercussions are not his responsibility and he only agreed to consider the so-called social tariffs.

In other words, there is a problem of "ideological" conception underlying the decisions.

That is why the problem of tariffs is a political issue that is addressed technically.

2.13.2 Something About Natural Gas and Electricity Tariffs

(i) **Natural gas tariffs**

Natural gas tariffs are composed of three items:

The cost at the wellhead, which represents approximately 40% of the tariff.

The cost of transport and distribution, which represents approximately 32% of the total cost.

The taxes, essentially VAT, which represent approximately 28% of the total.

The production of natural gas at wellheads is subsidized because it is recognized, at least by several producers, a price of US $7.5 per million BTUs, which then falls by US $0.5 per year to 6.0 and then, will be guided by the "market" (estimated at no less than 4 dollars) above the development cost which is assumed to be 1.7 dollars and the Henry HUB which oscillates at 3 dollars.

The first observation for producers is that subsidy is not a bad word.

Transport and distribution of natural gas are monopolistic activities, unlike production, and the consumer is captive to the company from which he buys the natural gas. In other words, he cannot change his company.

That the current rates are very high is proven by the enormous profits of the distributors as shown in paragraph 4.2.2 point xv.

There were public hearings and in them, ENARGAS presented a report justifying the increases.

But who audited the costs of the producers (we only know that the "marginal cost" is US $7.5 per million BTU), transporters and distributors?

Who verified and is controlling their future investment plans to support the new tariff levels?

The legislation of the 1990s that culminated in the privatization of Gas del Estado established a periodic tariff review with the intervention of the regulatory body (ENARGAS).

Table 2.53 presents information in selected years for natural gas price and natural gas tariffs. In the latter case, the tariff structure for extreme years and for a type of residential and industrial consumers is included.

We can see the enormous price increase after 2015 carried out in order to reduce the large subsidies, reduce the fiscal deficit and favor the profits of the companies that, as it was said, did not make the investments that these price increases facilitated and instead increased the payment of dividends to the shareholders.

Electricity rates

Electricity rates are composed of four items:

Generation, which involves approximately 65%.

High voltage transport, representing approximately 4%.

Distribution Added Value, in low voltage that brings the EE to the homes, approximately 31%.

The fourth item is represented by taxes, essentially VAT, but also gross income and sometimes the invoice include municipal charges and may be 30% of the invoice.

The generation is diverse and CAMMESA pays the producers according to the marginal cost of the most expensive machine that came into service. This is interesting because the renewable and hydraulic machines that have low or no cost of fuel benefit from the high cost that they do have, for example, the gas turbines if they come into service.

Table 2.53 Average natural gas prices in argentine basins

	Unit	Dec-14	Dec-15	Dec-16	Dec-17
Weighted average price Argentine basins	USD/MMBTU	3.22	2.85	5.2	5.2

Residential and industrial natural gas rates ($m³)

	Nov-15	Apr-17	Dec-17	Jun-18
Residential tariff 4ªCat, CABA894 m^3/year	1.18	4.42	4.42	10.1
Residential tariff 8ªCat, CABA1920 m^3/year	4.3	8.11	8.11	13.81
Industrial tariff Province Bs As annual consumption 3 million m^3 and 100% load factor	2.14	4.5	4.5	8.64

Composition of natural gas tariffs (%)

Residential rate 4ªCat, CABA894 m^3/year	Nov-15	Dec-17	Jun-18
Cost of gas	0.34	1.6	2.83
Transport	0.15	0.75	1.9
Distribution	0.29	1.15	2.68
Taxes	0.21	0.73	2.59
Trust fund	0.19	0.19	0.1
Total	1.18	4.42	10.1

Industrial rate Province Bs As annual consumption 3 million m^3 and 100% load factor	Nov-15	Dec-17	Jun-18
Cost of gas	1.18	3.14	5.23
Transport	0.12	0.26	0.59
Distribution	0.03	0.14	0.32
Taxes	0.31	0.96	2.4
Trust fund	0.5		0.1
Total	2.14	4.5	8.64

Source Monthly Price Report, MONTAMAT and Associates, various issues

The Distribution Value Added is remunerated according to the cost of the networks, the operation and maintenance costs and the commercial costs.

The cost of transport is very small and is carried out by TRANSENER, a state-owned company that the government wants to privatize even though it is profitable. They are governed by the principle of "what the private sector can do that the state cannot."

As with natural gas, the users are captive to the distributor who sells them the electricity. That is to say, they cannot opt for another one either. It is a monopolistic activity.

As in the case of natural gas, non-binding public hearings were held to explain the tariff increases.

Also here, who audited the costs mentioned by the companies? Who is controlling the realization of the investments, supposedly committed, which are the basis of the increases?

What can be verified is that in the city of Buenos Aires and in the suburbs the cuts have continued, essentially due to distribution and transformation problems (responsibility of the distributors).

For example, the Central Port, the main private generator, has earned 310 million dollars in the first quarter of 2018.

Table 2.54 includes electricity rates by type of user for the companies that provided the service in the autonomous city of Buenos Aires and Greater Buenos Aires suburbs.

Table 2.54 clearly shows the enormous increase in electricity tariffs between December 2017 and December 2015, an increase that has increased during 2018.

(iii) **Some effects of tariffs.**

Other effects of the tariffs are given by the analysis of the balance sheets which are eloquent with respect to the brutal change of relative prices and the transfer of income that, via tariffs, is made in favor of the companies that make up the privatized energy sector, to the detriment of the productive sector and the households. Thus, it can be seen that within 53 companies eight firms stand out that have a direct relation with the policy of tariffs of the National Government: EDESUR, EDENOR, Metrogas, CAMUZZI, Gas Pampeana, TGS, Pampa Energía, Endesa Costanera and Central Puerto.

Table 2.54 Electricity rates by type of user (thousandths of USD/KWH)

User type	Consumption	Feb-15	Dic-15	Abr-17	Dic-17	Jun-18	Growth June 2018–December 2015
Residential	150 KWH/month	11.2	8.7	68.8	84.4	64	10.6 times
Residential	300 KWH/month	10.3	8	68.3	83.1	62.8	11.4 times
Residential	600 KWH/month	6.5	5.1	96.8	114.5	84.1	24.7 times
Industry low voltage	Maximum demand 30 KW 40% usage factor	15.9	16.2	89.8	115	86.1	7.8 times
Industry medium voltage	Maximum demand 300 KW 80% usage factor	15.9	12.1	81.2	92.3	61.4	8.3 times

Source MONTAMAT and Associates Monthly Energy Price Report various issues
Note 1 The difference between June 2018 and December 2015 was calculated taking into account the devaluation of the peso between December 2017 and June 2018 which was 45%
Note 2 The rates correspond to the companies EDENOR and EDESU that serve the city of Buenos Aires and Greater Buenos Aires
Note The tariffs correspond to the companies EDENOR and EDESU that serve the city of Buenos Aires and Greater Buenos Aires

For these companies, the growth of their turnover was 99%, their gross profits expanded by 209% and their final net profits by an even greater percentage, 772%. Compared to 2016, a company like EDENOR expanded its gross profits by 2273.3%.

(iv) **Other alternatives**

What should have been done or what remains to be done?

I assume that a tariff increase was necessary after the freeze.

First other concepts.

Tariffs have to take into account the foreseeable expansion of the system.

In other words, no tariffs can be set without first having designed a medium- and long-term energy plan, from which an investment plan can be deduced.

It is impossible to set tariffs without taking into account the costs of the production chain.

That is: costs of producing natural gas in the different basins and formations.

That is: at least in the basins, Neuquén, Austral and Comodoro Rivadavia; the costs of the tight and shale gas and the forecasts of the prices of imported natural gas from Bolivia and LNG, in each case with their respective participation, to deduct the average costs that must be assigned to each basin.

The costs associated with the transport and distribution stages and as these are private companies a profit rate that should not exceed a reasonable dollar value.

Since the cost data are held by the companies, it is not acceptable to deduct them from secondary information.

This information should be known to the Secretary of Energy and Mining and the ENRE and ENERGAS as regulatory and control bodies.

But if the companies do not have expansion plans and the Secretariat does not have a coherent energy plan, it will be very difficult to ensure that the tariffs proposed are correct.

The comparison with the cost of alternative or imported energy is another path that does not represent local costs.

The companies refuse to give their costs because they move as if we were in a free market situation and not in a regulated market. The neoliberals consider as cost the border price or international price. As for the natural gas, there is not yet an international price they take the most expensive import price (i.e., today, one of the LNGs after gasification) or the marginal cost that would be one of the most expensive local natural gases, that is, one of the shale gases. In this way, those who produce other cheaper types of gas would be appropriating a differential income resulting from the difference between the much higher cost of shale and conventional natural gas.

The right thing to do would seem to be managed with average costs and not the highest cost. But the reasoning would be acceptable if there was only one state-owned company like YPF producing and importing the natural gas needed to supply the domestic market.

In the current situation, private companies acting in the market, if a price was established based on average costs, those producing the shale would "lose" money, because their costs would be above the price, and in this case, they would be close or should be subsidized as long as the cost of shale was below the imported natural gas.

Currently, most of the natural gas that is produced "unconventionally" is tight which is less expensive than shale, according to information from YPF S.A.

This reasoning leaves aside the consideration of the character of a necessary good, with captive consumers of a seller (the distribution company that sells the fluid, since they cannot choose another one. In the case of residential and most commercial consumers, they do not have an alternative energy source).

All this tangle is the result, as was said, of the policy implemented in the 1990s, which between 2003 and 2015 was not wanted, was not known or could not be changed.

Now, some proposals.

Do not aim for the disappearance or abrupt fall of subsidies (I do not think this is the cause of inflation, but this would be a good idea for another document).

First to verify the investment plans of the companies that should arise from the requirements of the internal market of the country in short, medium and long term, product of the indications of the energy planning.

Monitoring of the companies' commitments by the regulatory bodies.

Based on existing tariffs as of December 2017, generate an increase in average tariffs similar to the projected increase in the average wage of workers.

Implement a policy of cross-subsidies so that in the residential sector, users with high consumption have higher rates than those with lower consumption (the social rate should emerge from a statistical analysis of consumption and income); that industrial consumption sectors have lower rates than commercial ones; that the regions of the country that require more consumption for climatic reasons have lower rates than those that do not have such reasons. The objective would be to achieve an increased average rate. This would imply fewer subsidies to businesses and a reduction in subsidies to users, which would be much slower than what was implemented.

It is essential to coordinate energy tariff policy with that implemented for other production sectors in order to generate a new relative price structure.

Greater participation by Congress in decision-making and by users.

2.13.3 Fuel Tax

(i) Before and after deregulation.

Prior to the deregulation of the 1990s, there were four types of specific taxes levied on activity at different points in the oil chain. These were:

The tax on the processing of crude oil (equivalent to 10% of the FOB value of the crude oil);

That tax applied to the transfer of liquid fuels;

Table 2.55 Fuel tax as a percentage of the final price of each product (%)

Period	NC	NE	GO	KE	DO	FO
1984–1988	65	56.7	26	22	25	29
1992 Ley 23966	50.11	55.2	19.6	5.2	43.1	17

The one for the social welfare funds; and
VAT.

This tax mechanism was in place until the end of the 1990s and was modified when oil deregulation was implemented.

National Law 23,996 established a wide range of reforms in line with the Convertibility Plan of late 1991.

Table 2.55 shows the fuel tax as a percentage of the final price of each product before and after convertibility.

Table 2.56 includes the destination of fuel tax funds before convertibility and Table 2.57 after (1996 example).

A Comment:

As it can be seen, the National Energy Fund, the Fund for Chocón and Large Electric Works and the crude oil tax ceased to exist after convertibility.

The fact is that, with the privatization of practically the entire energy sector, the private companies would be in charge of making the investments, and consequently, these funds would not be necessary.

We only have to remember that, thanks to the energy funds, it was possible to build the extensive network of gas pipelines, hydroelectric and nuclear power stations in Argentina.

For more details on this period, see: "The Policy of Argentine Oil Deregulation," R. Kozulj, V. Bravo and N. Di Sbroiavacca, Centro Editor de América Latina, 1996."

(ii) Immediately before and after the March 2018 amendment.

Returning to the period analyzed in this document, the fuel tax was modified in March 2018 by introducing changes in Title IV of National Law 23,966. The two most relevant aspects introduced are that the taxes were changed from a variable amount

Table 2.56 Destination of fuel tax funds before convertibility (%)

Social security	30.0
National treasury	21.5
National roads	17.2
National energy fund	12.3
FFCC	6.7
Fondo chocón -grand electric works	6.3
Provincial roads fund	6.1

to a fixed amount and a new tax was added: the tax on carbon dioxide emissions (discounting biofuels, biodiesel and bioethanol) added to diesel and naphtha.

The fixed amount will be updated quarterly according to the variation of the Consumer Price Index published by the INDEC.

The immediate previous composition of the fuel tax was made up of three components:

Tax on the transfer and import of liquid fuels (ITC) and compressed natural gas (CNG), according to National Law 23.996/91, which applies differential rates for each type of fuel: unleaded naphthas 70–62%, leaded naphtha 70–62%, virgin naphtha 62%, natural gasoline 62%, solvent 62%, turpentine 62%, diesel 19%, diesel oil 19%, and kerosene 19%.

Tax for the hydric infrastructure fund (hydric rate), which reached with a 4% rate to the naphtha with and without lead and 9% to CNG.

Tax on diesel for automotive use, with a rate of 22%.

At the same time, fuels were taxed by other charges as, for example, the surcharge of $0.004 per m^3 of natural gas distributed to finance the program of social carafes and finally municipal rates that taxed the local sale of fuels, the tax to the gross income and the tax to the added value.

The distribution of these taxes was as follows: in relation to the ITC collected by Naftas, 23% was allocated to the National Treasury, another 23% was allocated to the provinces, 21% to ANSES and 33% to FONAVI. While the ITC collection corresponding to diesel, diesel and CNG were totally destined for ANSES.

On the other hand, the water rate was applied to the development of the sector's infrastructure and the diesel tax was allocated 8% to finance transport subsidies and 92% to the transport infrastructure trust.

Table 2.58 shows the distribution of the fuel tax before the March 2018 reform.

Table 2.57 Destination of fuel tax funds after convertibility (1966) (%)

VAT national treasury co-payable	32.3
National treasury	23
Provinces	22.3
VAT national treasury co-payable	22.3

Table 2.58 Distribution of fuel taxes

		%
ITC gasoline	National treasury	23
	Provinces	23
	ANSES	21
	FONAVI	33
ITC diesel/CNG	ANSES	100
Water rate	Public works	100
Gas oil tax	Grants	8
	FF Infrastructure	92

Table 2.59 Fixed amounts of tax according to the amendment of Law 23,966

	Monto fijo (en $)	Unidad de medida
(a) Lead-free petrol, up to 92 RON	6726	Liter
(b) Lead-free naphtha, over RON 92	6726	Liter
(c) Virgin naphtha	6726	Liter
(d) Natural or pyrolysis gasoline	6726	Liter
(e) Solvent	6726	Liter
(f) Turpentine	6726	Liter
(g) Diesel oil	4148	Liter
(h) Diesel	4148	Liter
(i) Kerosene	4148	Liter

Source National Law 23,966

Table 2.59 shows the fixed amounts applied at the time the law was amended in March 2018.

In the case of biodiesel and bioethanol fuel, the tax will be totally satisfied with the payment of the tax on the naphtha, diesel and diesel component or another taxed component. Biofuels in their pure state are not covered.

In the provinces of Neuquén, La Pampa, Río Negro, Chubut, Santa Cruz, Tierra del Fuego, Antarctica and the South Atlantic Islands, the county of Patagones in the province of Buenos Aires and the department of Malargue in the province of Mendoza are not taxed on unleaded and leaded gasoline, and diesel and kerosene pay $2246 per liter.

The fixed amounts consigned will be updated by calendar quarter, based on the variations of the Consumer Price Index (CPI), provided by the National Institute of Statistics and Census, considering the accumulated variations of such index since January 2018, inclusive.

Table 2.60 shows the carbon dioxide tax.

Fixed amounts will be updated by calendar quarter, based on the variations of the Consumer Price Index (CPI), provided by the National Institute of Statistics and Census, considering the accumulated variations of such index since January 2018, inclusive.

The destination of the collection of the mentioned taxes is the following:

National Treasury: 10.4%.
National Housing Fund: 15.07%.
Provinces: 10.40%.
Single Social Security System for national social security obligations: 28.69%.
Water Infrastructure Trust 4.31%.
Transportation Infrastructure Trust: 28.58%.
Public Transport Compensation: 2.55%.

Table 2.60 Carbon dioxide tax

	Fixed amount (en $)	Unit of measurement
(a) Lead-free naphtha, up to 92 RON	0412	Liter
(b) Lead-free naphtha, over RON 92	0412	Liter
(c) Virgin naphtha	0412	Liter
(d) Natural or pyrolysis gasoline	0412	Liter
(e) Solvent	0412	Liter
(f) Turpentine	0.412	Liter
(g) Diesel oil	0.473	Liter
(h) Diesel	0.473	Liter
(i) Kerosene	0.473	Liter
(j) Kerosene	0.519	Liter
(k) Petroleum coke	0.557	Kilogram
(l) Coal	0.429	Kilogram

Source National Law 23,966

(iii) **Comments**

As of March 2018, the share of fuel taxes and carbon dioxide charges represented 22 and 20% of the retail prices of super and regular diesel fuels, respectively. This is almost 40% of what they represented historically for gasoline and almost 30% of what they represented for diesel.

This change of modality favors the oil companies that distribute the fuels because now the increases in the prices of these products will go entirely to them, the State will lose and the consumers will assume the totality of the increases.

There are two reasons for this; the prices of oil derivatives (essentially naphtha and gas oil) increase much more than the Consumer Price Index (e.g., in the first nine months of 2018, the prices of derivatives increased by nearly 65% and the index by 35%) and the relative delay is one quarter.

As for the destination of the taxes, nothing goes to the energy sector (for the same reasons mentioned when analyzing the change caused in the 1990s by oil deregulation) and an important part will go to the trust funds that will be derived from the Public–Private Participation Programs.

2.13.4 Prices of Crude Oil and Its Derivatives

First, the prices of crude oil will be presented and then the prices of oil derivatives and then the comments (Tables 2.61 and 2.62).

Table 2.61 Crude oil prices. (US $/Barrel)

	Nov-14	Dic-15	Abr-16	Dic-16	Abr-17	Dic-17	Jun-18
MEDANITO (*)	84.58	76.21	67	62.75	57.2	66.01	68
Escalante (*)	68.21	50.84	54.9	48.44	47.7	56.58	62.48
Brent (+)	79.15	38.82	41.26	54.96	52.31	64.37	75.55
WTI (+)	75.14	37.28	41.17	49.97	51.22	58.21	69.07

Source
(*) MONTAMAT and Associates Monthly Price Report
(+) Own elaboration based on daily data from BLOOMBERG

Table 2.62 Oil derivatives prices—($/LT)

	Nov-14	Dic-15	Abr-16	Dic-16	Abr-17	Dic-17	Jun-18
SUPER naphtha without taxes	6.85	7.77	9	10.3	11.1	13.13	15.3
SUPER naphtha with taxes	12.31	13.8	15.86	17.33	18.7	23.02	26.65
Diesel oil without taxes	7.29	8.32	9.67	10.53	11.07	15.85	15.3
Diesel with taxes	11.17	12.72	14.53	12.7	16.67	19.15	23.34

Source MONTAMAT and Associates Monthly Price Report

In a letter, the former Minister of Energy and Mining, Mr. Juan José Aranguren, informed all the actors in the Hydrocarbon market that "in relation to the Agreement for the Transition to International Prices of the Argentine Hydrocarbon Industry," and depending on the dynamics of external quotations, it had been decided to free the price of fuels.

From then on, the oil companies would be authorized to modify the sales price of their fuels for consumption in the automobile market. In other words, the internal prices of oil derivatives would be freed up.

The winners of this measure were the refineries—those that sell naphtha, such as YPF, Shell, AXION, OIL, REFINOR and Petrobras—that established prices based on what they paid for the crude oil they bought from local producers (YPF, Pan American Energy, PLUSPETROL and Petrobras, mainly).

This will not favor the companies that extract the crude in the country as long as they receive a price for the crude that is lower than the international market price. In June 2018, this difference was 11%, as the international benchmark crude oil that had always been WTI had been replaced by Brent, which was approximately 10% more expensive.

In short, the oil companies in Argentina never lose.

From this decision of the authorities, the prices of oil derivatives in Argentina are freed, they are dollarized and it will be necessary to take into account the international quotation of Brent crude (with its equivalent in pesos); the value of the dollar; the price of Biofuels (since they are mixed in 10% with diesel and 12% with gasoline); and the quarterly variation in the INDEC's retail price index.

This change in rules put an end to a period of more than fifteen years in which local prices were regulated and independent of the international price of oil.

When crude oil reached over US $140 a barrel, fuels were priced low due to progressive export taxes. This situation began to be corrected, through small price adjustments.

From mid-2014, the price of crude fell to less than US $44 in nine months and forced the government to agree with the oil companies on a much higher sustained price ("criollo barrel" in pesos, i.e., "local barrel"), of US $73 on average. But with oil on the ground, fuel prices became the highest in the region behind Uruguay. That pattern was maintained by the current government, whose assumption coincided with the lowest Brent (US $36.2 per barrel) since 2009. The difference is that withholding taxes were eliminated and then a decreasing path was set for the "criollo barrel" which was actually a consumer subsidy to the oil companies, contrary to the years of expensive oil and low prices.

As of November 2018, gasoline and diesel fuel prices have already risen by an average of almost 70%, with the impact on freight and transportation costs adding to inflationary pressures. Despite the fact that during this period the Brent fell, and the dollar rose by almost 100%. In short, with free prices, fuels can rise when oil does not, but the dollar does, or vice versa. With the current volatility in both markets, the unknown is the frequency of adjustments. This paragraph is partly inspired by the article by Mr. Néstor O. Sibona, "Another era of changes in the oil sector," published in the newspaper La Nación, Buenos Aires, on February 25, 2018.

In the first days of November 2018, fuel prices increased again. The data correspond to YPF for the city of Buenos Aires and includes taxes:

Super naphthas: 37.59 $/liter increasing 41% with respect to June 2018.
Regular diesel: 33.76 $/liter up 45% from June 2018.
On the other hand, the price of Brent crude oil (a marker for the Argentine market and this was modified by the previous Energy Minister as it used to be WTI, which in November 2018 was 14% cheaper than Brent, which favors oil companies and harms consumers) in the same period fell by 5%.

In other words, it is verified that in the Argentine market the prices of oil derivatives are inelastic to the fall of the international price and elastic to the rise.

Another aspect that deserves to be taken into account is that with the current pricing policy, the value of, for example, naphtha and diesel oil is much more expensive in some provinces than in others. This means that prices are not fixed and homogeneous as they were before the nineties.

2.13.5 LPG Prices

Since the previous government, the so-called Plan Hogar has been in effect with the objective of making the carafes more accessible to the lower-income sectors. In June 2018, the price of the calorie paid by a user of LPG in 10 kg cylinders was 1.5 times

Table 2.63 Price of the 10 KGR Carafe ($)

	November 2014	December 2015	April 2016	December 2016	November 2017	December 2017	June 2018
Carafe of 10 KGR with VAT in commerce	16	97	97	97	135	185	195.47

Source Monthly Energy Price Report Montamat y Asociados

more expensive than that paid by a consumer of the R23 category of natural gas (Table 2.63).

2.13.6 Biofuel Prices

These prices are periodically updated by the Secretariat of Energy and Mining and are different for small and large producers. In the case of Bioethanol, the price is different if it is obtained from corn or sugarcane (Table 2.64).

The Secretariat of Energy and Mines publishes the so-called Reference Values. Thus, in October and November 2018, the values were as follows:

Biodiesel: 27,529 $/ton and 28,112 $/ton, respectively.
Bioethanol from sugarcane: 21,999 $/liter and 21,224, respectively.
Corn bioethanol: 18,318 $/liter and 19,846, respectively.
There is a difference between biofuel producers and oil refiners, who argue that this mixture increases their costs, but in this and other differences we must always consider the positive and negative impacts on the production network.

For example, there are experiences that increase the addition of biofuels over the current limits, reaching, in the case of biodiesel, the operation of buses with 100% of this fuel and the case of Brazil that generated a fleet of special cars that used all bioethanol. In other words, this is an issue that cannot be left to the producers of biofuels and oil derivatives alone.

Table 2.64 Price of biofuels ($/liter)

	November 2014	December 2015	April 2016	December 2016	November 2017	December 2017	June 2018
Bioethanol	8.61	8.62	10.65	13.06	13.72	13.90	16.58
Biodiesel	6.0	5.62	10.35	11.88	11.93	13.37	17.26

Source MONTAMAT and Associates Monthly Energy Price Report

In this sense, environmental impacts cannot be ignored, as is the case of biodiesel obtained from La Palma in Colombia. See: V.Bravo and D. Bouille, 2008. "Contribution of biofuels to the sustainability of development in Latin America and the Caribbean: elements for the formulation of public policies," March 2008-ECLAC.

2.14 Energy Subsidies

2.14.1 Some Concepts Related to Grants

Firstly, energy is a necessary good, and without it, the country's social and economic systems would not function.

Although this is a statement, it could be proved right if we divide the goods into those that are necessary and those that are dispensable. One can live without drinking beer but cannot do so without electricity or fuels (from wood to natural gas depending on where one lives, the level of income one has or the availability of one or another fuel).

An industrial or commercial activity or a service cannot function without these energies.

Another previous concept that the neoliberal economy does not accept is that energy is a public service.

It is like education and health are indispensable and the State has the obligation to supply them. It is discussed whether this should be done through public companies as was the case in Argentina in the seventies, or with private companies as was the case in Argentina in the nineties.

I am in favor of an entrepreneurial State, in this aspect, but with well-managed companies, as it happens in Electricity of France, where the managers enter by contest and must submit annually as many reports to the Congress about the fulfillment of the plan's contracts.

Another characteristic of the energy sector is that it is impossible for it to function without planning and without taking into account the short, medium and long term, since the works that the sector requires to function imply large investments that mature in the medium and long term.

Tariffs are used by companies to cover their operating costs and part of their investments. In the case of large investments that are usually made over several years (such as power plants, refineries, pipelines, exploration and exploitation), it would not be possible to charge the users present for all the investments that are displaced over time.

That is why, another tool of economic policy is financing (there is no energy company in the world that does not resort to financing to meet its investment needs).

In other words, investment needs cannot come entirely from tariffs, and a part of them must cover financial costs (it is like buying an appliance in installments. If you can't afford the one-time cost of an installment purchase, the cost is financed by a bank, or a vendor). In Argentina, before the 1990s, energy investments were financed by the National Energy Fund, which came from fuel taxes, as already mentioned in point 10.3, from Multinational Credit Banks and equipment suppliers.

Subsidies such as taxes are economic policy tools, and their goodness or timeliness will depend on how they are used.

Families that consume energy (e.g., electricity, natural gas and LPG) do not have the same level of income and consume in general in relation to their income level. A family living in a 20–40 square meter apartment will consume less than a family living in a 200 square meter house.

Then, you should take into account not only the average rate but implement a rate structure that can be progressive or regressive. It will be progressive when the price of the unit consumed (KWH or liter or kilogram) is lower, the lower the consumption and reciprocally. In other words, those who consume less will have a lower unit rate than those who consume more. This generates what is called a cross-subsidy, where those who consume more subsidize those who consume less. The important thing is that the average rate covers all the operating costs and a part of the investments (not all of them but the rate would be very high and the present generations would be financing future generations).

If the average tariffs were very low, future generations would either have no energy or they would have to finance companies with very high average tariffs if they want to have energy and they would be financing the past generations that had very low average tariffs.

The same should happen with the service and productive sectors. In this case, the final consumption sectors (commerce) should have higher unit rates than the intermediate consumption sectors (industry) and the public service providers (transport). Then, there would also be cross-subsidies between sectors.

This should lead to the design of a new tariff structure with different unit rates for low and high consuming households and for intermediate and final consumption sectors.

If the problem is analyzed as a whole, i.e., planned in the medium and long term, tariff structures will emerge that are equitable, socially, economically and generationally speaking.

This should lead to the stratification of household consumption and even determine a reasonable minimum consumption (which does not imply having a 40 W light bulb per room or gas consumption only for cooking) that would have a minimum tariff.

Of course, everything would be solved if all the inhabitants of the country had a stable, decent and fairly paid job. Then, the tariff structure could be different.

In summary, in recent years, in Argentina, there was an irrational tariff structure that led to the generation of a policy of indiscriminate subsidies (not for all provinces) that together with the delegation of much of the investment in the private sector (it should be remembered that since the 1990s almost all state enterprises had disappeared) led to problems of shortage of electricity and natural gas and the need to import fuels (this is responsible for the policy of energy exports in the 1990s and early 2000s and the late reaction, only in 2012, with the particular form of nationalization of YPF).

A last thought, the current distorted tariff structure that exists especially in the AMBA should not have been corrected, almost overnight, but gradually and elaborating a new tariff structure far from the neoliberal principle that says "everyone should pay according to their costs" because this led to a regressive tariff structure that seriously affected fixed-income family economies, such as salaried and retired workers, SMEs and Neighborhood Clubs.

The government that emerged from the December 2015 elections with its intention to cancel the primary fiscal deficit (increasing the secondary deficit with the increase in the interest on the contracted Foreign Debt) has implemented a strong policy of decreasing subsidies, in this case energy subsidies, with the objective that as soon as possible each user ends up paying the full rates for electricity, natural gas and the very high ones for LPG.

2.14.2 Subsidies in the Previous Government

To better understand the situation of the years 2016 to 2018, it is necessary to mention what happened in the previous government with the subsidies.

Energy subsidies are a relatively new phenomenon in the Argentine energy scene. In the last 25 years (1990–2015), it has been noted that they were very small until 2003 and that after that year they became a phenomenon of considerable magnitude and strongly increasing until the present.

In its current configuration, they have their origin in a set of measures adopted by the national government after the serious and traumatic economic, social and political episodes that occurred after the exit of convertibility at the end of 2001.

Economic Emergency Law 25,561, enacted on January 6, 2002, provided for the specification of gas and electricity utility tariffs from their original value in pesos convertible into US dollars at an exchange rate of pesos, 1 per dollar; all provisions relating to price adjustments and indexation mechanisms provided for in current contracts were revoked. At the same time, the law empowered the Executive Branch to renegotiate, within a reasonable period of time, the contracts of the public utilities and the tariffs corresponding to such services. This was done with the purpose of normalizing the services with tariffs that on the one hand could be faced by the population (strongly affected by the devaluation) and on the other hand that these tariffs would be sufficient to allow the companies to operate and adequately comply with the obligations established in the concession contracts.

The maintenance of the "Economic Emergency" through successive extensions of National Law 25,561 and with it the maintenance of the "tariff abnormality" until the present generated problems of significant magnitude to the energy sector, such as:

The tariff freeze; partial compensation; etc.

The production of crude oil decreases uninterruptedly since 1998 and that of natural gas from 2004 to 2014. Within the framework of the emergency and the price freezes imposed under its protection, the National State had to take over the higher costs of the energy products that had to be imported—fuel oil and gas oil and natural gas—to face the growth of domestic demand, in a context where the domestic production of natural gas and oil was decreasing.

On the other hand, the natural gas electric power distribution and transmission companies experienced a drop in their income in real terms as a consequence of the growing inflationary process that took place after the exit from convertibility. The combination of partially frozen revenues (tariffs) and increasing business costs—materials, personnel, expansion—led to a progressive deterioration of their equity situation and a fall in results.

The result was an evident drop in operating performance and loss of liquidity, which in turn caused many companies to postpone new investments in their networks, thus affecting the quality of services.

Finally, as a result of administrative claims and judicial presentations by some of the concession companies due to the delay in the implementation of tariff review agreements, the national government was forced to recognize and compensate through different mechanisms the increase in costs of these companies. For this purpose, very complex institutional mechanisms were used, which made the operation of the sector difficult, often mixing the responsibilities of the grantor and the concessionaire (taken from the document "Energy subsidies in Argentina" of the Argentine Association of Budget (ASAP) and the Instituto de la Energía General Mosconi (IAE) of December 2015").

During the period covering the years 2004 and 2014, subsidies to the energy sector amounted to about 342,000 million current pesos (equivalent to 2695,658 million pesos of the year 2018).

Energy subsidies are mainly channeled through two large companies, CAMMESA (55.8%) and ENARSA (30.7%), which are responsible, among other things, for subsidizing the generation of electricity (either through compensation to the generators or through the purchase of fuel) and the purchase of imported gas to supply the domestic market, respectively.

The transfers to gas-producing companies under the program for the injection of surplus natural gas (known as the "Gas Plan"), which was approved at the beginning of 2013 and is designed to increase production of this resource by granting subsidies to companies (5.0%), are also much lower.

Yacimientos Carboníferos Río Turbio (2.2%) for operating expenses, essentially salaries.

Trust fund for residential LPG consumption (1.5%) for the Plan Hogar for the so-called social carafe.

Yacyretá (1.4%) for payment of electricity to Paraguay.

Trust fund for residential natural gas consumption (0.2%) for tariff compensation for consumers in Patagonia and other areas of the country.

Other destinations (3.1%).

2.14.3 *The Current Government*

Table 2.65 shows the evolution of subsidies by recipient entity between 2016 and 2018.

In relation to 2016, subsidies would have decreased by almost 16% with strong falls in the Gas Plan and important increases in those destined for the former ENARSA and LPG consumers.

It can be seen that the enormous magnitude of tariff increases (NG and EE) has not been sufficient to significantly decrease subsidies.

So will the government continue to increase tariffs in real terms?

If this happens, a considerable number of low- and middle-class families will lose their right to consume energy and hundreds of SMEs and public good organizations will tend to disappear.

The solution would seem to be the one outlined in Point 11.1.

2.15 The Energy Matrix

Up to this point, the problems of each energy source or particular specific issues have been analyzed.

Now, we will deal with the energies as a whole.

Table 2.65 Amount of energy subsidies between 2016 and 2018 (millions of current pesos)

Receivers	11 months del 2016	8 months del 2017	11 months del 2017	8 months del 2018
CAMMESA	101,437	35,000	48,000	55,101
EX ENARSA	11,700	5000	11,347	20,557
Plan GAS	28,503	20,528	21,903	7882
YCRT	3030	2628	3346	2155
Residential LPG consumers	3946	3897	5034	5563
YACYRETÁ	1088	496	936	458
Residential consumers of NG	123	1270	3229	0
Others	563	2186	3206	146
Total	150,390	71,005	97,001	91,862

Source Trends Report, various issues GRL Energy Institute. MOSCONI (IAE)

Table 2.66 Final consumption plus own consumption (10^3 TEP)

Energy source	2014	2017
Oil and derivatives	19,248	20,506
Natural gas and derivatives	27,430	25,882
Coal and derivatives	1419	1446
LPG	1917	1816
Bioenergies	2500	2930
Wind power	129	129
Electrical energy	11,232	11,446
Total	63,875	64,155

Source Own preparation based on the Energy Balance Sheet of the Ministry of Energy and Mining of Argentina

2.15.1 Some Data

Although the observation of what has happened between 2014 and 2017 is too short a time to see robust structural changes, it can give an indication of trends.

The different energy sources have been grouped into the sets indicated below:

Petroleum and derivatives which comprises the following sources: crude oil; refinery gas; naphtha; kerosene; jet fuel; diesel; fuel oil; and waste petroleum coal.

Natural gas and derivatives comprising the following sources: natural gas and distributed gas.

Coal and derivatives comprising the following sources: mineral coal; coke; coke gas; and blast furnace gas.

LPG: It has been considered separately even though it comes from petroleum (refineries) and from oil fields (essentially natural gas).

Bioenergies that comprise the following sources: firewood; bagasse; waste; charcoal; bioethanol; and biodiesel.

Wind: only that destined for the pumping of water in the wind mills.

Electricity: which has been generated by the energy sources included in Table 12.1.3.

Solar energy used to heat water is not included due to lack of information.

Tables 2.66 and 2.67 present total energy consumption as the sum of own consumption and final consumption.

Finally, Table 2.69 presents some general indicators of the evolution of the Argentine energy system between 2014 and 2017.

2.15.2 Some Comments

First, it should be clarified that intermediate consumption, for example, those destined to produce electricity, is not included in the total consumption.

Table 2.67 Final consumption plus own consumption (%)

Energy source	2014	2017
Oil and derivatives	30.13	31.96
Natural gas and derivatives	42.94	40.34
Coal and derivatives	2.22	2.25
LPG	3.00	2.83
Bioenergy	3.91	4.57
Bioenergy	0.20	0.20
Electrical energy	17.58	17.84
Total	100.00	100.00

Source Own elaboration based on Table 2.66

Total energy consumption has hardly grown between the years 2014 and 2017. This is not as a result of energy saving measures but as a consequence of a drop in the country's economic activity and the income of its inhabitants.

There are few changes in the participation of sources: for example, the fall of natural gas and derivatives would be due to the strong tariff increases to and the mentioned lower activity of the productive sectors. On the other hand, the advance of oil and derivatives to the higher consumption of naphtha and gas oil, in spite of the price increases of the same.

Regarding bioenergy, the relative gain would have been due to the naphtha and gas oil since the bioethanol and the biodiesel are mixed with them.

As for the consumption of electricity, it would have practically stagnated and the commentary included in point (ii) of this section is valid.

Table 2.68 shows the impact of the income from Atucha II, how natural gas has continued to be used for conventional thermal generation and how the effects of the renewal plans are not yet observed in terms of the relative weight of wind and solar plants.

Table 2.68 Energetics for the generation of electricity of the public service (%)

Energy	2014	2017
Hydraulics	16.4	15.4
Nuclear	5.9	7.8
Coal	2.5	1.6
Wind	0.2	0.2
Solar	0.0	0.0
Natural gas	55.6	64.0
Diesel oil	7.2	5.4
Fuel oil	12.3	5.6
Biodiesel	0.0	0.0
Total 10^3 TEP	21,723	22,401

Source Own elaboration based on Table 2.36

Table 2.69 Energy matrix indicators

Concept	Unit	2014	2017
Existing energy (gross energy supply)	10^3 TEP	78,707	80,904
Energy available	10^3 TEP	78,458	79,088
Total net consumption	10^3 TEP	63,760	64,232
Net energy efficiency	(%)	81	79.4
Energy self-sufficiency	(%)	92.4	90.3
Energy intensity	TEP/million pesos 2004 GDP	90.8	88.16

Source Compilation based on the Energy Balance Sheets published by the Ministry of Energy and Mining of Argentina
Data from the INDEC GDP, Buenos Aires

Table 2.69 shows a slight drop in energy efficiency. That is, the system's losses would have increased (primary, secondary and transformer station losses); the level of self-supply would have fallen (obviously due to lower oil and natural gas production and higher liquid natural gas imports); and it would be noted that less energy was needed to generate one unit of GDP in constant currency.

Bibliography

Numerous documents, reports and news items have been consulted. Sometimes the opinions of third parties on different subjects have been transcribed verbatim and are in italics in the text.

References to relatively long documents have also been indicated in the text.

Unfortunately, the Secretariat of Energy and Mines discontinued years ago the very useful Fuel and Electric Energy Yearbooks that allowed the numerical data of the whole energy system to be collected and controlled. This made it possible to manage with a single source and saved a lot of time in searching.

Consequently, this type of data had to be obtained from the dynamic tables of the Secretariat of Energy and Mining, the Argentine Institute of Oil and Gas and CAMMESA, in addition to other sources that are also explained in the text.

Of course, National Laws, Decrees and Resolutions have been consulted where appropriate.

It has also been necessary to resort to information from magazines in general on line such as Nuevas Energías; Petroquímica; El Inversor Energético y Minero, ECOJOURNAL.

The material of the Argentine Energy Institute GRAL MOSCONI (IAE) and the Monthly Price Reports of MONTAMAT and Associates have been very useful.

Thanks to the daily reports of the Digital PAPERS, it has been possible to consult almost all the newspapers in the country (e.g., Naci6n, El Cronista; Ámbito Financiero; BAE; Página 12; Clarín).

Then, in the same document, it is included where the references and bibliographic citations correspond.

Chapter 3
An Opinion About Fracking

Victor Bravo

Abstract SE briefly describes the situation regarding the supply of oil and natural gas in the periods leading up to 2004 and shows the role of the state-owned energy companies. A very brief presentation is made of the so-called conventional exploitation techniques for the exploitation of hydrocarbons and then the same is done with the so-called non-conventional techniques (especially fracking), emphasizing especially the large volume of water that these non-conventional techniques require. Then the denomination as "Reserves" of what really are "Resources" of non-conventional hydrocarbons is questioned. In this way, the values mentioned by the International Energy Agency in June 2013 are taken in their true dimension. This opens up a question mark over the reasons why they want to exploit non-conventional hydrocarbons in Argentina through fracking techniques. The main environmental impacts that could be generated by the fracking technique are described. The situation at the end of 2015, in the rest of the world, of the use of fracking is analyzed. Next, a series of recommendations are made and some questions are asked that the application of the fracking technique may generate. As an annex, a critical analysis of the association contract signed between YPF SA and Chevron for the exploitation of hydrocarbon areas, within the Vaca Muerta formation, is included, since this contract started the investigation and exploitation of fracking in Argentina.

Keywords Conventional exploitation of hydrocarbons · The fracking · Unconventional hydrocarbons · Environmental impacts · Vaca Muerta · The YPF-chevron contract

3.1 Some Previous Elements to Better Understand the Topic

Argentina reached its self-sufficiency in petroleum energy and natural gas at the end of the 1970s of the last century due almost exclusively to the work of the state-owned YPF.

V. Bravo and N. Di Sbroiavacca, *Oil and Natural Gas Economy in Argentina*,
The Latin American Studies Book Series,
https://doi.org/10.1007/978-3-030-65520-4_3

That is to say that the Energy petroleum and natural gas that were consumed in Argentina were produced from Argentina's own deposits.

Exports were very small and so were imports. A small amount of natural gas was imported from Bolivia for geopolitical reasons.

In the state-owned company, it was said that Argentina was a country with oil but not an oil country.

Natural gas was replacing fuel oil in thermal power plants. So did water and nuclear fuels.

The idea was to lower petroleum consumption.

Even at the beginning of the 1980s, when democracy was recovered, alconafta (cane alcohol mixed with naphtha) began to be used, especially in the Northern Zone, NEA and the Center of Argentina.

But in the 1990s, with the Menem-Cavallo government, the energy policy changes completely, especially the petroleum and natural gas.

YPF, Gas del Estado and almost all the Electric Companies are privatized, except the binational ones (Salto Grande and Yacyretá and the nuclear companies Atucha and Embalse and some provincial ones).

It was said that the State was a bad businessman and that with the energy sector in private hands there would be a lot of energy, cheaper and at international prices.

In the matter of petroleum and natural gas the foreign companies, especially the Spanish REPSOL, that had bought very cheap to YPF, launched to increase strongly the production and constructed, even, gas pipelines to export the abundant natural gas to Chile, for example.

Exploration was stopped to replenish reserves and these decreased dramatically from almost 12 years, in the early 1990s to almost 10 years now (with a production almost 30% less) and from almost 20 years to less than 8 years for natural gas.

YPF drilled about 120–150 exploration wells per year, when it was state-owned, and then all the companies dropped to 60–50 and even 30 wells per year.

Without exploration, the reserves would run out and more and more had to be imported.

At the end of the 1990s, almost 40% of the petroleum energy produced and a good part of the natural gas was exported.

Exports amounted to US $22 per barrel of oil and less than US $3 per million BTUs of natural gas.

Today, petroleum (in the form of derivatives) must be imported at almost 120 dollars and Liquefied natural gas at almost 18.

That is 4–5 times more expensive.

This is because companies have dedicated almost all their efforts to exporting large quantities in order to recover as soon as possible what they have spent on buying YPF, to turn profits abroad and to increase dividends to their shareholders. The fact is that the foreign currency obtained by exports did not pass through the Central Bank and the oil was freely available (This situation lasted until 2011).

The local natural gas, at present, is not enough to satisfy the consumption needs of industries, power plants and even, sometimes, of residential users, despite the fact that more than 40% of the population lacks natural gas. It is that in addition hydroelectric and nuclear plants were stopped.

As it was not invested to expand the capacity of refineries, today gas oil must be imported, partly to replace natural gas in power plants, and sometimes naphtha and fuel oil.

These imports cause an important deficit in the Trade Balance and this added to the great subsidies that are given to the energy sector, essentially to buy LNG, and by the very low levels of the tariffs of GN and EE, contributes, also, to generate a deficit in the national budget creating a very difficult situation to the country that in part is reflected in the difficulties of the Provincial budgets, that sometimes are not even in conditions to pay the salaries of the public employees.

Faced with this situation, that of the fall in reserves and production of petroleum and natural gas, the government decided, rightly and belatedly, to recover the management of YPF by buying 51% of the shares from the Spanish group REPSOL.

ENARSA, which had been created at the beginning of the 2000s, could not contribute to the solution of this problem.

The other problem, the subsidies to the natural gas and electrical energy tariffs, is still delayed. The petroleum derivatives on the contrary gradually increase their price. This generates serious inconveniences to Camessa, the administrator company of the electrical system that buys the EE to the generators with those subsidies of the State since the distributors, that in turn buy the EE to CAMESSA, owe to this every day more money by the freezing of the tariffs.

That is the context.

3.2 Conventional Exploitation of Hydrocarbons

First we will see how petroleum and natural gas are extracted from conventional fields.

Structures containing petroleum and natural gas are in the subsoil at greater depths, generally at 1000 m and more, although in some cases they reach almost 4000 m.

They do not appear as underground petroleum and natural gas layers or rivers, but both are housed in capillaries, small pipes, sometimes not connected to each other, in solid rocks, but which have the property of being porous (i.e., there are spaces occupied by hydrocarbons) and permeable (i.e., hydrocarbons can flow and they can move).

They are deposited in what is called traps (you can imagine the interior part of a shade where the upper part is an impermeable rock that contains them and prevents them from escaping). Natural gas occupies the upper part of the capillaries, petroleum the intermediate part and water the lower part.

To extract them, as they are contained at high pressure (there are almost 1000 or 4000 m of rock layers above), it is necessary to reach the "hat" with a vertical well drilled from the surface. This well is drilled with a special piece called a trepan, placed at the end of a series of articulated bars. The well starts with a diameter of 50–60 cm and ends, in the trap, with 8–10 cm.

When the drill bit reaches the hydrocarbons, i.e., by piercing the cap, the petroleum comes out through the pipes of the vertical well toward the surface dragged by the natural gas and pushed by the water.

The wells that are drilled are vertical and for about 20 years, in Argentina, also by means of horizontal sections.

This means that the well reaches the area where the hydrocarbons are located vertically and then with a special tool it is made horizontal. In this way, many more hydrocarbons can be extracted than if it were only vertical.

In order to prevent the hole in the well-being drilled from being covered by crushed rock and to cool the drill bit, a fluid called mud is added through the center of the pipe. This mud is formed with water and a special mineral called baryte, which is non-polluting and sometimes contains chemicals, depending on the nature of the ground being drilled. These chemicals can be, for example, gas oil.

The sludge runs in a circuit from the wellhead to the drill hole, in the depths, and from there it returns to the wellhead dragging the bits of crushed rock.

The used sludge is recovered by removing the rock chips and reinjected.

At the end of the drilling, the excess mud is dumped into a pool next to the well.

This sludge usually contains oil and if the pool is not treated and the petroleum is removed, it becomes a contaminated area that ruins the soil and is very harmful to birds.

There are techniques for remediation.

But the history of well drilling in Patagonia has thousands of abandoned, untreated pools that are major environmental liabilities. In Chubut Province, more than 5000 untreated pools are reported.

The fact is that there has been very little control of the oil companies by the State.

A characteristic of the hydrocarbon fields is that their production declines over time. As one of the main forces that allow its extraction is exhausted which is the natural gas that is losing pressure.

For that reason, to extend the useful life, water and gas are usually injected under pressure from the surface.

For example in Argentina, the average production of a petroleum well is about 8–10 m^3 per day when at the beginning of its production can be at 80 or 100 and this decline usually occurs gradually over 10 or 15 years.

For example, the large producing countries in the Middle East produce about 500 m^3 per day per well, which is almost 50 times more than in Argentina.

This low productivity of the wells in Argentina, which produces only 0.2% of the world's petroleum, makes Argentina one of the 5 countries with the highest number of wells drilled.

In other words, what Saudi Arabia produces in one well, in Argentina requires 50 wells.

3.3 The Fracking

Petroleum and natural gas originate from what is called the bedrock. There, in hundreds of years, in beds of ancient seas, buried organic matter decomposed and generated petroleum and natural gas.

That is why it is called bedrock.

But from the bedrock, petroleum and natural gas migrate, moving until they are trapped in special formations that form a barrier of impermeable rock that prevents them from rising to the surface.

There they are housed, forming a deposit.

Today the technology recovers between 30 and 50% of the petroleum originating in the mother rock and the rest remains in the soil. Technical and economic reasons prevent the recovery of more.

In conventional deposits, the rock in which the petroleum is embedded sometimes has unconnected pores and to increase recovery, so-called hydraulic fracturing is performed, which is almost as old as the oil industry.

This technique consists of injecting water from the head of the well, with some chemical substance, under pressure that reaches the rock where the petroleum is located and increases the porosity and allows either to produce petroleum or to increase its production.

In other words, hydraulic fracturing is an old technique.

But where does fracking, which is a multiple fracturing technique, take place today?

Firstly, Fig. 3.1 shows the process and its impacts.

In formations where petroleum and natural gas are hosted in rocks of very low porosity and permeability, or in very compact clays, even in the same bedrock, they are called shale oil or shale gas (oil and natural gas from clays) or tight oil or tight gas (oil and natural gas from compact sands).

Fracking consists of an injection of water, sand and chemicals at high pressure but not in the vertical part but in the horizontal part of the well.

The horizontal section can be 1000 m and the vertical section 2000 or 3000 m long.

The walls of the horizontal section must be perforated by controlled explosions, which produce a series of holes along the horizontal section.

Water with sand and chemicals will pass through these holes at very high pressure causing multiple fractures in the rock area around the horizontal pipe.

This increases the porosity and permeability of these rocks that do not originally have it and this makes it possible for petroleum and natural gas to exit the pipe.

A lot of water is consumed almost 20,000 m^3 per well with the addition of about 400 tons of chemicals diluted in water so that the solution is 2% chemicals and 98% water.

The sand prevents the holes in the horizontal pipe from closing.

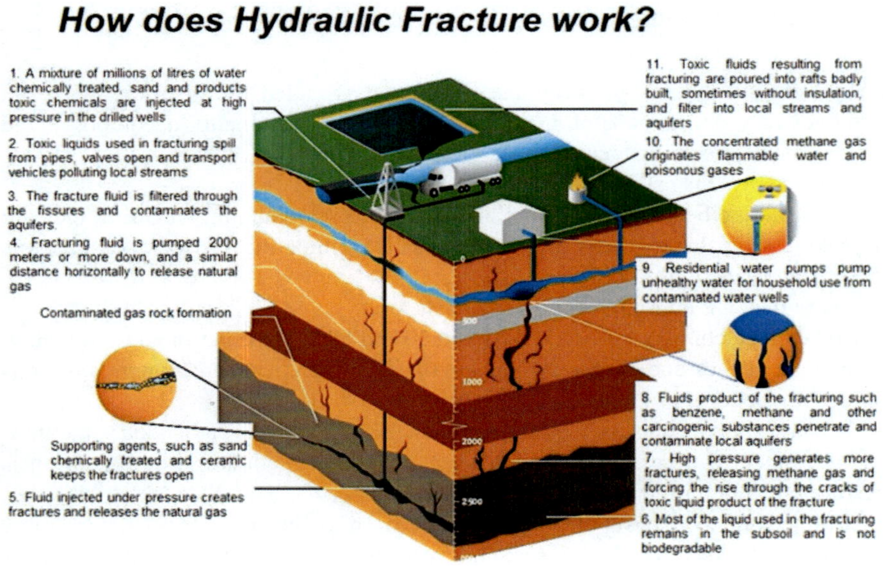

Fig. 3.1 How does hydraulic fracture work? *Source* Hidraulicano.info, May 14th 2012, Cantabria

The operations of this type consist of batteries of wells called platforms (2 or 3 per km^2) and from each one 6–8 wells are drilled, very close together. So the number of wells is enormous.

The production from each well is then collected on the platform. That is, the wells converge on the platform.

About 15 fractures are made in each well, which gives an idea of the amount of water and chemicals to be used.

The composition of the chemical products is not made known by the companies, as they say it is a commercial secret, but it is supposed to be about 500 chemical substances:

17: toxic to aquatic organisms
38: acute toxicants
8: proven carcinogens.

These are generally oxidants, biocides, aromatics, carbon disulfide, pyridines, etc.

In short, a large quantity of water, polluting and toxic chemicals and the expulsion of natural gas into the atmosphere, as not everything is recovered. This expulsion of natural gas increases the greenhouse effect, since the greenhouse effect power of natural gas is 23 times that of the main agent of this type which is carbon dioxide.

A large part of the injected fluid returns to the surface (between 30 and 50%) and this fluid is highly contaminating.

The one that remains in the subsoil can migrate toward water layers and contaminate them.

In Neuquén, the first 500 m of the vertical pipe will be required to be cemented in order to avoid the contamination of the underlying water layers.

However, due to the cracks generated in the rocks by the injections of water, sand and chemicals, the contaminated fluid can ascend to these layers.

Oil and gas wells that lose their structural integrity also leak methane and other pollutants through the liners and release them into the atmosphere and water sources.

Why do so many wells leak? Underground pressures, temperature changes, earth movement from drilling nearby wells, and shrinkage crack and damage the thin layer of cement that is supposed to seal the wells. Keeping the cement in perfect condition while drilling horizontally in the shale is extremely difficult. Once the cement is damaged, repairing it thousands of meters underground is expensive and often unsuccessful. Oil and gas industries have been trying to solve this problem for decades. See: "Shale gas towards a warmer future"; Engineer Anthony R. Ingraffe, researcher and professor at Cornell University, Ithaca, New York State, USA; source: Diario Clarín, Buenos Aires.

For the fluids that come out there are two alternatives:

Treat them in plants at the foot of each platform, but then the separate contaminants where they go, are they moved?

Take them out in trucks and deposit them in other places and then move them again.

In conventional farms, there are almost 5000 contaminated pools in Chubut Province alone.

The useful life of these unconventional wells is very low and the production is exhausted in 5 or 6 years. This then intensifies the drilling.

Once the life of the wells is over, they have to be abandoned and sealed with cement plugs.

The transport of water and materials and supplies causes a huge traffic of trucks, including cisterns, which circulate on rural roads.

In Neuquén, it is estimated that the water would come from Lake Mari Menuco or from the Colorado, Neuquén or Río Negro rivers. The consumption of water from the underground layers would be forbidden and only surface water could be used.

The Neuquén government has even announced that it will build a network of aqueducts in charge of the province to supply the needs of the oil companies.

The problem would not be water consumption but water contamination and the disposal of recycled sludge from the drilled wells.

Anyway, there are working techniques that can save water consumption.

According to the oil service company Schlumberger, the unconventional operations of five operators working in the Neuquén basin demand an annual average of 1,800,000 m^3 of water in an initial stage of exploration or semi-development. Of this figure, some 600,000 m^3 return to the surface (known as flowback water), a volume that must be treated and reinjected, and this is precisely the problem in the treatment and disposal of these waters.

The companies say that instead of 1,800,000 m^3, only 1 million, or perhaps less, could be used. But this would be a considerable technical challenge, requiring the help of new software. It would have to be taken into account that between 40 and 50% of the fractured wells do not end up producing. Therefore, more detailed analysis should be done at all levels until the operation is completed. And all these variables should be integrated into one platform in order to reduce costs and increase production.

According to business people, it is already totally feasible to model non-conventional fractures (which are irregular and have multiple ramifications) to better predict which points have the greatest production potential. By optimizing the fractures, water consumption can also be reduced.

The technique to be used would be that of fracture with channels, which was created in Argentina and is already being used effectively in many countries, including the USA. Thus, instead of injecting sand continuously during the entire fracture, intermittent pulses of sand and fracture fluids would be made, thus saving this raw material.

In short, the companies claim that it is possible to use up to 60% less water in the shale industry, combining two already proven techniques and taking advantage of the experience and knowledge available in the country. This saving would be achieved, in 20%, through the selection of the fractures, integrating the information and using it to make better decisions and the remaining 40%, by means of the mentioned technique of fracture with channels that besides promoting the hydric saving would have other advantages, among which a greater conductivity would stand out.

In any case, the volume of water used, which is still reduced by almost 60%, must be treated and reinjected, and this is precisely the problem, in the treatment and disposal of this water, known as flowback.

In other words, the main problem is not water consumption, which if reduced will be welcome, but the final disposal of polluted water and this is no longer a problem of quantity but a problem of quality.

3.4 Reserves of Resources? Of "Non-conventional" Hydrocarbons

According to the opinion of the International Energy Agency (IEA) in a document of June 2013, some countries have huge oil and gas reserves that until relatively recently could not be tapped, and it was not even in their plans to do so. Thanks to technological advances, these unconventional hydrocarbons can now be extracted more easily and at sufficiently reasonable prices. Shale oil and shale gas, their names in English (in Spanish they receive others such as bituminous sands, shale oil and gas, shale gas …) may end up revolutionizing the world energy map. According to the latest estimates from the U.S. Department of Energy, the world has deposits containing some 345 billion barrels of unconventional oil, 10% of the world's total crude reserves. And also with almost 7.3 trillion cubic feet of unconventional natural gas, which is 32% of the world's total reserves. These figures already represent a turning point in the conception of the future of fossil energies. They represent an increase in global reserves of 11% in the case of crude oil and 47% in natural gas and may also fall far short of the actual shale reserves the planet has. The report by the USA. Energy Information Administration (EIA) considers the reserves present in only 42 countries, only contemplates the resources that can be extracted by means of the technologies currently in use and, furthermore, leaves out other potential deposits that would be found under the large oil wells in the Middle East and the Caspian region, and which could be substantially larger than those already known. New hydraulic fracturing (fracking) and horizontal drilling techniques are being used to discover new oil and gas deposits trapped in the rock. But there could be much more. The huge reserves detected may be the gateway to an energy revolution, but it is still to be welcomed with caution that all of them may actually be exploited in the future. "The report shows significant international potential for unconventional oil and gas. But it is not yet clear to what extent technically recoverable resources are also exploitable in economic terms," says Adam Sieminski, director of the EIA. A new world energy map new reserves can drive a shift in the global energy status quo. At present, only the USA and Canada exploit their unconventional gas and oil reserves in truly commercial volumes. And they are called upon to be the protagonists of this new hydrocarbon boom. The USA seems to be the big winner. In fact, a few months ago the International Energy Agency (IEA) painted a new global scenario in which, thanks to its unconventional reserves, the USA would become the world's largest producer of natural gas in 2015 and in 2017 it would also be the leader in oil production. But other countries that are not yet exploiting the full potential of their unconventional fields may also become global giants in this new business. The great power of unconventional crude today is the USA, but Russia greatly surpasses it in shale oil reserves. The Russian giant holds one-fifth of all technically recoverable unconventional crude oil reserves in the world (75 billion barrels), followed by the USA (58 billion) and further behind by China (32 billion), Argentina (27 billion) and Libya (26 billion). These five countries account for more than 60% of all the world's shale oil reserves.

The battle for world leadership of unconventional natural gas will be much tighter. Estimates by the US Energy Information Administration recognize China as the country with the largest shale gas reserves (1.115 trillion cubic feet), followed by Argentina (802) and Algeria (707). The official data of the North American Government contemplates that the American reserves remain in the 665 trillion cubic feet of gas, which would leave it in fourth position. But the magnitudes handled by some consulting firms are pushing the USA's volumes up to the top of the world ranking. The Advanced Resources International group sets the US reserves at 1161 trillion cubic feet, which would surpass China as a major shale gas power.

However, other opinions, especially in Europe, and in some states of the USA and Canada, express well-founded concerns about the negative environmental impacts of the exploitation of these "non-conventional" hydrocarbons.

Finally, except in the USA and Canada, where they are already being produced commercially, in the remaining countries, the figures supplied are more similar to Resources than to Reserves, as there is still much work to be done in exploring and verifying the behavior, under an exploitation regime, of these structures (Fig. 3.2).

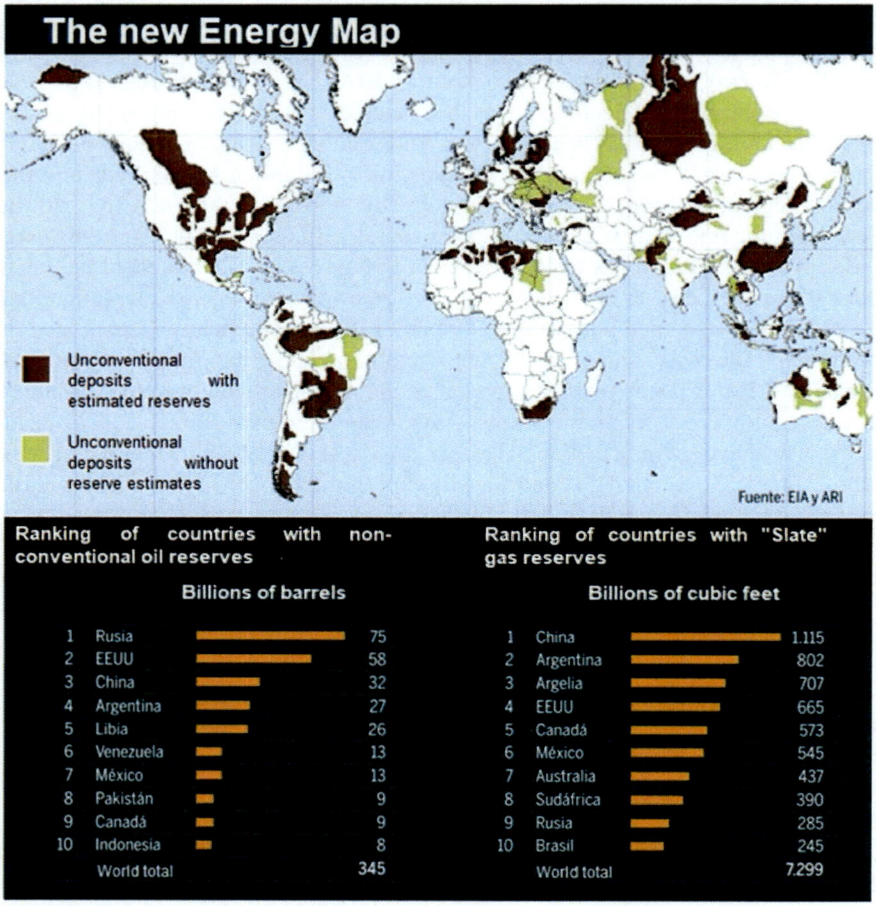

Fig. 3.2 The new energy map. *Source* Fracking a fracture that takes its toll, Aitor Ultrasti and Florent Marcellesi, September 2012, Political Ecology Journal N° 43, 16 September 2012

3.5 Why Fracking in Argentina?

Because more and more oil and natural gas are being imported and this affects the trade balance and there is practically a danger that the mentioned imports will absorb the trade surplus. And then where would the money come from for the imports that the country needs to make, for example, for the functioning of the industrial sector and to pay the foreign debt? Well, the Central Bank's reserves are not infinite and are partly committed.

Because the subsidies to the energy sector (to natural gas and electricity) together with those destined to the transport sector are one of the main causes of the budget deficit and the government inexplicably delays a change in the tariff policy. For example, if cross-subsidies were applied in the tariff scales, so that the residential users who consume more and the service sector would pay more per unit consumed than the lower-income residential users and the industrial sector, and making this gradually, companies could have an average rate that covers their costs and an adequate expansion of their investments.

Because exploration for conventional oil and natural gas was stopped and a report by the International Energy Agency places Argentina in third place, behind the USA and China as the country with the largest non-conventional natural gas resources. Resources that would be equivalent to almost 70 times the current proven natural gas reserves The resources would be located in: the Neuquén Basin (in the Los Moles and Vaca Muerta formations there are 14 deposits that are being investigated by YPF; Petrobras; Pluspetrol; Pan American Energy; Apache; Exxon; Shell; and the provincial company); in the San Jorge Gulf Basin (Aguada Bandera); in the Magallanic-Austral Basin and in the Chaco Paranaense Basin. Vaca Muerta oilfield is currently being exploited and the Apache Company has produced about 1000 m^3/day of oil, equivalent to 1.4% of the total oil production of the country. In the rest of the structure, Vaca Muerta YPF has placed its greatest hopes and with the production obtained, it estimates that it would reach in 4–6 years self-supply of oil and thus stop importing LNG. One of the most serious drawbacks, from the business point of view (YPF) is the magnitude of investment required to explore (resources must be converted into reserves) and exploit these structures. A well with fracking costs between 12 and 18 million dollars when a conventional exploration well does not exceed, exaggerating, 4 million. That is to say that the investment needs are considerable, mentioning no less than US $7000 million dollars/year for no less than 5 years and currently YPF does not have these figures. As for the availability of technology, without minimizing, although YPF does not have the necessary experience, it can obtain it quickly, since the two essential elements of it, horizontal drilling and hydraulic fracturing, have been handled by YPF for many years.

3.6 Main Environmental Impacts of Fracking

This point is taken verbatim from the Document: "Fracking: a fracture that takes its toll" by Aitor Utresti and Florent Marcellesi of September 2012.

Risks during drilling

As already mentioned, special drilling techniques are required to enable subsequent hydraulic fracturing. For all these reasons, in addition to the usual risks of drilling for hydrocarbons, there are also the specific risks of drilling for diversions. We are therefore talking about risks of explosion, gas leaks, hydrogen sulfide leaks (very toxic in low concentrations), and formation collapses on the pipe. The latter is much more common in the case of diverted boreholes such as those carried out in this case. Let us remember that an average of 6–8 wells are being drilled per platform, and between 1.5 and 3.5 platforms per km^2, so although a priori the risk of an accident of this type occurring per well is low, as the number of wells to be drilled increases the risk increases alarmingly.

Water pollution

One of the major concerns of hydraulic fracturing is the effect on underground aquifers. By fracturing the subsoil, there is a possibility that one of the induced fractures will reach an aquifer, contaminating the water with the fracturing fluids and with the formation gas itself. In addition to this risk, there is also the possibility that during fracturing it will connect to an old, poorly abandoned well, and hence the gas will communicate well with an aquifer, as it does with the surface. This type of accident has happened before, contaminating an aquifer through an abandoned well in the 1940s.

Chemical risk of additives

As mentioned above, about 400 tons of chemicals, most of which are highly polluting, are required in each drilling operation. When diluted to 2% in water, their level of toxicity is greatly reduced. However, these chemicals reach the platform unmixed. The risk of accident during transport must be taken into account. The number of truck transfers to be made for the density of wells to be drilled is high (which in turn leads to noise pollution and road safety). For each platform, the minimum movement of trucks is estimated to be 4000, a large number of them for the transfer of chemical products. Again, although the risk of an accident involving a chemical spill is low, the large number of operations to be carried out makes it a major risk.

Air pollution

Throughout the drilling and fracturing process, a large number of additives are used, many of which are volatile compounds. The same is true later in the production stage, where the extracted gas needs to be conditioned for injection into the pipeline. All these compounds pass to a greater or lesser degree into the atmosphere, and can generate ozone, or BTX (Benzene, Toluene, Xylene) among others.

Earthquakes

In those areas where the development of fracking is more advanced, an increase in seismicity has been observed, coinciding with periods of hydraulic fracturing. It must be taken into account that during fracking operations the subsoil is pressurized more than 100 times. This overstressing can be sufficient to cause displacement of underground faults, and therefore earthquakes, as has happened in Lancashire in the UK where the company Cuadrilla Resources has recognized that its drilling was the cause of two local earthquakes.

Greenhouse effect

Unconventional gas, because of the conditions in which it is found, is usually made up almost entirely of methane. This is a much more powerful greenhouse gas than CO_2 itself, specifically 23 times more powerful. This means that any leakage of it during drilling, fracturing, and production is much more harmful than the gases that are generated later during its combustion.

The added problem of fracking techniques with respect to gas leaks is the fracking water in its return. Having been in contact with the gas underground, it absorbs a quantity of gas, which on returning to the surface is emitted into the atmosphere. It has been estimated that in a well in which hydraulic fracturing has taken place, the increase in methane emissions is 2%. A report from Cornell University therefore estimates that shale gas results in an increase in greenhouse gas emissions of between 30 and 100% compared to coal.

Land occupancy

An added problem is the high land occupation of this type of exploitation. As mentioned above, a large number of wells need to be made in order to make good use of the resources. Usually, 1.5–3.5 platforms are drilled per km^2, with an occupation of 2 ha each. The visual impact of this accumulation of wells is very great.

3.7 What About the Rest of the World?

This point is based on the Document: "Fracking: a fracture that takes its toll" by Aitor Utresti and Florent Marcellesi of September 2012.

Unconventional gas fields are distributed throughout the planet with a novel character: they abound in countries historically poorer in hydrocarbons. While in the geopolitics of conventional gas—and energy (in)dependence—Russia, Iran, Qatar and Saudi Arabia concentrate more than 50% of world reserves, in the geopolitics of unconventional gas the list is headed by China, the USA, Argentina, Mexico, South Africa, Australia, India and Europe (central and eastern Europe, France, UK, etc.) and North Africa play an important role. Meanwhile, in South America, besides Argentina, it is Brazil, Chile, Paraguay and even Bolivia, who are far ahead of the traditional hydrocarbon giant, Venezuela. Given that on the one hand the exploitation of non-conventional hydrocarbons is a potentially very profitable business with a central geopolitical character, and that on the other hand it entails serious environmental and health problems, it is not surprising that more and more socio-ecological conflicts are arising at all the extraction points where this technique of hydraulic fracturing is applied.

USA, the guinea pig of fracking

The first alerts have come for years from the USA where, according to data from the US Energy Agency, non-conventional gas production has risen from 1.4% of total US gas supply in 1990 to 14.3% in 2009, and may reach 24% by 2035. This previous experience allows us to have more perspective when analyzing what has happened, given that the first wells were started in the eighties, with a boom in the 2000s, and there are currently 500,000 wells drilled, and a forecast rate of over 30,000 new wells per year.

It is clear that the anti-fracking movement has made its voice heard: documentaries like Gasland, made by US environmental activist Josh Fox, which can be downloaded from the Internet, or Fracking Hell and organized citizen platforms like "No Fracking" show the political and commercial complex hidden behind the exploitation of unconventional gas and its environmental and health consequences in the USA.

"The Promised Land," an anti-fracking film co-written by Matt Damon and John Krasinskipor, and directed by Gus Van Sant, is in preparation: https://www.huffingtonpost.com/2012/04/06/promised-land-matt-damon-fracking_n_1408501.html (it's pretty bad despite good intentions).

The main conclusions that can be drawn from the American experience are once again reminiscent of the classic patterns of conventional socio-ecological conflicts.

The collusion between political and economic powers: under the Bush Junior administration and after an incessant lobbying by the energy transnationals, it caused the repeal of several of the most important environmental protection laws in the US, among them the Safe Drinking Water Act, so that this law would not be applied to fracking. This provision has come to be called the "loophole" or the "Halliburton amendment," since the multinational Halliburton is one of the pioneers and one of the largest providers of hydraulic fracking services in the US, and whose previous executive director was none other than Dick Cheney, then Vice President of the US when this specific legal exemption was approved.

The economic power of the energy multinationals with checks to buy the land; with promises of new sources of employment around the wells. On the other hand, the very aggressive publicity campaigns, get the approval of both the owners of the land, where the gas is, and the economic, political and legislative actors. They also generate their own expert reports where they self-excuse themselves from any environmental pollution or health effects. In fact, due to the weak role of the federal Environmental Protection Agency, the federal states, with tight budgets, are defenseless to face interests and budgets in the millions, which in turn explains that "21 states out of 30, where there are wells, have no specific regulations and none require companies to report the amount of toxic fluid that remains underground".

Harmful effects on the environment and health: in addition to the environmental risks mentioned above, there have been cases of cancer, respiratory problems, brain damage, neurological disorders, hypersensitivity to chemicals, mainly due to water and air pollution. As stated by Grandoso, Duke University has carried out a study that shows that the drinking water wells near the extraction sites have a very high concentration of methane.

A study showed that drinking water wells near extraction sites have very high concentrations of methane, "an indoor asphyxiant and a fire and explosion hazard," while in the Texas town of Dish surrounded by wells, 61% of the recorded illnesses were associated with the contaminants used by the fracking. In November 2010, a study by the Environmental Protection Agency in Wyoming linked contamination of drinking water wells to fracking.

After fifteen years of trials, struggles and diverse lobbying, confrontations between anti-fracking movements and transnational energy companies have resulted in the temporary suspension of water fracking in New Jersey, New York and Pennsylvania until the risks of contamination of drinking water are better understood, while 16 municipalities have approved local bans (but without regulatory capacity over the gas and oil industry), and Vermont has become the first federal state to approve the ban in May 2012. In addition, in September 2010, the federal Environmental Protection Agency (EPA) requested information on chemicals used by extractive companies: eight responded, and a subpoena was needed for Halliburton to respond. As a symbol of this struggle—among other less publicized ones—Josh Fox is today on the US government's "Terror Watch List" and, denounced by the American Civil Liberties Union, he was arrested in February 2012 when he attended with his camera a public hearing of a US Congress commission dominated by the Republicans and where the harsh conclusions of the EPA about fracking were analyzed.

A new study reveals the contamination of drinking water wells near shale gas extraction sites in the USA, which could reignite the debate on the environmental impact of this controversial technique.

Researchers from Duke University in North Carolina (east) analyzed water samples from 141 private wells that supply homes in the Marcellus shale gas basin in northeastern Pennsylvania and southern New York State.

Methane concentrations in drinking water from homes within a kilometer of the drilling sites averaged six times higher than water from homes farther away, while ethane concentrations were 23 times higher.

The amount of methane in most of these wells far exceeded 10 mg per liter of water, the maximum level accepted by the US health authorities.

Propane was also detected in 10 water samples from household wells located less than one kilometer from the extraction sites.

"The results on methane, ethane and propane, as well as new indications of traces of hydrocarbon isotopes and helium, lead us to believe that the extraction of shale gas has affected the sources of drinking water in homes," said Robert Jackson, professor of environmental sciences at Duke University and lead author of this paper published in the Proceedings of the U.S. National Academy of Sciences (PNAS) Jun. 24–28.

The data on ethane and propane contamination "are new and difficult to disprove," he insisted.

"There is no biological source of ethane and propane in the region, and the Marcellus shale gas basin is rich in these two gases," said the researcher.

These scientists have considered all the factors that could explain the contamination, including the topography and geological characteristics of the site.

"Our research shows that distance from extraction sites, as well as variations in local and regional geology, are the main factors in determining the possible risk of groundwater contamination that should be considered before drilling," explained Avner Vengosh, professor of geochemistry and water quality, and co-author of the paper.

Previous studies by researchers at the same university had found evidence of methane contamination in water wells located near drilling sites in northeastern Pennsylvania (Source AFP-Terra.com, June 26, 2013).

However, a third study, conducted by scientists from the U.S. National Geophysical Institute, had found no evidence of drinking water contamination from shale gas extraction in Arkansas (center).

None of these investigations have detected contamination from the fluids—a mixture of water and chemicals—that are injected under high pressure to fracture the rock and release the shale gas.

Europe and the rest of the world

After the US trial, fracking has landed in Europe and many other parts of the world where, as in the US, several socio-ecological conflicts have been generated, with the dominant economic and political powers facing up to neighborhood, social and environmental groups that denounce the risks inherent in the exploitation of non-conventional gas.

At the level of the European Union, where the International Energy Agency estimates that Europe's reserves of non-conventional gas are 35 trillion m^3, equivalent to forty years of gas imports by current standards, the legislative complexity is once again apparent. So far, the recommendations of the European Parliament, which indicated in July 2011 that "the environmental risks and burdens [of fracking] are not offset by its corresponding potential benefit," are opposed. It recommends that they be regulated and that the components used in the wells be made public, as well as those of the European Commission, which has the capacity to promote a directive on the subject, and which considers that existing legislation can be applied to both conventional and non-conventional gas. In addition, the battle continues in the European Parliament where, on the initiative of a Polish MEP, a report promoting the expansion of fracking gas in Europe has been discussed in the Committee on the Environment, Public Health and Safety. On the other hand, at the end of April 2012, the European Union's Committee of Petitions studied 8000 signatures on the risks associated with this technique. For its part, the anti-fracking movement, through a coalition of 36 non-governmental organizations specializing in the fields of the environment and health, mainly from Europe but also from the USA, Australia and South Africa, has called on the European Union "to ensure that until these problems are properly addressed through a thorough scientific assessment (...) activities related to the extraction of gas and oil from shale, as well as methane from coal seams, should not be continued." They urge Member States to "suspend all ongoing activities, revoke permits and prohibit all new prospecting and exploitation projects." Finally, in May 2012 the European Green Party, very active in the European Parliament, and the social mobilizations with the MEP and alter globalist José Bové, spoke out for all of Europe against the exploitation of non-conventional gas by means of hydraulic fracturing.

The situation in Europe at country level

Germany

Moratorium in the state of North Rhine-Westphalia from March 2011.

Bulgaria

The Bulgarian Parliament passed a resolution banning water fracturing on its territory in January 2012 and provides for a fine of around 50 million euros and the confiscation of the equipment used by those entities that practice it.

Previously, Bulgaria withdrew Chevron's first license for slate gas extraction in North-Eastern Bulgaria.

Opponents of fracking are calling for the resolution to become law.

Czech Republic

The Minister of the Environment proposes a temporary ban for about two years so that new laws can be drafted to regulate the rules for surveys in which various foreign companies have taken an interest.

France

Exploitation of hydrocarbons by fracking prohibited by law since July 2011 for injecting "extremely aggressive products" and resulting in "destroyed landscapes, contaminated water" and "dubious safety."

There is a "National Coordination of Groups against Slate Gas and Oil."

The new French president, François Hollande, while opposed to the exploitation of unconventional gas, does not oppose research, exploration or the withdrawal of drilling licenses or permits.

Northern Ireland

At the end of 2011, he declared a moratorium until environmental studies are carried out.

The Netherlands

National moratorium until summer 2012 until the effects of the technique are known.

Poland

No specific regulation.

Called "fracking heaven," the Polish field would have a central geopolitical importance in counteracting dependence on Russian gas and Gazprom's interests.

After publicizing an estimate of reserves that placed it as the largest field in all of Europe, the latest news reports a volume of gas 10 times less than predicted.

Seven people, including government officials and businessmen linked to Petrol Invest, have been accused of offering or accepting bribes in the granting of licenses to search for unconventional gas in 2011.

UK

No specific regulation.

Some operation of Cuadrilla Resources in Lancashire has been suspended following the occurrence of seismic movements.

Romania

Despite the support of the Romanian President, the newly elected government is preparing a moratorium on fracking.

After the ban in Bulgaria, Romania represented Chevron's new hope.

Sweden

Sweden would allow small-scale fracking under an appropriate regulatory framework.

Switzerland

In April 2011, the canton of Freiburg suspended all authorizations to search for slate gas on its territory.

Ukraine

No specific regulation.

Although estimates differ, it would potentially have with Poland the most important reserves in Europe.

Auctions have started to grant the first exploration and exploitation permits.

Ukraine wants to reduce its dependence on gas imports from Russia.

Exxon, Chevron, Shell, BP, ENI have made offers for the first lots.

Spain

Due to the great lack of transparency or information on the part of the public authorities, it is not easy to know exactly the reality of non-conventional gas exploitation in Spain and to know which permits or concessions in force or requested are for conventional and non-conventional hydrocarbons. As far as non-conventional gas is concerned, the areas of greatest interest for political and economic interests and of greatest social conflict are currently to be found in the so-called Basque-Cantabrian basin (which includes mainly deposits in Cantabria, Álava, Burgos and to a lesser extent Bizkaia, Navarra, La Rioja) and, with a view to the future, also in Aragon, Seville and Jaén. While in Cantabria there is a permit with several wells under the name Arquetu and in Burgos some 20 research wells are planned, the most important deposit is in Alava in the Gran Enara permit where, according to the Basque Energy Agency (EVE), an estimated 184,500 Mm^3 of reserves are located, which would mean, ignoring the extraction ceiling of the wells, the consumption of the Basque Country of 60 years and of Spain for 5 years.

Therefore, although there is currently no state platform linking the different local struggles against fracking, different anti-fracking movements have been organized in each affected area. Whether in Cantabria, where the first permits were issued and a website was launched with the most information in Spain on fracking, in the Basque Country (mainly in Alava and then in Bizkaia) or in Burgos and Navarre more recently, the social and political struggle has reached a significant degree of conflict, making it difficult on the one hand for the political-economic roller to roll and on the other hand allowing a greater degree of awareness and information for society. Furthermore, thanks to this work, more than 15 municipalities in Alava have declared themselves free of fracking, while Vitoria-Gasteiz (European green capital 2012!) is demanding an Environmental Impact Study for all the wells or in Cantabria several town halls have appealed for permits in the Arquetu area.

It is also of great interest in the construction of more global alternatives that the anti-fracking struggle once again demonstrates the confluence and unity of action that is increasingly normal and powerful among movements for environmental, social and democratic justice. For example, in the Basque Country, the anti-fracking collective includes ecological associations (Ekologistak Martxan, Eguzki, Gaia, Mendialdetik, etc.), political parties (Bildu, Equo, Izquierda Unida, Aralar), the 15M movement and individuals, or in Cantabria, where they include people affected by the issue and organizations concerned with it, such as the 15M movement, Democracia Real Ya, Ecologistas en acción, ARCA, Asamblea contra el TAV, Agitación Rural or Regüelta (You can see the situation in Spain in the documentary "The Shadow of the Fracking" that can be downloaded from internet https://vimeo.com/46871495#t=4).

In the rest of the world

In the rest of the world, whether in South Africa, Quebec, Australia, for example, citizen mobilizations have achieved moratoriums. In contrast, in China, where due to the lack of opposition, the Chinese government signed an agreement with the USA in 2009 and has already begun to drill with the promise that its reserves will be the most important in the world.

Something else about the USA

The exploitation of "unconventional" hydrocarbons through the technique of fracking (especially Permian) in the USA has changed the world's hydrocarbon landscape.

To demonstrate this, Table 3.1 includes the situation of the main concepts referring to hydrocarbons in the USA, comparing what was happening in 2008 (before the fracking boom) with what happened in 2018.

The growth in all concepts has been remarkable, becoming in 2018 the first world producer of oil and Natural Gas and the third world exporter of LNG after Qatar and Australia.

In 2008, the USA imported the equivalent of 66% of the oil it consumed and in 2018 it will import 48.5%.

In 2008, the USA exported 29% of the oil it produced and in 2018 46.6%.

In this way, the USA has reduced its dependence on imported oil, especially from the Middle East, and has practically limited it to that which it buys from Canada and Mexico.

The OPEC countries and Russia have been the most affected by the appearance of the USA as a competitor in the world hydrocarbon market.

The question is whether this situation can be maintained in the medium and long term, beyond the processes of substitution of fossil fuels that are taking place in the world.

But the boom of fracking in the USA has as its counterweight the enormous investment effort required to contain the natural decline of the productive structures,

Table 3.1 US hydrocarbon data

	Units	2008	2018
Oil proved reserves	Thousand million barrels	28.4	61.2
Oil production	Thousands barrels per day	6783	15,311
Oil consumption	Thousands barrels per day	19,490	20,456
Oil imports	Thousands barrels per day	12,872	9929
Oil exports	Thousands barrels per day	1967	7131
Natural gas proved reserves	Trillion cubic meters	6.8	11.9
Natural gas production	Billon cubic meters	546.1	831.8
Natural gas consumption	Billon cubic meters	628.9	817.1
Liquefied natural gas imports	Billon cubic meters	9.7	2.1
Liquefied natural gas exports	Billon cubic meters	1	28.4

Source BP Statistical Review of World Energy

which at present would seem to be sustained almost entirely by the exploitation of the Permian and which has led to the bankruptcy of numerous small- and medium-sized oil companies. The fact is that international oil prices and LNG prices have not accompanied the growing evolution of production, despite the efforts that companies have made to reduce their costs.

That is why it is pertinent to show the work included below:

Fracking in the U.S. all that remains is the permic and that will always last (taken from HFI Search, Seeking Alpha, February 19, 2020)

Summary: In the coming months, the story that U.S. oil production growth will slow down materially will become the main reality. Using an estimated 11,728 wells to be completed this year, we have U.S. shale oil growth of ~424 k barrels per day (b/d) year over year. So, although the overall figure is still positive, all the growth will come from the Permian.

The declining productivity profile of the wells at Bakken, Eagle Ford, Niobrara and Anadarko, combined with the completion of lower wells in 2020, will result in a decline in extraction. This will be the first annual decline in these basins since 2016.

But to make matters worse, the Permian will not be able to sustain the growth of US oil production much longer either. If we assume ~5500 completed wells in the Permian per year, the growth rate begins to plummet after 2020.

In the coming months, the story that U.S. oil production growth will slow down materially will become the mainstream. We are already seeing this in the real-time data with US oil production. Now falling to ~12.6 mb/d, but this will become more evident in the coming months as the EIA 914 reports double-check the data we see.

Most important, however, is the breakdown of US oil production growth this year (Table 3.2 and Fig. 3.3).

Table 3.2 Shale oil production

Shale Oil Production	Average Production	Growth
2017	5,831,751	516,180
2018	7,383,498	1,551,747
2019	8,599,148	1,215,650
2020e	9,023,224	424,076
Total Growth	3,191,473	

Source EIA, HFI Research

Using an estimate of 11,728 wells to be completed this year, we have a growth in US shale oil ~424 kb/d year on year

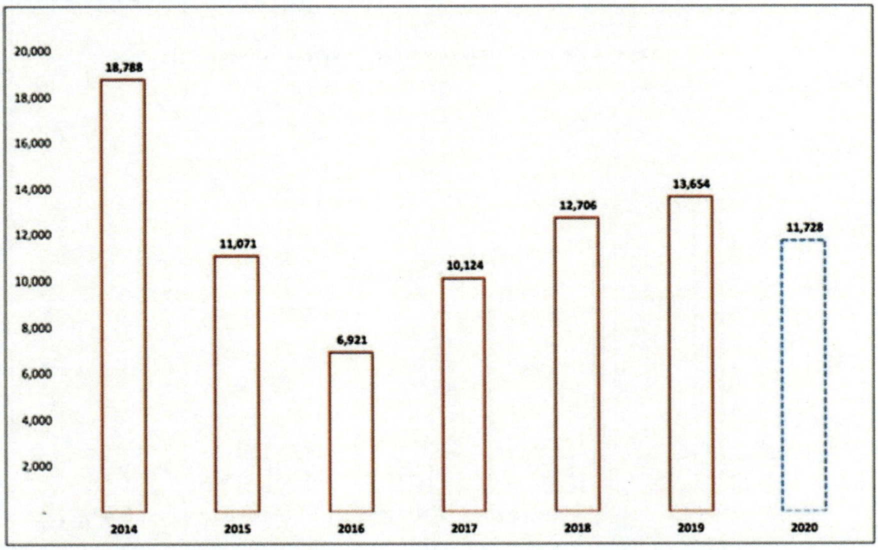

Fig. 3.3 Total wells completed. *Source* EIA, HFI Research

So, although the overall figure is still positive, all the growth will come from the Permian (Fig. 3.4).

The declining productivity profile of the wells at Bakken, Eagle Ford, Niobrara and Anadarko, combined with the completion of lower wells in 2020, will result in decreased production. This will be the first annual decline in these basins since 2016.

But to make matters worse, the Permian will not be able to sustain the growth of US oil production much longer either. If we assume ~5500 completed wells in the Permian per year, the growth rate begins to plummet after 2020 (Figs. 3.5 and 3.6).

This is just the nature of shale. As the base production increases, so does the base rate of decline. The proverbial treadmill increases with pace, so without a corresponding increase in productivity or well completion rates, growth inevitably stops and becomes absolute decline. According to our estimate, the total production gain in the Permian from 2020 to 2025 is only a small ~426 kb/d.

So where does that leave the world's oil supply in the future? (Fig. 3.7).

Goldman estimates that by 2025, OPEC needs to increase oil production by ~5.4 mb/d. But that is under the assumption of +2.3 mb/d of US oil shale. If U.S. shale can only add ~1 to ~1.5 mb/d, the deficit increases to ~6.2 mb/d.

We do not think this scenario is feasible at all with Brent below $60/bbl. One of these will have to give way. Either oil prices will skyrocket to begin destroying demand, or the supply deficit is unavoidable.

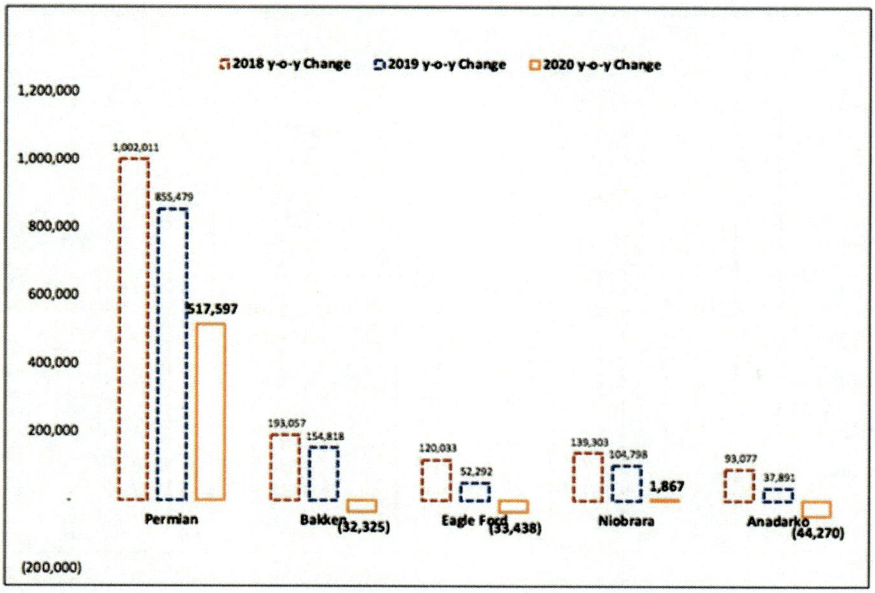

Fig. 3.4 US shale production growth by basin. Fuente: EIA, HFI Research

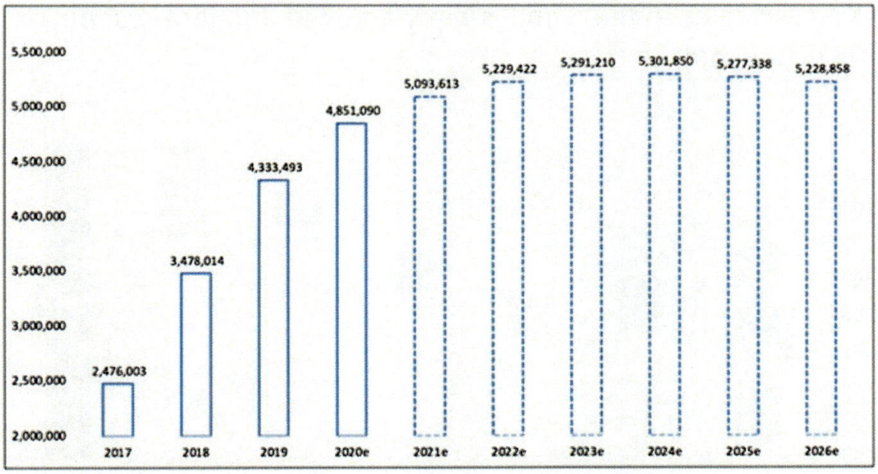

Fig. 3.5 HFI research Permian oil forecast average production. *Source* EIA, HFI Research

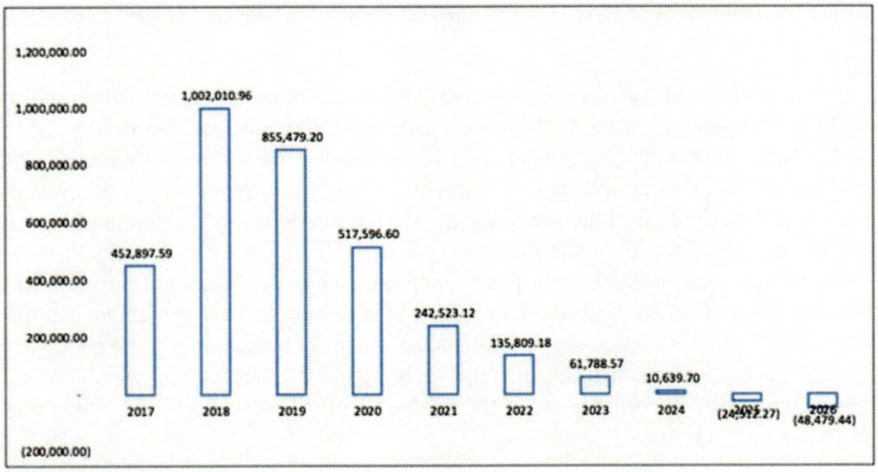

Fig. 3.6 Permian production growth year to year. *Source* EIA, HFI Research

3.8 Conclusions

At the global level, we must reduce the consumption of hydrocarbons by changing the consumption patterns of today's society, using energy savings and gradually replacing these energies with other renewable energies (hydroelectric, solar, wind) and nuclear, in the knowledge that they will not disappear anyway.

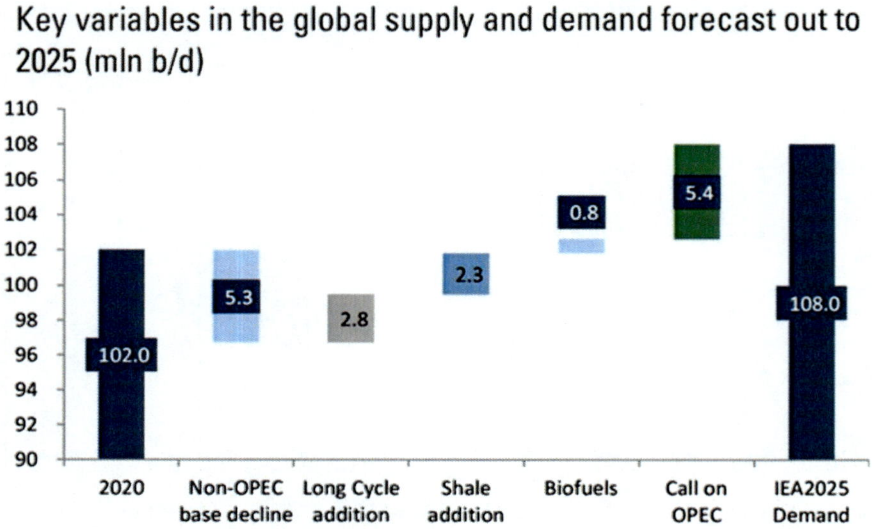

Fig. 3.7 WE estimate a 5.4 mln b/d callus n OPEC between 2020 and 2025 to meet IEA "current policies" demand forecasts *Source* IEA, Goldman Sachs Global Investment Research

For Argentina, in addition to what was expressed in the previous paragraph, it would be necessary to insist on the search for conventional oil and natural gas, since for almost 15 years it has been practically unexplored in the mature basins and in the continental territory or explore the continental shelf whose potential is unknown or increase the production of the mature deposits, the old deposits. One example is the long life of Cerro Dragón in Chubut.

Given the environmental impacts generated by the exploitation of non-conventional hydrocarbons through the use of the fracking technique, declare a moratorium while the corresponding specialists carefully analyze and with the necessary time all the implications that this technology brings with it and then obtain the social license granted by the inhabitants of the affected regions, as well as those who carry out productive activities in them.

Annex: The YPF-Chevron Contract

First of all, it must be said that the content of the agreement is secret so that what is known has actually transcended from authorities and specialists who are supposed to be well informed.

1. **Concise content**

It is signed between YPF and the international company Chevron.

Chevron is one of the main oil companies in the world, with private capital, of North American origin, and which has behaved poorly in terms of respect for nature and the rights of native peoples (in reality the same could be said of the large international oil companies).

In the case of Argentina, the State YPF, although it was not an example in terms of care for the environment, it was the creator of towns, roads and harbors that continued to exist even with less production of petroleum.

Investments and tasks of the agreement

Amount Chevron	Amount TPF	Period	Area	Wells
10^6 US $	10^6 US $	Years	Km^2	N°
1240	260	1.5	20	100
4500	4500	5	375	900
8250	8250	33.5	375	1677
13,990	13,010	35	375	2677

10 million dollars per well
Source Private information

A concession is given, that is to say, take a certain area of the Province of Neuquén, 395 km^2, which is given to take out the hydrocarbons for 35 years.

In exchange, Chevron invested 1240 million dollars in 18 months to drill, in 20 km^2, 100 wells, in what they call a "pilot" job.

If things go well, that is, if the wells produce interesting amounts for the companies, the other investments to develop the field in the remaining 375 km^2 would be put in equal parts by YPF and Chevron. In the next 5 years, they would put 4.5 billion dollars each to drill 900 wells, and in the years remaining until completing the 35 years, they would put 8.25 billion dollars each to drill 1677 more wells.

YPF is the one that operates the area, that is to say, it is in charge of the responsibility of the works with the contribution of the technical knowledge of the personnel of Chevron.

The petroleum and natural gas at are taken out are distributed in equal parts.

When the country is able to produce enough so that petroleum and natural gas do not have to be imported, Chevron, after the fifth year of operation, will be able to export the petroleum or natural gas, having the freedom to send abroad up to 20% of the US dollars product of that sale.

If the country continues to import, Chevron can sell this 20% to the country as if it were exporting it and send the corresponding dollars abroad.

For the Agreement to be effective, the owner of the hydrocarbons from Vaca Muerta, which is Neuquén province, must give their approval. The governor agreed to a Decree that must be approved by the provincial Congress. This Decree essentially extends the concession in one of the areas, so that it also expires in 2048. In exchange,

Neuquén seems to receive some special benefits. These are: a canon and something called Responsibility, which add up to 65 million dollars; in addition, the natural gas is removed from the subsidium for a Methanol plant, from which it is will be in charge of the National Government, which will also deliver 1 billion pesos for works in several locations near Vaca Muerta oil field. Additionally, Neuquén will receive 5% of the profits left by the YPF-Chevron association after 2027. To cover itself the provincial Decree includes a reinsurance clause that declares the concession null and void, and the area flies to the Province, if the companies do not comply with their investment plan. That is to say, it receives concrete promises in money if the project, the reason for the Agreement, reaches a successful conclusion.

In order not to think that this agreement was only for the benefit of Chevron, the National Executive issued Decree 929 of July 15, 2013, which provides for similar conditions for companies that invest no less than one billion dollars during the first five years of operation.

2. **Observations and comments**

(i) Firstly, the potential environmental damage implied by the type of operation, the "multiple hydraulic fracture." This technology is highly questioned in North American countries (where it has been most applied), has been banned in some European countries and has merited a negative statement by the European Economic Community's Commission on the Environment. The environmental damages refer, essentially, to the contamination of the water used for the operations (it has some aggregates of chemical products that are dangerous for the health of living beings), more than to the amount of water used. A part of it returns to the surface and it is not defined where it will be deposited and how it will be treated on the surface to be "clean." In addition to possible contamination, water used under high pressure to make hydraulic fractures can escape through cracks in underground rocks into fresh groundwater layers even if these are far away from the place where the fractures were made. Natural gas can also escape to the outside, and it has happened in North America, and this is very harmful to the environment because it raises the temperature of the planet. Of course companies have their answers saying that they can solve these problems.

That is, the main problem is not water consumption, which if reduced will be welcome, but in the final disposal of contaminated water and this is no longer a problem of quantity to become a quality problem.

Productivity of "non-conventional" and "conventional" wells

Oil

Concept	Non-conventional EP	Conventional EP
Initial production	350 Bl/day	350 Bl/day
Production at 5 years	50	205
Additional wells	4 times	

(continued)

(continued)

Concept	Non-conventional EP	Conventional EP
Investment per well	10–12 million US $	1–4 million US $

Source Tecpetrol, Views & Strategies for Long-Term Development of Unconventional Resources in Argentina July 17, 2013

For conventional estimates.

Natural gas

Concept	Non-conventional EP	Conventional EP
Initial production	255,000 m^3/day	255,000 m^3/day
Production at 5 years	11,300 m^3/day	180,000
Additional wells	15 times	

Source Tecpetrol, Views & Strategies for Long-Term Development of Unconventional Resources in Argentina July 17, 2013

For conventional estimates.

(ii) The lack of serious consultation with and information for the people living in the areas where the operations are to be carried out, as indicated by the International Labour Organization (Convention 169 June 1989) and the Argentine Environmental Law (National Law 25,675 November 2002).

(iii) The fact that the Convention is secret and not of public knowledge, at least for the legislators, opens questions as to what clauses it may not contain that are beneficial for the country. They say it is because of the commercial secret.

(iv) The association of State companies with international or national private companies is now common practice in almost all countries (including Venezuela, Ecuador and Bolivia), and everything depends on how these associations are made and how the national interest is protected.

(v) The fact that the companies can keep 20% (in this case, after the fifth year of operation) is better than what happened in the 1990s with decree 1589 of 1989, which allowed the transfer abroad of 70% of the foreign currency obtained and did not respect the condition that before exporting the country was self-sufficient in petroleum.

(vi) If what is wanted is to stop importing by increasing production, it must be taken into account that today YPF does not have all the country's deposits but only produces 36% of the country's petroleum and 24% of the natural gas.

(vii) With this agreement, assuming that the pilot plan on the 20 km^2 area satisfies the companies, in 2048 a quantity of petroleum (10,000 m^3 per day) would be produced that would not be enough to compensate for the normal drop in petroleum production between 2013 and 2048 which would be (being very conservative) of 16,000 or 20,000 m^3 per day. In other words, this investment of almost 28 billion dollars between 2013 and 2048 would not even be enough to cover the natural decline of the current YPF deposits.

(viii) On the other hand, in the "non-conventional" petroleum fields, production decreases year after year (e.g., wells start producing 350 barrels per day and after 5 years they produce 50 and in the natural gas fields they start producing 255,000 M^3/day and after 5 years they produce 1300) and this forces, as it happens in North America, to drill many wells annually and to spend a lot of money. On the other hand, the "conventional" petroleum fields start producing 350 barrels per day and after 5 years they produce 205 and the natural gas fields start producing 255,000 m^3/day and after 5 years they produce 180,000. This requires spending less money and drilling fewer wells.

(ix) This is an agreement fundamentally to produce petroleum and additionally some natural gas, when the fuel that the country needs most is natural gas and that represents, between what is bought from Bolivia by pipeline and what is brought in liquefied by ship, almost 7 out of every 10 dollars spent on importing fuels.

(x) The proposed investment would not be far from the possibilities of the country and YPF, if it were convenient to exploit today Vaca Muerta oil field, because it would involve 620 million dollars a year until 2015; 1800 million per year between 2016 and 2020 and 600 million dollars a year between 2021 and 2048.

(xi) The knowledge to carry out the exploitation today is not available but can be "bought," hiring and paying foreign specialists until ours learn it.

(xii) In my opinion, given the characteristics of this type of exploitation, and this goes beyond the Chevron-YPF Agreement, those who really win are the companies selling equipment and specialized services, all of them foreign.

3. **Why this agreement is made**

(i) **The "ideological" framework**

Throughout history, since the beginning of the twentieth century, in Argentina there have been, roughly speaking, two types of policies with respect to hydrocarbons (petroleum and natural gas). One of national content (governments of Irigoyen, Alvear, Illia; first and third presidency of Perón and partly of Alfonsín's) that thought in the country and in the use of the natural resources for the well-being of the people before the private gain.

Another one of favorable content to the national and foreign private capital that alienated this wealth (all the other governments) and that was unfavorable for the energy companies of the State.

In 2012, it seemed, with the expropriation of 51% of REPSOL shares, that the State YPF would reappear, but many doubts remain and the YPF-Chevron agreement does not contribute to clarify them.

(ii) The production of petroleum and natural gas in Argentina has been falling year after year, while consumption is growing.

The main causes of this situation are:

The reduction of the production capacity of the State YPF first and then the sale at a rather low price of the State YPF to REPSOL, a Spanish company.

This company was dedicated to strongly produce petroleum and natural gas, already discovered by YPF State-owned, to export it and recover as soon as possible the money from the purchase of YPF State-owned.

REPSOL stopped investing, that is, dedicating money to explore, to replace the petroleum that it exported and the country was left without petroleum and natural gas reserves. The petroleum and natural gas were sold abroad at a time when prices were almost four times lower than what it costs today to bring them from abroad.

(iii) As a result, today the country must spend many dollars to import the petroleum and natural gas that are needed for consumption. Fuel imports, mainly natural gas, take 15 out of every 100 dollars that come in for exports and there is less money left to buy outside things that the country needs for industries and the countryside to work.

The other fuels that are imported are the fuel oil and the gas oil that are consumed to produce electricity and for the industrial ones and that before used natural gas.

In addition, in the last 20 years the necessary investments were not made to build new refineries that convert petroleum into naphtha, gas oil and fuel oil, so more and more of the last two must be imported, as there is not enough natural gas to replace them, especially to produce electricity. Neither is the abundant wind used, nor new hydroelectric plants, nor nuclear plants, which would avoid consuming more natural gas.

(iv) The third cause mentioned, to sign the Agreement with Chevron, is that, according to the international experts Argentina has a lot of petroleum and natural gas in deposits called "non conventional" hydrocarbons. They say that the reserves (that is to say what can be extracted from what is in the rocks) can be, and only in the zone of Vaca Muerta oil field; 10 times greater than the reserves of petroleum and 70 times natural gas that the country has today. This is only a hypothesis that has not been verified yet.

(v) But to take out this non-conventional petroleum and natural gas from the rocks underground is very expensive (almost 3 or 4 times more than the petroleum and natural gas that is taken out today from other places in the country) and it is said that YPF does not know how to take it out, it does not have the knowledge to do it, which is called technology.

In other words, it is said, you have neither the money nor the technology to produce this "non-conventional" petroleum and natural gas without the input and knowledge of foreign companies.

(vi) In summary, the reasons given for signing the Convention are: the fall in production, the increase in imports, the lack of money in YPF to invest and the lack of technology to produce in the area of Vaca Muerta oil field.

4. **Possible alternatives**

The energy problem of the country is not only of petroleum and natural gas but of the set of all the forms of energy that can be used (electricity produced by the water, by the nuclear fuels, by the wind, by the sun; and what can be obtained with a good handling of the vegetables). That is to say that we have to see what country we will have in 30 years.

We have to modify the structure of electricity and natural gas rates and increase them little by little for high- and medium-income family consumers and for the commercial sector and to a lesser extent for industries. In this way, we will be able to make gas prices in carafes more similar to natural gas and the amount, today very large, of energy subsidies that in good measure are received by the rich who can pay higher rates will decrease.

Instead of putting almost all the eggs in the basket of petroleum and natural gas from fields like Vaca Muerta oil field, allocate an amount of money, about US $1000 million per year, to make between 200 and 250 exploration wells in areas where they can find petroleum and "conventional" natural gas. The cost of drilling wells in "non-conventional" areas, such as Vaca Muerta oil field, is 3–4 times higher than in conventional areas. Furthermore, in the last 20 years, since the privatization of the state-owned YPF, the exploratory effort is almost nil. If it is not explored it is impossible to discover petroleum and natural gas. But more "conventional" petroleum and natural gas must be found and instead the "non-conventional" ("Vaca Muerta" type) has already been "discovered" and the big question is how much petroleum and natural gas will finally be extracted and at what environmental, social and monetary cost. More information on this will be available after investing US $1240 million dollars in the first stage of Vaca Muerta oil field and verify how much the production of petroleum and natural gas is lowered annually.

Ideally, a group of specialists would have studied the energy, environmental, social and economic aspects of exploiting areas such as Vaca Muerta oil field, before launching a massive drilling plan. The results would then be communicated to the population, especially to the native peoples living in the area, and only then would measures be taken, such as the signing of contracts with private companies or whether the state could do it alone. That is to say, to decree a moratorium for the development of these areas as Vaca Muerta oil field until all the information required to face the exploitations is available.

Bibliography

BP Statistical Review of World Energy, June 2019

"Fracking a Fracture that takes its toll", Aitor Ultrasti and Florent Marcellesi, Political Ecology Journal N°43, September 2012

Fracking in the US All That Remains is the Permic and that will always last, taken from IEA, HFI Search, seeking Alpha, February 10, 2020

IEA, Goldman Sachs Global Investiment Research, 2019

"Reserves or Resources? Of Uncoventional Hidrocarbons" International Energy Agency, June 2013
"Shale gas towards a warmer future"; Engineer Anthoni R. Ingraffe researcher and professor at Cornell University—USA, Clarín Newspaper, Buenos Aires, 2012
Source: Hidraulicano.info, May 14th 2012, Cantabria
Source: "How does Hydraulic fracture work" Hidraulicano.info, May 14th 2012, Cantabria
Tecpetrol, Views and Strategies for long term development of Unconventional Resources in Argentina, July 17, 2013

Chapter 4
Shale Oil and Shale Gas in Argentina. State of Situation and Prospective

Nicolas Di Sbroiavacca

Abstract In the last years, a large number of articles, interviews, proposals and announcements have been published about topics on non-conventional hydrocarbons (mainly shale oil and shale gas), both at the international level and in Argentina. The high speed of circulation in which such information circulates, it is modified and renewed, does not sometimes allow to get a firm notion of the meaning of the shale phenomenon and even less, to establish which would be their impacts in the economy, the safety of the supply and the requirements of investments for its development. This document aims to systematize and analyze information on the subject of unconventional hydrocarbons (shale oil and shale gas), mainly its resources and reserves and also present prospective exploratory scenarios for Argentina (of its own elaboration), where the supply and demand is analyzed for oil and gas, with the aim of finally obtaining some conclusions and finally proposing a possible road map.

Keywords Vaca Muerta · Non-conventional · Shale oil · Shale gas · Scenarios

4.1 Introduction

In the last years, a large number of articles, interviews, proposals and announcements have been published about topics on non-conventional hydrocarbons (mainly shale oil and shale gas), both at the international level and in Argentina. The high speed of circulation in which such information circulates, it is modified and renewed, does not sometimes allow to get a firm notion of the meaning of the shale phenomenon and even less, to establish which would be their impacts in the economy, the safety of the supply and the requirements of investments for its development.

The present document pretends to systematize and analyze the more recent information about resources and reserves and also to present prospective exploration scenarios of our own elaboration where the offer and the demand of oil and gas are analyzed, with the objective of finally obtaining some conclusions, finally presenting a possible route map.

V. Bravo and N. Di Sbroiavacca, *Oil and Natural Gas Economy in Argentina*,
The Latin American Studies Book Series,
https://doi.org/10.1007/978-3-030-65520-4_4

4.2 About Resources and Reserves

It may be said that in Argentina, after June 2010, the epoch of non-conventional hydrocarbons has been finally and officially initiated, starting from the exploration well performed by the YPF-REPSOL company, of the first shale well at the locality of Loma La Lata, province of Neuquén, northern Patagonia. It may be noted that the existence of this resource has already been known since the 1960 and 1970 decades, when YPF-State owned discovered the deposits of Puesto Hernández and Loma La Lata and they started the perforation of the Vaca Muerta and Los Molles formations (which are rich in non-conventional hydrocarbon resources). In those times, neither the existing hydrocarbon prices, nor the lack of appropriate technology would have permitted their exploitation. Non-conventional resources correspond to a hydrocarbon classification that comprises those accumulations present in the subsoil, either of petroleum and/or natural gas that are not extracted by the usual techniques, but instead they require other extraction processes, which generally make these resources more expensive than the conventional ones.

As a consequence of the declination of the oil and gas reserves in the USA, observed during the last decades, a series of projects oriented were started in this country since the 1980s to design and develop a viable technique, from the economic and technological points of view, that would allow the hydrocarbon extraction from shales and similar rocks. Shale is interpreted as a fine-grained sedimentary rock, composed of many layers that were deposited one on top of the other in marine environments, mainly composed of clays, known as "mother rock," where the kerogene that originated the hydrocarbons was accumulated.

As a result of this process, in 1997, the first hydraulic fracture that combined the technique of "fracking" (water injection at high pressure with sand and chemical additives) took place together with development of horizontal wells, thus formally initiating the shale era in the whole world. Since 2005, with the beginning of production in a commercial scale of the first non-conventional oilfield, named as the Barnett Shale in the state of Texas, started the development of shale in the USA.

Simultaneously with this new scenario, in April 2011, the Energy Information Administration (EIA) of the USA published a preliminary evaluation of the shale resources in other regions outside the USA, in which information about Argentina is included. In such report, it is suggested that Argentina would have around 774 Tcf of non-conventional resources of natural gas which would be technically recoverable. "Tcf" means Tera cubic feet, equivalent to 27,000 million m^3.

This report was later on up-dated and, in June 2013, EIA published a new document where the hydrocarbon resources of the non-conventional type were estimated, but in this case comprising information from 41 countries, providing information about new resources for oil and natural gas.

The in-situ resources with risk adjustment are obtained multiplying the quantity of in-situ identified gas by a success factor. This last factor represents the probability that a portion of the studied rock formation may have commercially attractive production rates of natural gas and oil. Such factor cited in the EIA study was determined for all countries mentioned in the report with geological formations similar to those present in the USA.

In a new report, EIA mentioned that Argentina would possess 802 Tcf of unproven hydrocarbon resources technically recoverable of natural gas (equivalent to 21,654 thousand million m^3) and 27,000 millions of petroleum barrels (equivalent to 4293 million m^3). This would represent, respectively, 67 and 11 times the present proven reserves of conventional hydrocarbons (estimated in 323 thousands of million m^3 of natural gas and 394 million m^3 of oil of December 2011, data from the Secretary of Energy of Argentina).

Based upon the information provided in such report, the resources available for each of the analyzed basins are detailed as follows, according to the EIA.

4.2.1 Shale Gas

Neuquén Basin: In situ gas resources adjusted for risk (the Los Molles and Vaca Muerta formations): 2184 Tcf (93,083 thousands of million m^3). Out of this figure, the unproven, technically recoverable resources (those that in the EIA Report were estimated using a recovery factor of 27%) would be placed in the order of 583 Tcf (15,741 thousands of million m^3), equivalent to 49 times the presently proven gas reserves of the entire country. Out of this total, 53% corresponds to the Vaca Muerta formation (8308 thousands of million m^3).

San Jorge Gulf Basin: In situ gas resources adjusted by risk (the Aguada Bandera and Pozo D-129 formations): 438 Tcf (11,831 thousands of million m^3). Out of this value, the unproven, technically recoverable resources (estimated with a recovery factor of 20%), would yield a total of 86 Tcf (2322 thousands of million m^3), equivalent to 7.2 times the present proven gas reserves of the whole country.

Magellan Austral Basin: It is only considered here the portion of the basin pertaining to Argentina, since this basin is shared with Chile: in situ, risk adjusted in situ gas resources (the Inoceramus and Magnas Verdes formations): 606 Tcf (16,348 thousands of million m^3). Out of this value, the unproven, technically recoverable resources (estimated with a recovery factor of 21%), would be found in 130 Tcf (3510 thousands of million m^3), equivalent to 11 times the present proven gas resources of Argentina.

The Paraná-Chaco Basin: It is considered here only the part of the basin pertaining to Argentina, since this basin is shared with Paraguay and Bolivia: in situ, risk adjusted, in situ gas resources (the Ponta Grossa Formation): 16 Tcf (440 thousands of million m^3). Out of this value, the unproven, technically recoverable resources (obtained from a recovery factor of 20%), it would provide a total of 3.2 Tcf (86.4 thousands of million m^3), equivalent to 0.3 times the present proven gas reserves of Argentina.

Total for Argentina:

In situ, risk-adjusted shale gas resources:

Neuquén Basin: 2184 Tcf (58,968 thousands of million m^3)
San Jorge Gulf Basin: 438 Tcf (11,831 thousands of million m^3)
Magellan Austral Basin: 606 Tcf (16,348 thousands of million m^3)
Paraná-Chaco Basin: 16 Tcf (440 thousands of million m^3)
Gran Total: 3244 Tcf (87,587 thousands of million m^3).

Unproven, technically recoverable shale gas resources

Neuquén Basin: 583 Tcf (15,741 thousands of million m^3) (72.6%)
San Jorge Gulf Basin: 86 Tcf (2322 thousands of million m^3) (11%)
Magellan Austral Basin: 130 Tcf (3510 thousands of million m^3) (16%)
Paraná-Chaco Basin: 3.2 Tcf (86.4 thousands of million m^3) (0.4%)
Gran Total: 802 Tcf (21,659 thousands of million m^3).

As it may be observed, in the case of Argentina a recovery factor in the order of 25% was used, and for this reason the unproven, technically recoverable shale gas resources would be equivalent to 67 times the present proven reserves of conventional gas that Argentina has identified.

However, according to the most recent literature where precise cases developed in the USA are evaluated, it is indicated that the observed recovery rates in the case of shale gas are found around an average of 6.5%, with a range from 4.7 to 10%. It should be remembered that the boom of this activity in that country was achieved during the 2005–2008 period in such oil districts as Barnett, Marcellus, Haynesville, Fayeteville and Eagle Ford, the latter being compared with the Vaca Muerta Formation in respect to the behavior of oil production. Such estimations were made by the U.S. Geological Survey in the USGS Model Document for undiscovered conventional oil, gas and liquefied natural gas Resources, in the Seventh Approximation (2000), where the last resources contained in the 5 sedimentary basins of Argentina, actually in production, are estimated, compared with 20 basins that potentially may contain hydrocarbons.

Considering then this new mean recovery factor, the values of the technically recoverable shale gas resources that might be obtained from the subsoil were calculated:

Unproven, technically recoverable, shale gas resources with a recovery rate of 6.5%

Neuquén Basin: 142 Tcf (3833 thousands of million m^3)
San Jorge Gulf Basin: 29 Tcf (769 thousands of million m^3)
Magellan Austral Basin: 39 Tcf (1064 thousands of million m^3)
Paraná-Chaco Basin: 1 Tcf (28 thousands of million m^3)
Gran Total: 211 Tcf (5694 thousands of million m^3).

Based upon this recovery rate, the values of the unproven, technically recoverable, shale gas resources are equivalent to 18 times the present proven gas reserves of Argentina.

As it may be appreciated, with these new recovery rates, one of the formations that would be mostly affected is the Vaca Muerta Formation, since when passing from a recovery factor of 27% (applied by the EIA), to 6.5%, the recoverable shale gas resources of that formation are reduced to 3833 thousands of million m^3, equivalent to 12 times the present proven gas reserves of Argentina, instead of 49 times (using the recovery factor proposed by EIA).

If the unproven, technically recoverable, shale gas resources, obtained with a recovery factor of 6.5%, are compared with the conventional natural gas resources that would remain in the country still to be discovered according to the U.S.G.S. (with a value up to 1217 thousands of million m^3), it may be observed that the relationship between these values is of only 4.6 times.

Based from these estimations, a first conclusion may be obtained. Even though the magnitude of the unproven shale gas resources in the subsoil seems to be of a magnitude much higher than the proven gas reserves that Argentina actually possess, the degree of uncertainty is very high. Several very important variables as the success factor and the recovery factor (among many others) may challenge the concept that it is today firmly installed at the level of public opinion, concerning the magnitude of the shale gas resources. Available shale gas would be so abundant, that Argentina would be changed into a gas super-power. In fact, the pathway of shale in Argentina is just beginning and we are at the beginning of the learning curve. Higher investments in this area would permit to acquire own information about the behavior of the recovery rates, level of initial well production, well declination rates, environmental impacts and their mitigation, as well as the maximum reserves that may be extracted from them. All these would allow to improve the quality of the information of these estimations because they would be determined, based upon local parameters, the hydrocarbon recoverable potential volumes.

Although in the oil industry environment, there is a principle that establishes that the real reserves of a hydrocarbon deposit are known exactly when it is abandoned (because all possible resources have been extracted), the acquisition of more and better information permits to diminish the degree of uncertainly of such estimations and thus orientate in a better way the resource management strategies, such a lowering costs, improve the technique of exploitation, reduce the potential environmental impacts, create economic conditions to make the extraction profitable, etc.

To carry on this process may require between 5 and 10 years, and starting from the perforation of more new wells and the interpretation of such information, it would be possible to extract robust conclusions about the technical, economic and environmental feasibility of these type of enterprises. In relationship to this latter aspect, the analysis of topics such as the use of water, together with the management of the fracking returned waters and the production water, the carbon and water track, socioeconomic aspects, the analysis of the EROI of the non-conventional resources, etc., would allow to provide elements for the development of integral studies of site-specific environmental impacts, that should be carried on previously to the exploitation of the non-conventional resources, to determine precisely if it is truly convenient or not to use this extraction technology.

In conclusion, and because of the risk that would imply to follow just one unique pathway in the search of the recovery of the gas self-sufficiency, by means of the shale, for instance, it would be not appropriate to abandon the exploration of conventional resources, because there is still a great potential, since there would have been extracted only 57% of the last resources since gas was discovered in Argentina until today, still remaining the detection and extraction of 1217 thousand of million m^3 of the five basins still in exploitation today, equivalent to 3.8 times the present proven reserves. If this exploratory effort is also performed in high risk areas, such as the Argentine marine basin or the deepest waters of the Salado Basin, the perspective could be even better.

The strategy would be then to diversify the risk opening three processes that should be done in parallel: to start the shale, the exploration of the remnant conventional resource in known basins and the exploration of new basins. This would allow to transit a more conservative pathway, because if the country goes into shale only and if this is not viable then for its commercial exploitation, the country would have not only lost resources but it would also see seriously delayed in time the original objective of reaching again the gas self-sufficiency, with the expected impact that this generates onto the economy in general.

So as to provide more data to this analysis, an evaluation study about the Vaca Muerta Formation is presented here, which was developed by the Ryder Scott Company, as requested by the REPSOL Company and published at the beginnings of 2012.

It may be noted that of the 30,000 km^2 that are covered by the Vaca Muerta Formation, the cited company analyzed geological information of a smaller surface of 8071 km^2 (equivalent to the 27% of the total). Of the whole analyzed surface, this study concluded that 77% of the area would contain oil and the rest is distributed between dry and wet gas.

In Table 4.1, the principal results of the study may be observed.

It is observed herein that the level of shale oil reserves 3PA, YPF would count on 81 million barrels, equivalent to 13 million m^3, that is, the 3.3% of the proven oil reserves of Argentina. It should be noted that the concept Reserves 3P is preferred to those reserves that have been discovered, but only the proven reserves are in

Table 4.1 Resources and reserves of the Vaca Muerta Formation

	Gross (100%)				Net YPF			
	Petroleum (Mbbl)	Condensed (Mbbl)	Gas (Mbep)	Total (Mbep)	Petroleum (Mbbl)	Condensed (Mbbl)	Gas (Mbep)	Total (Mbep)
Prospective resources	5.732	396	15.038	21.167	3.966	224	8.161	12.351
Contingent resources	1.115	0	410	1.525	883	0	330	1.213
Reserves 3P	81	0	35	116	81	0	35	116
Possibles	33	0	15	48	33	0	15	48
Probable	25	0	11	35	25	0	11	35
Proven	23	0	10	33	23	0	10	33

Source Ryder Scott—1 bep = 5615 thousands of cubic feet of gas conditions to be extracted, taking into consideration the present economic conditions and of geological certainty.

In the case of oil, the Reserves 3P represent 1.2% of the total of recoverable resources estimated by Ryder Scott for such surface in the Vaca Muerta Formation. The total recoverable resources come out of the sum of Reserves 3P, plus the contingent resources. The latter are concentrations of hydrocarbon that have been discovered, but the present market conditions or the lack of appropriate infrastructure for their extraction do not allow them to be included as reserves, plus the prospective resources, that is, those that have not been discovered yet, but the detected geological conditions allow to infer their existence.

It is then noted that, for oil, 1.2% are Reserves 3P, 16% are contingent resources and 82.8% are prospective resources. In consequence, there is yet much to be done to know with better precision the available resource, the recovery percentages and the extraction rates, among other aspects.

In the case of natural gas, the Reserves 3P reach a 0.2%, the contingent resources are 9.8% and the prospective resources expand to 90%. In this case, the uncertainty about this resource is even greater.

It is then observed that a good portion of the shale resources must still be discovered, what implies that it is needed a significant exploration effort and a strong vocation to risk propensity.

4.2.2 Shale Oil

Neuquén Basin: In situ oil resources risk adjusted (Los Molles and Vaca Muerta formations): 331,000 million bbl (52,623 million m^3). Out of this value the unproven, technically recoverable resources (those that in the EIA report were estimated using a recovery factor of 6%), would be placed in the order or 19,880 million bbl (3160 million m^3), equivalent to 8 times the present proven oil reserves of the country. Out of this total, 80% correspond to the Vaca Muerta Formation (2528 million m^3).

San Jorge Gulf Basin: In situ oil resources adjusted for risk (Pozo D-129 Formation): 16,700 million bbl (2655 million m^3). Out of this value, the unproven, technically recoverable, resources, estimated with a recovery factor of factor of 3%, would yield a total of 500 million bbl (79 million m^3), equivalent to 0.2 times the present proven oil reserves of Argentina.

Magellan Austral Basin: It is herein considered only the portion of the basin that belongs to Argentina. In situ, oil resources, adjusted by risk, the formations Inoceramus and Margas Verdes): 131,200 million bbl (20,858 million m^3). Out of this value, the unproven, technically recoverable resources, estimated with a recovery factor of 5%, would be around 6560 million bbl (1043 million m^3), equivalent to 2.7 times the present proven oil reserves of the country.

Paraná-Chaco Basin: It is herein considered only the portion belonging to Argentina: In situ gas resources adjusted by risk (Ponta Grossa Formation):

300 million bbl (48 million m^3). Out of this value, the unproven, technically recoverable oil resources, obtained from a recovery factor of 3%, would provide a total of 10 million bbl (1.6 million m^3), equivalent to only 0.004 times the present proven reserves of Argentina.

Total for Argentina:

In situ oil resources adjusted for risk

Neuquén Basin: 331,000 million bbl (52,623 million m^3)
San Jorge Gulf Basin: 16,700 million bbl (2655 million m^3)
Magellan Austral Basin: 131,200 million bbl (20,858 million m^3)
Paraná-Chaco Basin: 300 million bbl (48 million m^3)
Gran Total: 479,200 million bbl (76,184 thousands of million m^3).

Unproven, technical recoverable shale oil resources

Neuquén Basin: 19,880 million bbl (3160 million m^3) (73.7%)
San Jorge Gulf Basin: 500 million bbl (79 million m^3) (1.8%)
Magellan Austral Basin: 6560 million bbl (1043 million m^3) (24.3%)
Paraná-Chaco Basin: 10 million bbl (1.6 millones de m^3) (0.04%)
Gran Total: 26,950 million bbl (4285 thousand million m^3).

As it is observed in the EIA report for the case of Argentina, a recovery factor of the order of 5.6% was used, and with it, the unproven, technically recoverable shale oil resources would be equivalent to 11 times the proven, conventional oil reserves that Argentina possesses.

Considering the last conventional oil resources estimated by the U.S.G.S., there would be to be discovered in Argentina around 350 million m^3 in the 5 productive basins. This means that the relationship between the unproven, technically recoverable unconventional oil resources is 12 times greater than the conventional oil resources.

As it has been exposed when discussing the case of natural gas, to diversify the exploration strategy in search of unconventional and conventional oil, both in the known basins as in promising basin, would be the strategy to follow.

Finally, the estimations of gas and conventional and unconventional oil reserves, herein presented, have been taken to a common unit (m^3 of equivalent oil), based upon the information published by the EIA and the U.S.G.S.

From Table 4.2, the great potential in terms of natural gas that Argentina possesses is exposed, since 82% of the potential reserves and resources are referred to this hydrocarbon. Besides, 92% of the reserves and resources would be of the unconventional type. Likewise, the proven reserves of Argentina represent only 2.5% of the total, exposing the great challenge that Argentina has, it is pretended that in the near future these resources would be converted in reserves. This objective would be reached based upon important investments in exploration and development, the implementation of an oil politics that would accompany such process and that nature would confirm the preliminary estimation of resources.

Finally, and as a summary, it should be noted that 6 shale deposits have been discovered so far in Argentina:

Table 4.2 Reserves and resources of oil and natural gas (in million m^3 oil equivalent)

	Oil Convencional	Oil No Convencional	*Sub-Total*	Natural Gas Convencional	Natural Gas No convencional	*Sub-Total*	**Total**
Proven Reserves	393	4.285	393	323	21.659	323	716
Probable Reserves, Posible and Potencial Resources	350		4.635	1.217		22.876	27.511
Total (millon m3 oil eq.)	743	4.285	**5.028**	1.540	21.659	**23.199**	**28.227**
Total (%)	3%	15%	**18%**	5%	77%	**82%**	

Source N. Di Sbroiavacca, elaborated on the basis of information of EIA, U.S.G.S. and the Secretary of Energy of Argentina

3 of them in the San Jorge Gulf Basin (Cañadón Yatel, Los Perales and Lomas del Cuy). With 100% of YPF participation, being basically shale oil deposits.

3 more in the Vaca Muerta district within the Neuquén Basin (Orejano, Loma del Molle, Loma La Lata/Loma Campana):

Loma del Molle (45% YPF, 45% Exxon-Mobil and 10% G&P). It contains shale oil and shale gas.
Orejano (50% YPF and 50% Dow). This area would produce mainly shale gas.
Loma La Lata and Loma Campana (YPF 50% and 50% Chevron). Discovered in 2010. It is in the window of shale oil.

4.3 About Production

As it was previously exposed, the production of shale oil and shale gas in Argentina is only a few years old as it started in 2010. For this reason is that very little is known about well performance, with such short productive history behind, the well performance, their EUR, the recovery rates and the declination rates, among other central aspects related to reservoir and production engineering are still poorly known.

Until the end of 2012, only 72 non-conventional wells had been drilled in Argentina, 67 of them in the Vaca Muerta district (Neuquén Basin), with the following sequence: in 2010, 2 wells were completed, in 2011 32 wells were done and in 2012, around 33 wells were finished. Toward the end of 2012, YPF had drilled 49 wells in the Vaca Muerta shale. In 2013, it was estimated that around 130 wells would be drilled.

Concerning the distribution of the wells reported by the 8 oil companies operating in Vaca Muerta (toward the end of 2012), 59 wells (88%) has as a main objective the oil production and 8 wells were interpreted as gas yielding perforations (12%). As it is observed, the objective of drilling unconventional wells has been a criterion mainly of oil resources search. Besides, most of the wells have been of the vertical type and only 10 of the 67 wells drilled in Vaca Muerta have been horizontal wells, that is, a 15% of the total.

Due to the characteristics described when analyzing reserves and resources, where it is observed that Argentina would possess a larger quantity of gas non-conventional resources, and considering as well the present lack of sufficient supply of this resources at the national level, the need of a hydrocarbon policy that would support the search and production of natural gas is evident. A revision of the prices at the mouth of the well would clearly generate a strong signal in this direction.

In spite of its very short productive history, several papers have been published based upon the experiences in the development of the Neuquén deposits, mainly in the Vaca Muerta and Los Molles formations, whose conclusions allow having a preliminary approach of the productive scenario of the non-conventional resources (principally in terms of oil).

In this sense, it is observed that several articles are coincident in the range of EUR that the Vaca Muerta non-conventional oil-producing wells could contain. These values are found between a minimum of 15,600 m^3 in 25 years (for vertical wells) and a maximum of 130,000 m^3 in 25 years (for horizontal wells). The mean values of EUR of the vertical wells were estimated in 28,000 m^3 in 25 years, and that of horizontal wells in 62,000 m^3 in 25 years, what implies a relationship between the potential extractable volumes from such formation is in the order of 2.2 times. This means that in the same formation the EUR of a horizontal well doubles that one of a vertical well. According to other studies, the range of EUR in the Vaca Muerta district would be between a maximum of 115,000 m^3 in 25 years, with a mean of 17,800 m^3 and median (that is, 50% of the cases) of 12,600 m^3.

Another central element that comes from these preliminary analyses is referred to the declination rate of the oil wells. The productive profile of these wells follows a hyperbolic trajectory at the beginnings of their useful life, which is derived toward an exponential curve when the minimum declination value is reached. This value would be achieved between the years 9 and 10 since their production period. This implies that, until this moment, the declination is in the order of 40–45% per year, being placed later on around 6% per year. These very high declination rates in the first years of production are coincident with those observed in the USA (40% per year in the case of oil and between 29 and 52% in the case of natural gas).

With the objective of estimating these EUR values and the declination rates, the present authors have proposed a series of hypotheses related to several parameters that are used for their calculation, based upon the existing literature and the empirical experience of the field data on which these estimations have been performed. For these reasons, it is very important to continue the acquisition of experience based upon the development and the analyses of production of the first wells, since this would allow obtaining local parameters that may be useful to estimate the EUR and the declination rates. This would permit to establish with a better precision the economic viability of exploiting this resource and based on the learned lessons to orientate toward the best productive techniques that would allow the feasibility of the extraction of the non-conventional resources.

In relation to natural gas, national publications with productivity estimations for these types of wells have not been found yet, assuming that this is due to the scarce impulse that non-conventional gas exploitation has received so far in Argentina.

4.4 About the Possible Scenarios of Offer and Demand

Based upon recent estimations with respect to the prospective of oil and natural gas demand in Argentina (in the period 2013–2050), prepared by the Bariloche Foundation, various scenarios of hydrocarbon offer have been developed in the present project, from which a series of conclusions may be extracted.

Concerning the scenario of oil demand, it has been proposed the incorporation of a set of oil refineries along the period, so as to accompany with local production the increasing demand of oil derivatives. To simulate the growth of oil demand, this parameter it has been represented in a step function which reflects the starting of activities in the proposed refineries along this period. Starting from a refinement capacity of 32 million de m^3 of oil per year, for 2012, then generating a distillation capacity of 62 million m^3 around 2050.

Taking in consideration the natural gas demand, this would evolve at a rate of 2.4% a.a. along this period, what implies to move from a gas consumption of 135 million m^3/day for 2012 to 347 million m^3/day in 2050. The growth projected for the gross national product (GNP) is of 3.0% per year, during the 2013–2050 period.

From the point of view of the hydrocarbon offer, three exploration scenarios have been considered. One of them is of minimum magnitude, named as the Chevron scenario, where it is expected that only this project, related to the non-conventional oil extraction, is started in Argentina up to 2050, producing an average of 3.8 million m^3 per year of non-conventional oil. The Conservative scenario estimated that the starting of several non-conventional oil projects, but in this case accumulating along the forthcoming 35 years the extraction of roughly 440 million m^3 of non-conventional oil, equivalent to the 10% of the unproven, technically recoverable resources oil non-conventional oil, with a mean annual production of this resource of 12.5 million m^3, whereas in the case of the natural gas the production accumulated for the period reached only the 8% of the unproven, technically recoverable resources of non-conventional gas. Finally, a self-sufficiency scenario, where the effort is determined in the number of wells that should be completed in such a period to reach the self-sufficiency in oil and natural gas.

In all scenarios, it is discussed the extraction during such period of the proven reserves of conventional hydrocarbons (cited in Table 4.2), considering a declination of oil and natural gas production, based upon the observed rates during the last years, which are placed at 6% per year in the case of oil and of 2.5% per year in the case of natural gas. Besides, it has not been considered the inclusion within the category of proven reserves of the following concepts: probable, possible and potential resources (as seen in Table 4.2), with the objective of demonstrating the contribution to the future supply of the non-conventional resources, under the hypothesis that all the effort of investments in exploration and development would be concentrated in shale oil and gas.

4.4.1 The Chevron Scenario

To represent this case, it has been considered the press information recently exposed by the YPF Company, in which they project the oil volumes to be extracted from the Vaca Muerta district during the 2015–2050 period.

During such period, 124 million m^3 would be extracted. This project is considered the most important in the country.

In Fig. 4.1, it is observed that the contribution of this project to the oil offer would be marginal. Although it would reduce the oil deficit recorded in Argentina between 2014 and 2017, it is concluded that under the hypotheses of this scenario, Argentina would present a growing crack between offer and demand. Therefore, the contribution of the Chevron Project would be more oriented to improve the quality and quantity of information that might be acquired along the exploitation process of these resources, more than to reduce the supply deficit. The accumulated oil deficit is found around 1400 million m^3.

In spite of the low impact that this project would have over the total oil production, the effort in terms of investment and operation costs to reach these increments are important. These values were estimated by the YPF Company in 25,000 million US $ and 1562 wells should be drilled, what implies a mean investment per well of 16 million US $ (in which CAPEX and OPEX are included).

Based on the accumulated oil production that is planning to extract 124 million m^3 following this project, a EUR of 80,000 m^3 per well is estimated. This value is 30% higher than the mean value observed in horizontal wells in the Vaca Muerta area (see the item about production in this document). This implies a certain optimism from the company side concerning the EUR of the formation or a high proportion

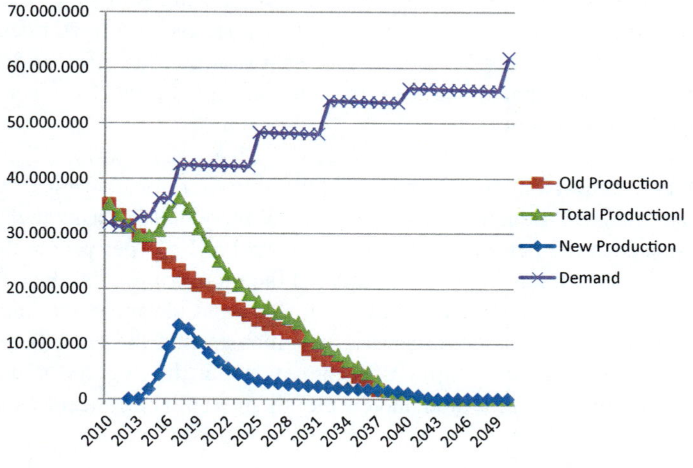

Fig. 4.1 Offer and demand of oil. Chevron scenario (in m^3). *Source* N. Di Sbroiavacca, own elaboration

of horizontal wells over the total of proposed wells, since they possess almost twice EUR than one vertical well.

This pilot try implied by the Chevron Project would provide more experience to the area operator, but also, if it is successful, it may give a positive signal to other investment groups to launch their activities to the search and extraction of the shale resources. It also may show the great challenge that Argentina has in its future, in terms of the volume of requested investments, in the case of willing to recuperate the self-sufficiency as a result of putting shale into production, as it will be seen later on in this document.

4.4.2 The Conservative Scenario

In this case, it is planned to perform around 25,000 wells of shale oil and shale gas between the years 2013 and 2050, that is, almost 700 wells per year in average, along the entire period. The estimated investment to fulfill this scenario is around 250,000 million US $ (equivalent to 6800 million US $ per year).

The EUR used in this study for each oil well reached 42,000 m^3 (around 50% above the mean value of a vertical well and 35% below the mean cost of horizontal well). Therefore, to achieve this EUR value it is proposed that a somewhat greater participation of horizontal and vertical wells with respect to that presently being used in the Vaca Muerta district. Besides, the initial production per well and the production profile of the oil wells was calculated considering the mean parameters for declination rate proposed in the already mentioned papers and taking in consideration a mixed declination (both of the hyperbolic and exponential types). In the case of shale gas, the used EUR was of 124 million m^3 (similar to that of such areas as those of Haynesville, USA). Following these hypotheses, a model was elaborated which permitted to estimate the production evolution.

In Fig. 4.2, it may be observed that, with the entrance in production of such quantity of wells proposed in this scenario, the oil and gas self-sufficiency may not be achieved during this period. Although the gap between offer and demand is smaller from 2018 onwards (compared with the Chevron scenario), Argentina would not be in condition to recuperate the self-sufficiency, even though a great economic effort is exposed in this scenario. The accumulated oil deficiency is found around 1100 million m^3. It may stress that under the categories of probable, possible and potential resources, according to the United States Geological Survey (USGS), Argentina would keep a total of 350 million m^3 of oil. If there were investments along this period in the search and set in production of these conventional resources, the deficit would be reduced to 750 million m^3. It is then concluded that with the aim of reducing even further such imbalance, it would then be required the set in production of a larger proportion of non-conventional resources and/or the exploration and development of hydrocarbons from other geological basins still unproductive.

In the case of natural gas, something similar happens, because although the natural gas production would be increased, such production would not reach the point of

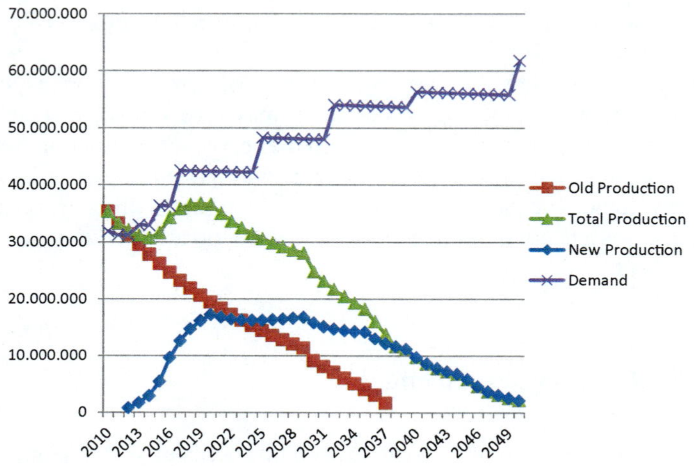

Fig. 4.2 Oil offer and demand. Conservative scenario (m^3). *Source* N. Di Sbroiavacca, own elaboration

covering the local needs. Therefore, Argentina would still be a natural gas importing country and in 2050 (Fig. 4.3).

However, it may be noted that in terms of natural gas, this scenario is more beneficial that in the case of oil. Nevertheless, the growing requirements of natural gas with national gas are not fully satisfied in this model. The accumulated deficit of natural gas in this scenario adds up to 540,000 million m^3. In that sense, it may be highlighted that under the categories of probable, possible and potential resources, according to the USGS, there would remain in Argentina up to 1,200,000 million

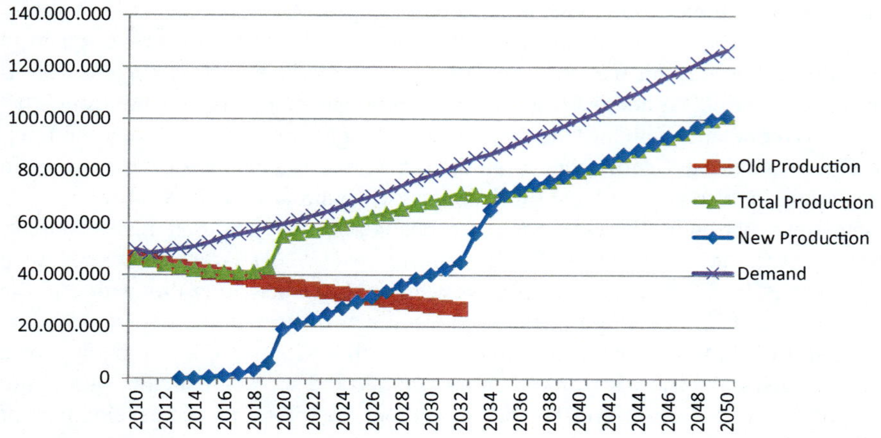

Fig. 4.3 Natural gas offer and demand. Conservative scenario (in thousands of m^3). *Source* N. Di Sbroiavacca, own elaboration

m^3 of natural gas (see Table 4.2). If there were investments in the search and set in production of these resources, the deficit would be eliminated.

It is then concluded that within the conservative scenario, in the case of natural gas Argentina would be in better conditions to provide the needed supply due to increasing demand, whereas in the case of oil, in spite of incorporating the conventional resources, the deficit would be still standing, unless there would be investment in exploration of new geological basins and that these investments would be successful.

4.4.3 Self-sufficiency Scenario

In this scenario, the proposal was reversed, due to that the objective was posed in reaching the self-sufficiency and with the model elaborated, the number of requested well was determined.

Based upon this, the requested investment to reach a sustained growth in oil and natural gas production would be around 600,000 million US $ and close to 57,000 wells of shale oil and shale gas. This is equivalent to the development of 1500 wells per year.

In the case of oil, the self-sufficiency may be achieved with the extraction of an accumulated production of shale oil of at least 1700 million m^3, equivalent to 40% of the unproven, technically recoverable non-conventional oil resources (Fig. 4.4).

In the case of natural gas, the self-sufficiency may also be sustained during the entire period. This implies an accumulated production between 2013 and 2050 of

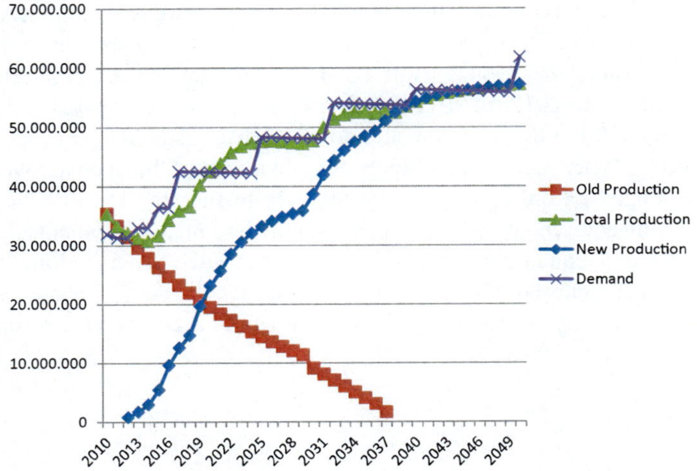

Fig. 4.4 Oil supply and demand. Self-sufficiency scenario (in m^3). *Source* N. Di Sbroiavacca, own elaboration

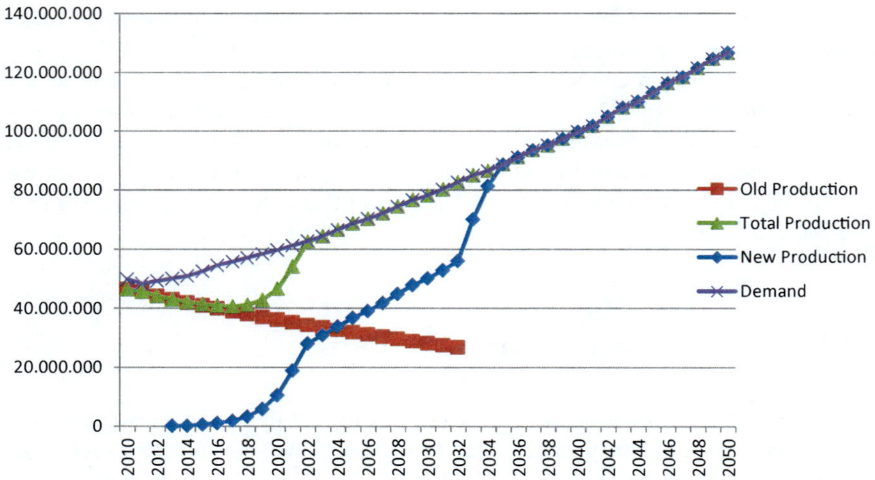

Fig. 4.5 Natural gas offer and demand. Self-sufficiency scenario (in thousands of m^3). *Source* N. Di Sbroiavacca, own elaboration

2600 thousands of million m^3, equivalent to the 12% of the unproven, technically recoverable, non-conventional natural gas resources. However, it may be observed that during the 2013–2021 there would be an accumulated deficit of natural gas in the order of 110,000 million m^3 (equivalent to 33 million m^3 per day). This is due to the maturation period that the industry usually requests to be able to set in production the shale deposits (Fig. 4.5).

It is then concluded that to reach this scenario a mean annual investment of 16,000 million US \$ is required. This amount is similar to the payments that Argentina would have to do in 2013 to buy the badly needed oil resources overseas, mainly natural gas, a value that would triple the investment in the Argentine upstream recorded in 2011 and 2012 (in 2011, 5200 million US \$ were invested and in 2012, approximately 4500 million US \$ would do so as well, a sum that was used to drill around 1000 wells per year). Undoubtedly, the concretion of this scenario would pose a great challenge, not only for the requested investments, but also for the requirements in drilling equipment, sandy materials for the fractures, equipment appropriate for fracking, management of the water resources associated to the exploitation of the oil resources, the control of the environmental impacts, the human resources needed and the development of a proper legislation that would accompany and support the entire process.

4.5 Conclusions and a Possible Road Map

Based upon the information presented along this document, the situation of non-conventional, hydrocarbon resources and reserves, the possible production profiles and the diverse prospective scenarios of offer and demand were explored.

A central conclusion that emerges from this analysis is that the development of non-conventional oil resources is feasible. However, to do it so more information based on local data is needed, that would allow better and more reliable estimations about recoverable hydrocarbon volumes, initial well productivity, rates of extraction and declination, water well production, possible environmental impacts, thus permitting to generate a better knowledge about the management of these resources, and from this, adapt and/or design regulatory and promotion frameworks that may stimulate the development of the shale deposits.

To reach this objective would require a span of time not less than 5–10 years, with an open end until obtaining concrete and tangible results in the exploitation of non-conventional oil resources. Therefore, to diversify the widening of the hydrocarbon offer is one of the pathways to follow. In this sense, the National state, by means of the YPF National Company, should participate in the shale process, but it also should be the priority to continue with the exploration of the already known geological basins and even in those basins that are still unproductive. The State should also implement better production improvements, tending to increasing the recovery rate in the presently operating wells, reactivate the secondary and tertiary exploitation (which would provide a good portion of the conventional resources still unexploited) and to intervene in the hydrocarbon exploration/exploitation overseas, with the aim of making oil and natural gas provision of hydrocarbons produced in other countries so as to cover the offer deficiencies.

These measures implemented from the side of the offer would permit to reduce the impact upon the foreign debt payment from now onwards, when the self-sufficiency would be achieved, both in natural gas and oil resources, if this scenario would be followed. Notwithstanding, to achieve this goal and maintain it, an investment of around 200,000 million US $ should be done until the year 2030 and during such period (140,000 million US $ in the case of oil and 60,000 million US $ in the case of natural gas).

It is for these reasons that, in addition to these actions, other ones should be implemented from the viewpoint of the demand. In this sense, the national state should have a more pro-active role in the field of energy efficiency, massively putting at operation and sustained with time, such programs as PRONURE or other concrete initiatives oriented to diminishing the specific consumption of the transportation sector or promoting the use of massive transportation media or promoting the use of railway in the transportation of heavy loads. In all these cases, aims, objectives and performance indicators should be stated. There should also be promoted the energy substitution, both in the final energy demand, by means of the penetration of renewable means in heating uses, for instance, as in the offer as well, with a greater participation of hydro-energy, nuclear energy and renewable sources of energy, and creating conscience among the citizens about the importance of energy, creating plans and media campaigns that would promote the rational use of energy.

All these actions would allow the diminution of the pressure on the hydrocarbon demand and, at the same time, enhance the offer. Argentina should be able to convert such weakness present today from the viewpoint of the hydrocarbon supply, in a great strength, defining and designing energy politics that would allow overcoming the present situation and generating the bases to build and propose a diversified and sustainable energy matrix.

4.6 Present Situation

From June 2010 to December 2019, there have been important advances in the development of the non-conventional, oil and gas resources of Argentina. During this period, 1169 wells have been drilled in the Vaca Muerta Formation, of which 929 wells were effectively in production on December 31, 2019. During the year 2019, 909 wells were completed all around the country, of which 819 were finally devoted to production. Off these 819 wells, 532 of them were considered as of the conventional type and the remaining 287 as of the non-conventional category, of which 194 were of the Shale kind, whereas 93 are of the Tight type. In the year 2016, 110 wells were drilled as non-conventional in shale, whereas in 2015, 199 wells were drilled, in 2017 104 wells were completed and in 2018, 194 wells more. The 70% of the total were performed by YPF (the oil company with 51% of State participation). About 56% of them are of the horizontal type (652 wells) (Fig. 4.6).

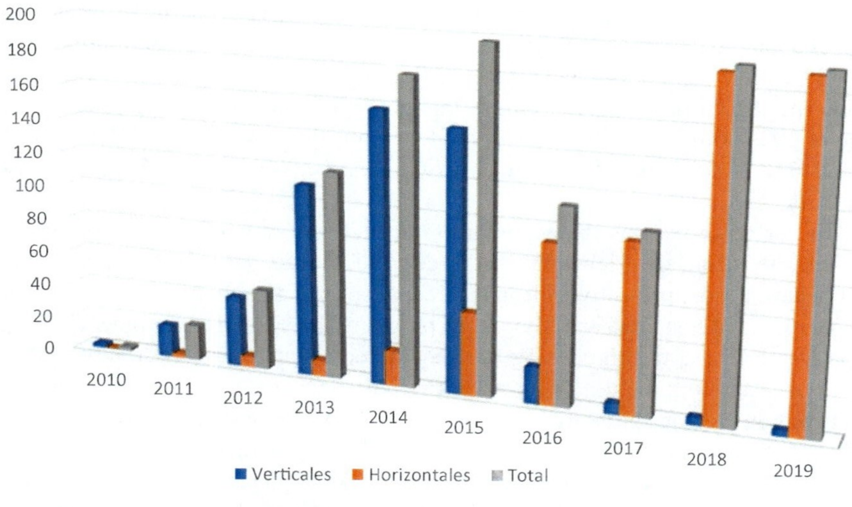

Fig. 4.6 Evolution of wells drilled in Vaca Muerta. *Source* Own elaboration based upon GiGa, "Tri-Monthly Report – Non Conventional resources – Neuquén and Austral basins – Q4 2019"

The observation of this graph allows the recognition that there has been a change concerning the drilling strategy. During the first 5 years of exploitation, the vertical wells were predominating, but after the year 2016, the drilled wells are mostly of the horizontal type. This change made possible to account for the highest rates of resource recovery and, therefore, a better profitability of the wells. Likewise, there has been an increment concerning the length of the horizontal branches, which were originally of 1500 m, presently moving into a mean length of 2200 m, with some wells that have branches that reach more than 4000 m. Concerning the number of fracturing phases, the present mean value is of 25 phases per well, with exceptional cases of 40 fracture phases per well.

Besides, since up to 10 years of experience have been accumulated in the exploitation of the Vaca Muerta Formation, comparisons between the final oil and natural gas recovery per well, checked against other non-conventional oil and gas fields all over the world.

Considering the extracted resources accumulated from an average oil type well of the Vaca Muerta district, along their first two years, it may be observed that the accumulated production is about 300 Mbbl, against 200 Mbbl from the Permian fields (USA) or 150 MMbl in the Eagle Ford district (USA).

In the case of natural gas, the accumulated production of a typical well in the Vaca Muerta fields, after 12 months, is 3 BCF, when in the Marcellus district is of 3.1 BCF (USA) and in the Haynesville fields are of 4.2 BCF (USA). This would place the Vaca Muerta district as geological formation with competitive productive with productive yields. From the economic point of view, high CAPEX values are still observed, in comparison with those found in the USA, thus allowing still for space to reduce these values and improve their economic competitiveness.

Concerning the present production of the Vaca Muerta Formation, and its representativeness in the national context, in December 2019 the production reached a rhythm of 110.7 Mbbl/d (17.6 Mm^3/d) of oil, which represents the 21% of the Argentine national total (528 Mbbl/d).

In relation to natural gas, in December 2019 the production achieved 1044 MMscft/d (29.6 MMm^3/d), around 23% of the Argentine total production (4.480 MMscft/d).

In consequence, it is found that in the last 10 years there have been advances in the knowledge of the oil and gas resources, in the improved and adapted techniques to the Vaca Muerta district conditions. Increments in the recovery rates and in the cost reduction should still be studied and discussed.

In synthesis, Argentina, along these years, advanced in the detailed knowledge of the non-conventional resource, starting from a central role performed by the YPF Company. Besides, exploration areas in the Argentine continental platform have been recently offered, where 18 new areas were adjudicated, with the idea of exploring the Argentine Sea. To continue the study of the technical aspects and the diversification of the hydrocarbon supplied should be part of the national strategy in the long run.

Bibliography

Energy Information Administration (2011) World shale gas resources: an initial assessment of 14 regions outside United States

Energy Information Administration (2013) World shale gas and shale oil resources assessment

Giampaoli H (2012) Vaca Muerta: Dos años de shale en Argentina. Análisis estadístico de producción a noviembre de 2012. Rev Petrotecnia

Gutiérrez Schmidt J, Alonso J, Giusiano A (2013) Subsecretaría de Minería e Hidrocarburos. Dirección de Estudios, Provincia de Neuquén Evaluación del Shale Oil de la Formación Vaca Muerta. Análisis de la declinación de la producción. Rev Petrotecnia, Buenos Aires

Kokogian D (2010) Perspectivas exploratorias de la plataforma continental argentina. Rev Petrotecnia 51(5)

Newsletter diario Inversor Energético & Minero (2013) YPF confía en aumentar 30% su producción de petróleo con Vaca Muerta

Rogers D (2013) Shale and wall street: was the decline in natural gas prices orchestrated? Energy Policy Forum

US. Geological Survey (2000) USGS model for undiscovered for conventional oil, gas and NGL resources—seventh approximation

Final Summary

This book is composed of four interrelated chapters. The central focus is the "unconventional" exploitation of hydrocarbons called "Fracking."

In Argentina and in the rest of the world, this technique has been questioned. Its positive and negative impacts are analyzed throughout this book.

This is reflected in the two chapters that specifically mention the topic: "A Technical Opinion on Fracking" and "Shale Oil and Shale Gas in Argentina: State of Play and Prospects".

"A Technical Opinion on Fracking" describes the issue and warns of the possible environmental and socioeconomic impacts of its application, using as an example what is happening in other parts of the world.

In "Shale oil and Shale Gas in Argentina: State of affairs and perspectives," the resources, reserves, production, demand and supply of hydrocarbons in Argentina are described, particularly those of the "non-conventional" ones, pointing out the importance of them for Argentina as long as they are properly exploited.

But dealing with the issue of fracking in Argentina implies analyzing the policies on hydrocarbons applied by the Kirchner and Macri governments, since this technique began to be applied in the first of them and tried to be consolidated during the second.

The economic and social situation of Argentina in 2020 is really delicate, essentially because of the enormous weight of its foreign debt and the peremptory terms granted by the creditors (International Monetary Fund and private creditors), which are impossible to comply with.

These two types of policy are referred to in the documents: "Analysis of the Argentine National Law N° 27,007, called the "Hydrocarbon Law,"" "Hydrocarbon Policy for the period 2003 to 2014" and "Argentine Energy Policy for the period 2014 to 2018."

In summary, this book deals with a hydrocarbon technological subject called "Fracking" which has generated a very important impact on the World hydrocarbon market, especially in the USA.

V. Bravo and N. Di Sbroiavacca, *Oil and Natural Gas Economy in Argentina*, The Latin American Studies Book Series,
https://doi.org/10.1007/978-3-030-65520-4